DESOLATE LANDSCAPES

THE RUTGERS SERIES
IN HUMAN EVOLUTION

Robert Trivers, *Founding Editor*
Lee Cronk, *Associate Editor*
Helen Fisher, *Advisory Editor*
Lionel Tiger, *Advisory Editor*

David Haig, *Genomic Imprinting and Kinship*

John F. Hoffecker, *Desolate Landscapes: Ice-Age Settlement in Eastern Europe*

John T. Manning, *Digit Ratio: A Pointer to Fertility, Behavior, and Health*

DESOLATE LANDSCAPES

Ice-Age Settlement in Eastern Europe

John F. Hoffecker
Foreword by Richard G. Klein

RUTGERS UNIVERSITY PRESS
New Brunswick, New Jersey, and London

Library of Congress Cataloging-in-Publication Data

Hoffecker, John F.
   Desolate landscapes : Ice-Age settlement in Eastern Europe / John F.
Hoffecker ; with a foreword by Richard G. Klein.
     p. cm. — (The Rutgers series in human evolution)
   Includes bibliographical references and index.
   ISBN 0-8135-2991-3 (cloth: alk. paper) — ISBN 0-8135-2992-1 (pbk. : alk. paper)
   1. Neanderthals—Europe, Eastern. 2. Prehistoric peoples—Europe,
Eastern. 3. Human beings—Effect of climate on. 4. Human evolution.
5. Europe, Eastern—Antiquities. I. Title. II. Series.

GN285 .H64 2002
569.9—dc21   2001019802

British Cataloging-in-Publication information is available from the British Library.

Copyright © 2002 by John F. Hoffecker
All rights reserved
No part of this book may be reproduced or utilized in any form or by any means,
electronic or mechanical, or by any information storage and retrieval system,
without written permission from the publisher. Please contact Rutgers University
Press, 100 Joyce Kilmer Avenue, Piscataway, NJ 08854–8099. The only exception
to this prohibition is "fair use" as defined by U.S. copyright law.

Manufactured in the United States of America

For Lilian

Contents

Figures   ix
Tables   xiii
Foreword
By Richard G. Klein   xv

Preface and Acknowledgments   xvii

CHAPTER 1: Human Evolutionary Ecology and Eastern Europe   1

CHAPTER 2: Environmental Setting   15

CHAPTER 3: Middle Pleistocene Settlement   36

CHAPTER 4: Neanderthal Adaptations   55

CHAPTER 5: The Transition to Modern Humans   138

CHAPTER 6: People of the Loess Steppe   192

CHAPTER 7: Retrospective   249

Bibliography   257
Index   285

Figures

1.1. Mean annual temperature and brachial index in selected modern populations   7
1.2. Effective temperature and the percentage of meat and fish in the diet among recent hunter-gatherers   8
1.3. Effective temperature and average distance of residential moves among recent hunter-gatherers with low dependence on aquatic foods   9
1.4. Contribution of hunting to diet and territory size among recent hunter-gatherers   9
1.5. Effective temperature and technological complexity among selected recent hunter-gatherers   10
1.6. Patterns of adaptation to northern environments among recent hunter-gatherers   11
2.1. Map of Eastern Europe, illustrating modern political boundaries and major geographic features   16
2.2. Temperature in northern Eurasia expressed as number of days per year with temperatures above 0 degrees C   18
2.3. Oxygen-isotope stages and East European loess-soil stratigraphy   20
2.4. Paleolithic open-air site buried in loess-derived colluvium: Korman' IV in the Middle Dnestr Valley   21
2.5. Late Pleistocene loess and buried soil profiles for the East European Plain   29
2.6. Environments of the Last Interglacial climatic optimum   29
2.7. Environments of the Last Glacial Maximum   33
3.1. Map of Middle Pleistocene sites in Eastern Europe   43
3.2. Stratigraphy of Treugol'naya Cave   45
3.3. Stone artifacts recovered from Layer 4 at Treugol'naya Cave in the Northern Caucasus   46
3.4. Stratigraphy of Mikhailovskoe and Khryashchi localities on the Northern Donets River   49
3.5. Flake cores recovered from the lower buried soil at Mikhailovskoe on the Northern Donets River   50
4.1. Map of Late Pleistocene Mousterian sites in Eastern Europe   62
4.2. Topographic setting of Mousterian sites in the Middle Dnestr Valley   68

4.3. Geomorphic context of Mousterian sites in the Middle Dnestr Valley 70
4.4. Stratigraphic profile of Khotylevo I on the Middle Desna River 73
4.5. Stratigraphic profile of Betovo on the Middle Desna River 75
4.6. Topographic setting and excavations at Sukhaya Mechetka (Volgograd) on the Lower Volga River 76
4.7. Map of Mousterian sites in the Crimea 78
4.8. Stratigraphy of Zaskal'naya V 80
4.9. Map of Mousterian sites in the northwestern Caucasus 84
4.10. Stratigraphy of Mezmaiskaya Cave in the Northern Caucasus 86
4.11. Chronology of Mousterian sites in Eastern Europe 87
4.12. Crural index for modern human populations and mean annual temperature showing crural indices for Neanderthals 92
4.13. Mousterian cores, including discoidal, Levallois, and proto-prismatic, from Khotylevo I on the Middle Desna River 99
4.14. Mousterian unifacial tools, including side-scrapers, points, knife, and denticulate 102
4.15. Mousterian bifacial tools from Sukhaya Mechetka 105
4.16. Arrangement of mammoth bones and associated occupation debris from Mousterian level at the open-air site of Ripiceni-Izvor in the Prut Valley 108
4.17. Age mortality profile for steppe bison based on molar crown-height measurements from Il'skaya I in the Northern Caucasus 117
4.18. Occupation floor mapped from Horizon 6 at Rozhok I on the coast of the Sea of Azov 118
4.19. Early and Middle Pleniglacial sites in the northwestern Caucasus: variations in elevation, artifacts, and large mammal remains 124
4.20. Occupation floor mapped from the western excavations at Sukhaya Mechetka (Volgograd) on the Lower Volga River 130
4.21. Effective temperature and percentage of meat and fish in diet among recent hunter-gatherers, showing projected values for Neanderthals 134
4.22. Effective temperature and technological complexity among selected recent hunter-gatherers, showing projected values for Neanderthals 134
4.23. Contribution of hunting to diet and size of territory among recent hunter-gatherers, showing projected values for Neanderthals 134
5.1. Map of early Upper Paleolithic sites in Eastern Europe 144
5.2. Stratigraphy of Molodova V in the Middle Dnestr Valley 145
5.3. Geomorphic setting of the Kulichivka site in the northern Volyn-Podolian Upland 147

5.4. Geomorphic setting and stratigraphy of the Kostenki sites on the Middle Don River   149
5.5. Topographic setting of the Kostenki sites and ravine systems on the Middle Don River   150
5.6. Geomorphic setting and stratigraphy of the Sungir' site on a tributary of the Oka River in northern Russia   151
5.7. Chronology of early Upper Paleolithic sites in Eastern Europe   154
5.8. Brachial index for modern human populations and mean annual temperature showing brachial indices for Upper Paleolithic specimens   157
5.9. Early Upper Paleolithic artifacts from Layer 10 at Molodova V in the Middle Dnestr Valley   165
5.10. Early Upper Paleolithic artifacts from Layer 3 at the rockshelter of Brynzeny I (tributary of the Prut River) assigned to the "Brynzeny Culture"   166
5.11. Artifacts of the early Upper Paleolithic "Spitsyn Culture" from Layer 2 at Kostenki XVII on the Middle Don River   170
5.12. Artifacts of the early Upper Paleolithic "Strelets Culture" from Layer 5 at Kostenki I, Kostenki VI, and Layer 3 at Kostenki XII on the Middle Don River   170
5.13. Artifacts of the early Upper Paleolithic "Gorodtsov Culture" from Layer 2 at Kostenki XIV (Markina Gora) on the Middle Don River   172
5.14. Ornaments from Layer 2 at the site of Kostenki XVII on the Middle Don River   175
5.15. Decorated bone objects from Layer 2 at Kostenki XIV (Markina Gora) on the Middle Don River, assigned to the "Gorodtsov Culture"   176
5.16. Decorated ornament and possible art object from Layer 3 at the rockshelter of Brynzeny I on a tributary of the Prut River   177
5.17. Figurative art from the site of Sungir' located on a tributary of the Oka River in northern Russia   178
5.18. Early Upper Paleolithic occupation floor of Layer 9 at Molodova V in the Middle Dnestr Valley   186
6.1. Map of late Upper Paleolithic sites in Eastern Europe   196
6.2. Stratigraphic profile of Ataki I in the Middle Dnestr Valley, where occupation layers are buried in loess-derived colluvium deposited during the Last Glacial Maximum   199
6.3. Radiocarbon dates from late Upper Paleolithic sites in the central East European Plain   201
6.4. Generalized stratigraphic context for the Middle Desna River   203
6.5. Stratigraphic profile for Eliseevichi I on the Sudost' River   205

6.6. Stratigraphic profile of Avdeevo on the Seim River in the Central Russian Upland, illustrating the effects of frost disturbance and rodent burrowing   208
6.7. Depositional context of the bison bone bed at Amvrosievka, which is located on the southern margin of the Donets Ridge   213
6.8. Chronology of late Upper Paleolithic sites in Eastern Europe   216
6.9. Kostenki shouldered points and Kostenki truncation from Layer 1 at Kostenki I (Eastern Gravettian) on the Middle Don River   223
6.10. Nonstone implements from Kostenki I and Gagarino (Eastern Gravettian) on the Middle Don River   224
6.11. Mattocks of mammoth ivory from Layer 1 at Kostenki I (Eastern Gravettian) on the Middle Don River   227
6.12. Backed bladelets from Mezin (Epigravettian) on the Desna River   231
6.13. Mammoth-bone structure at Mezhirich in the Dnepr Basin   232
6.14. Figurative art of the Eastern Gravettian: "Venus" figurine of mammoth ivory from Layer 1 at Kostenki I on the Middle Don River   234
6.15. Mammoth bones decorated with geometric designs from Mezin on the Desna River   236
6.16. Arrangement of pits and hearths of the first feature complex ("longhouse") in Layer 1 at Kostenki I on the Middle Don River   244
6.17. Arrangement of mammoth bone structures and associated hearths and debris scatters on the occupation floor of Dobranichevka on the Supoi River in the Dnepr Basin   247

Tables

2.1. Late Pleistocene and Modern Climates in Eastern Europe   23
4.1. Neanderthal Skeletal Remains from Eastern Europe   90
4.2. Stone Tools in Mousterian Sites of the East European Plain: Southwest Region   94
4.3. Stone Tools in Mousterian Sites of the East European Plain: Central and Southern Regions   95
4.4. Stone Tools in Mousterian Sites of the Crimea   96
4.5. Stone Tools in Mousterian Sites of the Northern Caucasus   97
4.6. Large Mammal Remains in Mousterian Sites of the East European Plain   114
4.7. Large Mammal Remains in Mousterian Sites of the Crimea   120
4.8. Large Mammal Remains in Mousterian Sites of the Northern Caucasus   122
5.1. Modern Human Skeletal Remains from Early Upper Paleolithic Sites in Eastern Europe   156
5.2. Artifacts in Early Upper Paleolithic Sites of the East European Plain: Southwest Region   164
5.3. Artifacts in Early Upper Paleolithic Sites of the East European Plain: Central and Northern Regions   168
5.4. Large Mammal Remains in Early Upper Paleolithic Sites of the East European Plain: Southwest Region   180
5.5. Large Mammal Remains in Early Upper Paleolithic Sites of the East European Plain: Central and Northern Regions   182
6.1. Skeletal Remains in Late Upper Paleolithic Sites in Eastern Europe   218
6.2. Artifacts in Late Upper Paleolithic Sites of the East European Plain: Eastern Gravettian   222
6.3. Artifacts in Late Upper Paleolithic Sites of the East European Plain: Epigravettian   230
6.4. Large Mammal Remains in Late Upper Paleolithic Sites of the East European Plain: Southwest Region   238
6.5. Large Mammal Remains in Late Upper Paleolithic Sites of the East European Plain: Central Region   240

Foreword

Human beings originated in equatorial Africa, and their physiology remains ill suited to cooler climes. Today, their geographic range would probably resemble the chimpanzee's, if it weren't for one crucial human trait—the ability to extend anatomy with culture. It is sometimes said that chimpanzees also have culture, but if so, it involves only a handful of inventions that could probably be lost with minor impact on chimpanzee existence. In contrast, human culture involves the accumulation of inventions that build vitally on earlier ones, and their loss would produce a dramatic reduction in human numbers, if not extinction. In short, unlike chimpanzee culture, human culture evolves, it is vital to human survival, and cultural evolution explains how human beings can inhabit environments that differ so dramatically from the environment of origin.

Early on in human evolution, people developed the cultural means to expand their range within Africa, and sometime between 2 million and 1.5 million years ago, they succeeded in reaching its northern and southern bounds. From its northeast corner, in what is today Egypt, they dispersed eastward to southern Asia, where environmental conditions often resembled those in Africa. Their expansion farther north in Eurasia occurred later, and it was perhaps only 500,000 years ago that they gained a permanent foothold in Europe. Even then, they were apparently confined to the warmer, more maritime western and southern provinces, and during episodic glacial periods, they repeatedly retreated yet farther west and south. It was only after 300,000 years ago that they developed the cultural capacity to remain more widely in place when the glaciers advanced, and it is only then that we see the first tentative colonization of the eastern, most continental part of Europe. In modern political terms, this is the part that was long governed by the Soviet Union and that still lies within the sphere of Russian political and cultural dominance. Anyone who has visited Moscow or Saint Petersburg in January will testify that by ordinary human standards, Eastern Europe presents a forbidding environment to humans.

Desolate Landscapes chronicles the early prehistory of this harsh region, beginning with initial human penetration perhaps 300,000 years ago and ending with widespread habitation at the end of the last glaciation roughly 12,000 years ago. The title reflects the exceptionally adverse conditions that Ice-Age occupants had to overcome. The author, John Hoffecker, is an American archaeologist who has specialized in early East European prehistory for nearly two decades and who has published seminal journal articles on the subject. He bases *Desolate Landscapes* partly on his comprehensive reading and partly on repeated trips to Russia, during which he conferred with Russian archaeologists about their findings, he studied important artifact and animal bone collections, and he participated in some key excavations.

The primary literature that the author synthesizes is mainly in Russian, but there is no Russian source that treats the same material so thoughtfully and thoroughly. Equally important, Hoffecker is not interested in East European prehistory for its own sake, but for what it can tell us about global issues in human evolution. He focuses particularly on the fate of the Neanderthals, a distinctive group of humans whose evolutionary roots lie in more archaic European populations dating to at least 300,000 years ago. By 80,000 years ago, the Neanderthals had extended their range to western Asia, and they prospered in both Europe and western Asia until between 50,000 and 40,000 years ago. Then, over a period of 10,000 years or less, they were replaced by anatomically modern humans. Human fossils and genetics indicate that these modern people originated in Africa, and Hoffecker details archaeological discoveries that show that they exploited East European environments far more effectively than their Neanderthal predecessors. This helps to explain why they replaced the Neanderthals so quickly and completely. The archaeology the author summarizes indicates that early modern East Europeans still suffered when the last glaciation reached its peak 20,000–18,000 years ago, but they bounced back quickly, and the cultural adaptations to cold that they developed anticipate those that allowed modern humans to become the first occupants of northernmost Asia and to expand from there to the Americas. By 10,000 years ago, modern humans had come to live almost everywhere that people live today, and *Desolate Landscapes* illustrates the central role that culture played in this dispersion.

Most English-reading archaeologists know that East European prehistory can illuminate cultural evolution over tens of thousands of years and that East European archaeology is crucial to understanding the fate of the Neanderthals and to documenting the history of cultural responses to extreme cold. However, until now, English speakers had to rely mainly on brief partial summaries in technical journals. Now, for the first time, they have a readily available, highly readable, authoritative, and comprehensive source to inform both themselves and their students.

Richard G. Klein
Stanford University
December 2000

Preface and Acknowledgments

This book is an attempt to summarize the Ice-Age settlement of Eastern Europe within the broader context of human evolution. Because the occupation of Eastern Europe took place relatively late in prehistory, that broader context lies primarily in the more recent phases of human evolution—especially the transition from the Neanderthals to our own immediate ancestors. However, it is the lateness of human settlement in this part of northern Eurasia that makes it such an interesting place in which to view the past.

Isolated from the milder oceanic climates of the North Atlantic, most of the lands that lie east of the Carpathian Mountains are colder and drier than those of Western Europe. And during the Ice Age, they were often colder and drier than they are today. The harsh climates and their effects on local plant and animal life seem to have been a formidable challenge to the genus *Homo*. As a consequence, Eastern Europe was settled late and remained a marginal zone of human occupation, which continued to fluctuate throughout most of the Ice Age. The story of the formidable people who met this challenge tells us much about the evolution of human adaptations during the last quarter of a million years. Above all, it shows us—as clearly as any place on earth—how the capacity for creating a mental, material, and social structure of symbols underlies the triumph of modern humans.

Because the Ice-Age settlement of Eastern Europe was so thoroughly shaped by cold and arid climates, the evolutionary ecology of humans in northern environments is a fundamental frame of reference for its study. Humans are tropical creatures, and cannot set foot in higher latitudes without confronting a number of environmental realities. The pattern of recent hunter-gatherer adaptations to cold and dry landscapes provides a baseline for interpreting the Ice-Age record of Eastern Europe.

My fascination with the Ice-Age settlement of Eastern Europe began with the writings of Richard G. Klein on this subject between 1965 and 1974. Several years later, I had the good fortune to complete my graduate studies under his guidance at the University of Chicago, and I have continued to benefit from that guidance to the present day. At the time of his departure from Chicago in the early 1990s, Professor Klein kindly bequeathed to me his extensive collection of books, papers, and notes on the Russian Paleolithic, which were invaluable to writing this book. During my years at the University of Chicago, I also benefited considerably from the advice and encouragement of Karl W. Butzer, for which I will always be grateful.

My northern perspective on Eastern Europe is derived in no small measure from my studies and travels in Alaska, and I am indebted to many friends and colleagues there. I am especially thankful to W. Roger Powers, G. Richard Scott, and R. Dale Guthrie of the University of Alaska in Fairbanks, all of whom taught me a great deal about Ice-Age peoples and environments.

I am also immensely grateful to many colleagues in Eastern Europe who have made this book possible. I would particularly like to thank those with whom I have worked in the Northern Caucasus, including G. F. Baryshnikov (Zoological Institute, Russian Academy of Sciences), L. V. Golovanova and V. B. Doronichev (Laboratory of Prehistory, Saint Petersburg), and G. M. Levkovskaya (Institute of Material Culture History, Russian Academy of Sciences). Two other specialists in the Caucasus region who have been generous with their hospitality and knowledge are V. P. Lyubin and E. V. Belyaeva (Institute of Material Culture History, Russian Academy of Sciences).

Many other members of the Institute of Material Culture History in Saint Petersburg have shared their collections and thoughts with me, and I am especially grateful to M. V. Anikovich, N. K. Anisyutkin, N. D. Praslov, A. A. Sinitsyn, and L. M. Tarasov. Much of this book is based on the results of their excellent research on the East European Plain.

My own research in Eastern Europe has been funded from a variety of sources, but I would especially like to thank the L.S.B. Leakey Foundation, which has consistently supported my field and laboratory studies of Paleolithic sites and collections in Russia. I am also grateful for financial support from the National Academy of Sciences, National Geographic Society, and the International Research and Exchanges Board.

Drafts of one or more chapters were reviewed by V. B. Doronichev, L. V. Golovanova, R. G. Klein, and A. A. Sinitsyn, and I thank them for their helpful comments.

Finally, my thanks to Helen Hsu, Suzanne Kellam, and Adaya Henis at Rutgers University Press, and to my wife, Lilian, who prepared the illustrations.

DESOLATE LANDSCAPES

CHAPTER 1

Human Evolutionary Ecology and Eastern Europe

Spoken language . . . was, in fact, the supreme human technology.
—Robert McC. Adams (1996)

The Great Leap: An East European Perspective

The most important event in human evolution occurred not two or three million years in the past, but only about 50,000 years ago. At this time, hominids—anatomically modern humans—rather suddenly began to leave traces of the use of symbols in the archaeological record. Many anthropologists believe that this marks the birth of fully modern human language, for which there is some supporting anatomical evidence. In any case, the appearance of symbols coincides with a transformation in technological skills—a quantum jump in human ability to manipulate the natural environment. It also seems linked to fundamental changes in social behavior, although this is less obvious in the archaeological record of the time.

The consequences of the transition to "behaviorally modern humans" were only fully apparent with the end of the Pleistocene (roughly 10,000 years ago). From that point onward, many human societies began to capture increasing amounts of energy through plant and animal domestication and other new technologies. These societies increased in size and organizational complexity. The latter generated new technological advances—some of which entailed new uses of symbols—further increasing their size and complexity. Within ten millennia, technological change has accelerated to the point at which major innovations and their effects occur within a single generation, and organization has become increasingly global. This is the legacy of the transition that took place 50,000 years ago.

One of the most striking developments associated with the transition is the more or less simultaneous dispersal of modern humans throughout most of the Old World. Only the extreme conditions of the Bering Land Bridge temporarily barred their invasion of the Americas. Although living humans are clearly derived from one or more populations of *Homo sapiens* that evolved in Africa during the late Middle Pleistocene, the process of their dispersal remains a controversial issue. Some paleoanthropologists perceive genetic continuity between local archaic and modern human groups in various regions, and postulate a complex process of gene flow from Africa to other areas (e.g., Wolpoff 1999). Many researchers now believe that archaic *Homo* groups outside Africa were effectively replaced by behaviorally modern humans (e.g., Stringer and Andrews 1988; Wilson and Cann 1992; Klein 1999).

Without question, the most impressive aspect of the dispersal of modern humans was their rapid colonization of the cold regions of northern Eurasia. Having evolved in tropical and subtropical environments, modern humans were adapted to warm climates and productive habitats. Their morphology was particularly unsuited to temperatures at latitudes above 40 degrees North during the Last Glacial. Moreover, the archaic humans who occupied these latitudes were the Neanderthals—the only true northern representatives of *Homo*—specially adapted to cold climates and environments poor in available plant foods (Stringer and Gamble 1993). And as modern humans dispersed northward, they were squeezed into an increasingly narrow hominid niche that probably left them little alternative to competing for the same large mammal resources.

Despite their handicaps, modern humans successfully replaced the Neanderthals within ten millennia, and even colonized new areas of northern Eurasia that their predecessors had been unable to occupy. In fact, the case for outright replacement of an archaic population—with minimal genic exchange at best—is stronger in the Neanderthal region than other parts of Eurasia (Klein 1999, 477–491). It has recently been reinforced with analyses of fossil Neanderthal DNA, which indicate wide divergence between the two lineages (Krings et al. 1997; Ovchinnikov et al. 2000).

The triumph of modern humans in northern Eurasia was undoubtedly a consequence of the behavioral transformation that accompanied the use of symbols. Modern humans quickly developed an array of novel technological solutions to the challenges of northern habitats that more than compensated for their warm-climate morphology. Their technology reflected a fundamental advance in ability to manipulate the environment (e.g., Mithen 1994). They probably also used symbols to create new forms of organization—like the networks of modern hunter-gatherer peoples—that gave them the flexibility to exploit widely dispersed resources in marginal environments (e.g., Gamble 1986; Whallon 1989).

The combined advantages of technological innovation and organizational flexibility linked to the use of symbols were perhaps critical to modern human dispersal throughout the northern and southern hemispheres. But the importance of these new forms of behavior in the dispersal process—including an almost certain contest with the existing population of archaic humans—is most clearly evident in northern Eurasia (i.e., above 40 degrees North). And within the latter, the contrast between archaic and modern humans is sharpest in the cold and dry landscapes of Eastern Europe and Siberia (Hoffecker 1999a).

In northern Eurasia, the climate gradient runs from west to east, as well as from south to north. Isolated from the moderating influence of the North Atlantic, climates become increasingly continental east of the Alps. Although mean annual temperature is not affected, winter temperatures at midlatitudes fall dramatically on the East European Plain. Low rainfall on the southern plain creates an open steppe that is unknown in Western Europe. The east-west climatic gradient was present throughout the Pleistocene, and it is significant that the cold-adapted Neanderthals—who were the first humans to settle widely across Eastern Europe—came from the west and not the south (Hublin 1998).

In the early 1930s, archaeologists working in Eastern Europe described the transition from the Neanderthals to modern humans as the "great leap" (*bol'shoi skachok*) (e.g., Boriskovskii 1932). Like that of archaeologists and historians in other times and places, their perspective on the past reflected the current issues of their society, which was experiencing its own wrenching transformation under Soviet authority, and one that also involved the power of symbols and technology. But the young revolutionary archaeologists of this period were inspired by the very real contrast between the paleoanthropological record of modern humans and that of their predecessors. In Eastern Europe, these contrasting records of settlement reflect the formidable challenges that northern continental environments posed to humans. This is the basis of the East European perspective on the transition to modern humans.

Human Evolutionary Ecology in Northern Environments

Principles of Evolutionary Ecology

Like other aspects of hominid evolution, the colonization of northern latitudes and transition to modern humans are best understood within the theoretical framework of evolutionary ecology (e.g., Turner 1984; Foley 1987; Gamble 1994). Evolutionary-ecology concepts are employed throughout this book to explain the paleoanthropological record of Eastern Europe, and some of them are discussed below with specific reference to archaic and modern humans.

Adaptation is the core concept of evolutionary ecology, and it is central to explaining the changes that occur in the paleoanthropological record as hominids colonized higher latitudes during the Middle and Late Pleistocene. Adaptation has been defined simply as "the condition of showing fitness for a particular environment" (Mayr 1970, 413). Although it has morphological and physiological components, as well as a behavioral component, archaeologists have often ignored all but the last of these—particularly in the context of the Late Pleistocene. However, both morphology and physiology have always played a role in hominid adaptations (e.g., Coon 1962), and the archaeological record cannot be properly explained without reference to the morphology (and inferred physiology) of the people who produced it.

Although defined in simple terms, adaptation is a very complex and dynamic phenomenon in the natural world (Pianka 1978). To begin with, organisms are adapted not to their environment as a whole, but to a small spatial/temporal piece of it. While simple life forms during the early history of the earth were primarily adapted to abiotic aspects of their environment, the growing diversity and complexity of the biotic world increasingly compelled organisms to adjust to each other, creating a vast web of interrelated adaptations. As applied ecologists know, a small change in one aspect of an environment may generate a chain reaction among many plant and animal species. Furthermore, there is much short-term and long-term instability in the abiotic component of most environments (and this instability was especially pronounced in northern latitudes during the Pleistocene [Frenzel 1968]).

In recent years, "something of an anti-adaptationist backlash" has arisen in biology and anthropology (Eldredge 1985, 141). The causes of this backlash are understandable, and lie in the wretched excess of past applications of the adaptation concept. Organisms have often been subdivided into a collection of traits, and adaptive explanations concocted for each trait. The literature abounds with such "just-so stories," which ignore the possibility that organisms reflect not only selection for adaptive characters, but other effects (e.g., drift, pleiotropy), and that selection can favor nonadaptive characters (Gould and Lewontin 1979). Adaptation should be invoked as an explanation "only where it is necessary" (i.e., when all other plausible explanations have been exhausted) (Williams 1966, 4). Nonadaptive characters are almost certainly present among both archaic and modern humans in northern Eurasia (Howell 1957; Hublin 1998; Holliday 1999).

Another central concept in evolutionary ecology is the *niche*, which has been defined as the "profession" of an organism—as opposed to its habitat or "address" (Odum 1975, 44). A niche may comprise an intricate four-dimensional component of the environment. In broad terms, it is useful to distinguish between the narrow niche of a "specialist" and the wide niche of a "generalist" (Pianka 1978, 253–256). The niche is closely linked with models of competition between species, and the principle of *competitive exclusion* (i.e., that no two species can occupy the same niche) is an important element of ecological theory (MacArthur 1972, 21–58).

Although evidence of competition between closely related species in natural environments is rare, this may be the consequence of past competition that led to niche separation, and/or limiting factors other than competition that restrict populations to below a competitive threshold (Krebs 1978, 207–238). The concept of the niche and its apparent exclusivity are essential to understanding the relationship between Neanderthals and modern humans, and the transition from the former to the latter (e.g., Hoffecker and Cleghorn 2000).

The complex nature of adaptation and the niche is underscored by the need for an organism to allocate time, matter, and energy to a variety of potentially conflicting demands (e.g., foraging versus reproduction). The morphology, physiology, and behavior of an organism should reflect optimal solutions to these demands (Pianka 1978, 257–260). Although the use of *optimality models* has been criticized as part of the "anti-adaptationist backlash" (Gould and Lewontin 1979), the problem again lies in the application and not the concept. If used properly (i.e., without the assumption that all traits necessarily represent part of an adaptive solution), optimality models help to avoid the tendency to consider traits in isolation (Stephens and Krebs 1986, 206–215). They are predicated on the assumption that an organism is an integrated whole comprising mutually constraining parts.

Optimality modeling has been applied with particular vigor to *foraging theory*, which assumes that selection will favor an efficient foraging strategy that maximizes its energy and material benefits over its costs (MacArthur 1972, 59–69). Foraging costs may include the time and energy expended searching for food items, as well as the effort (and possibly risk) of capturing a food item. A fundamental distinction is made between the exploitation of patches (or "clumps of food") and prey (Stephens and Krebs 1986, 13–37). Optimal foraging models provide insights to recent hunter-gatherer behavior (e.g., Winterhalder and Smith 1981), and are helpful in understanding the paleoanthropological record. Because humans often hunt and gather in groups, balancing their foraging choices with social and reproductive benefits/costs is especially important.

Social behavior in an evolutionary-ecological context (or *socioecology*) presents a fundamental theoretical dilemma, because it entails selection for behavior that contributes to the fitness of another individual. As Darwin (1859, 236) observed, the presence of sterile casts among the social insects creates a "special difficulty" for the natural selection model. The problem was solved by recognizing selection at the level of the family ("kin selection") (Hamilton 1964; Williams 1966, 193–203). Selection should favor cooperative (or "altruistic") behavior among individuals who share a high degree of genetic relatedness (just as it favors parental care that contributes to the long-term reproductive success of the parent). Kin selection helps explain social behavior among the social invertebrates, as well as among some vertebrates (Wilson 1975).

However, the kin selection model cannot account for cooperative interactions in many higher vertebrates where these interactions occur among individuals that are not closely related. There are numerous examples of such behavior, which include the cooperative rearing of offspring between mating pairs of birds and—less commonly—mammals. Mutually beneficial interactions among nonrelatives seem to have evolved through selection for the capacity to establish and maintain reciprocal social relationships (or "reciprocal altruism") (Williams 1966, 92–124; Trivers 1971). The capacity for such relationships would appear to be confined to organisms with well developed neocortical areas and the ability to store and manipulate large amounts of information (i.e., higher vertebrates). Because selection should favor individuals who maximize the benefits and minimize the costs of reciprocal relationships, highly complex and dynamic systems of alliances may develop among the latter (Trivers 1971; Maynard Smith 1982).

Human Ecology in Northern Environments

If evolutionary ecology provides a general theoretical framework for the study of human settlement in Eastern Europe during the Pleistocene, the ecology of northern environments—and especially of human hunter-gatherers in such environments—offers a more specific context. Northern environments place special demands on living organisms. The most obvious of these is temperature, which declines in the higher latitudes as a function of reduced solar energy per unit area of the earth's surface. Environments in higher latitudes also experience greater seasonal variations in climate, including increasingly severe winter temperatures (Pianka 1978, 19–34).

Cooler terrestrial environments tend to be drier, and the combination of reduced temperature and moisture lowers *primary productivity*, which is defined as the rate at which organic matter is produced by plant photosynthesis (Whittaker 1975, 193–200). While a tropical rain forest produces an average of 2200 grams per square meter each year, temperate deciduous forest and temperate grassland produce 1200 and 600 grams per square meter, respectively. Arctic and alpine tundra yield a yearly average of only 140 grams per square meter (Whittaker 1975, 224, table 5.2; Archibold 1995, 9, table 1.4). Cold and dry environments thus provide a low amount of energy (in the form of plant food) to their herbivore inhabitants (secondary productivity), which in turn support a reduced number of animals at higher energy or *trophic levels*, such as carnivores and humans.

Studies of recent hunter-gatherer peoples reveal that many aspects of their morphology, physiology, and behavior are correlated with latitude and temperature. Peoples who have lived in cold environments for at least a few thousand years have developed a variety of morphological adaptations to low temperature that conserve body heat and protect from the effects of cold (e.g., frostbite). Most notably, these include larger body mass and shortened distal limb segments, which reduce the ratio of exposed surface area to volume and conform to the predictions of Bergmann's and Allen's biogeographic rules, respectively. The latter may be illustrated by the relationship between mean annual temperature and the brachial index (ratio of radius to humerus length) (fig. 1.1). Northern peoples also exhibit some physiological responses to low temperature, such as increased blood flow to extremities (Harrison et al. 1977, 428–439; Trinkaus 1981, 208–213).

Hunter-gatherer diet is strongly influenced by latitude and temperature. To begin with, energy demands increase significantly in cold climates and caloric intake in arctic environments may be as much as 30 percent higher than it is in tropical regions (Harrison et al. 1977, 403, table 26.2). The percentage of meat and fish in the diet of recent hunter-gatherers increases as temperature, moisture, and primary productivity decline

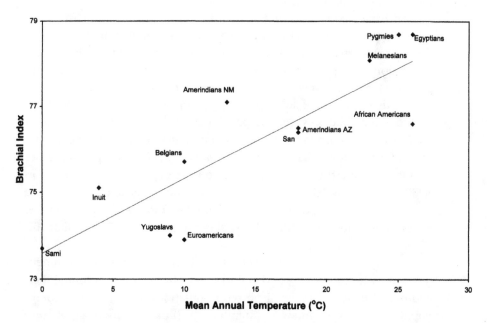

Figure 1.1. Mean annual temperature and brachial index in selected modern populations, illustrating the trend toward shorter distal limb segments in colder environments (based on data in Trinkaus 1981, 211, table 7).

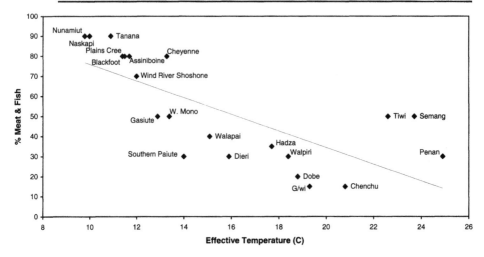

Figure 1.2. Effective temperature and the percentage of meat and fish in the diet among recent hunter-gatherers (based on data in Kelly 1995, 67–69, table 3.1).

(fig. 1.2), and equals or exceeds 80 percent among most peoples who live in areas with an effective temperature of 10 degrees C or less (Kelly 1995, 66–73). Defined by Bailey (1960), *effective temperature* reflects the duration and intensity of the growing season, providing an ecologically meaningful measure of warmth (Binford 1980, 13–14).

The high protein-fat diet and hunting and fishing subsistence of hunter-gatherers in northern environments has major implications for foraging strategy. Although cold maritime settings often provide rich concentrations of aquatic resources that require limited mobility, hunter-gatherers in northern continental environments who subsist on terrestrial mammals must forage across large areas in order to secure highly dispersed and mobile prey. Among peoples who rely primarily on nonaquatic foods, there is a correlation between temperature and the average distance of residential moves (fig. 1.3), and a related correlation between the percentage of hunted food in the diet and territory size (fig. 1.4) (Binford 1980, 1990; Kelly 1995, 65–160). Another consequence of low temperatures and a high meat diet is that males procure most or all food resources, generating a more pronounced sexual division of labor (Kelly 1995, 262–270).

The high mobility requirements of northern continental environments not only incur added time and energy costs, but also carry potential social and reproductive costs for dispersed populations in such environments. Populations must maintain a minimum threshold density in order to remain viable and avoid extinction, and it is estimated that the "minimum

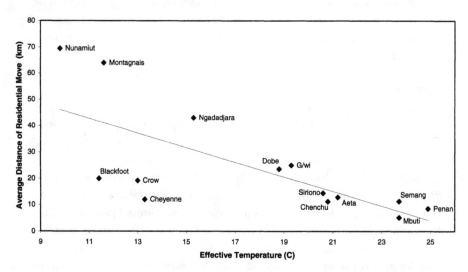

Figure 1.3. Effective temperature and average distance of residential moves among recent hunter-gatherers with low dependence (≤20 percent) on aquatic foods (modified after Kelly 1995, 128, figure 4.7).

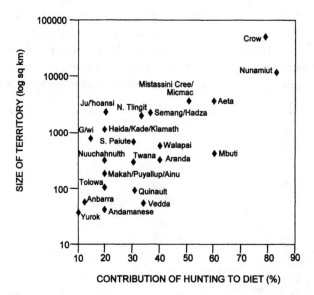

Figure 1.4. Contribution of hunting to diet and territory size (log square km) among recent hunter-gatherers (redrawn from Kelly 1995, 131, figure 4.8).

equilibrium size" for a *mating network* of modern hunter-gatherers is between 175 and 475 individuals (Wobst 1974, 163). The degree of dispersal of these individuals across the landscape (typically grouped into bands containing roughly 25 individuals) cannot exceed their ability to sustain a social network through at least periodic contact and aggregation (Kelly 1995, 209–221).

The technology of recent hunter-gatherers is also influenced by temperature and diet. Both the diversity of tool types and the complexity of individual tools and weapons (measured in terms of "technounits" [Oswalt 1976, 38]) increase as effective temperature and the percentage of plant foods in the diet decline (Oswalt 1976, 181–195; 1987; Torrence 1983, 17–20) (fig. 1.5). This apparently reflects the need for greater foraging efficiency in habitats where resources are available for limited periods of time. Recent hunter-gatherers in cold environments also tend to make increased use of storage technologies and untended facilities (e.g., traps and snares). The former represent another adaptive response to seasonal variations in resource availability, while the latter reflect an efficient approach (i.e., reduced mobility) to collecting unpredictable and widely dispersed resources (Binford 1980, 1990; Torrence 1983). Finally, modern hunter-gatherers in northern environments produce relatively complex technology for heat conservation and cold protection (e.g., tailored fur clothing).

The pattern of recent hunter-gatherer adaptations to cold environments (summarized in fig. 1.6) provides a frame of reference for the study of the human colonization of northern latitudes, and more specifically for

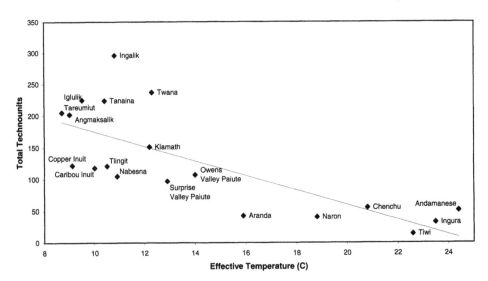

Figure 1.5. Effective temperature and technological complexity (measured in total "technounits" for food-getting technology) among selected recent hunter-gatherers (based on data in Oswalt 1976, 173, table 9.1).

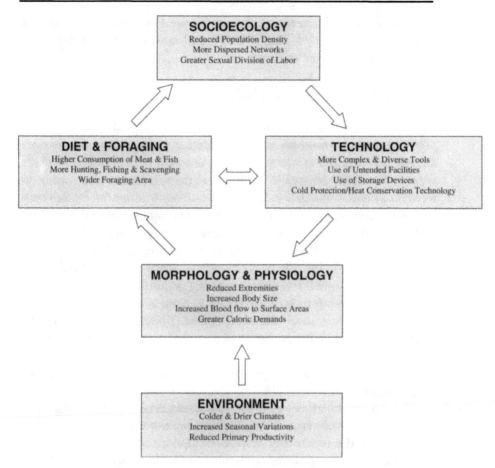

Figure 1.6. Patterns of adaptation to northern environments among recent hunter-gatherers.

the Pleistocene settlement of Eastern Europe. Virtually all of the aspects of hunter-gatherer adaptation discussed earlier can be measured to some degree in the paleoanthropological record. Morphological (and inferred physiological) adaptations to climate are evident among the skeletal remains of both archaic and modern humans in northern Eurasia (e.g., Coon 1962; Trinkaus 1981). Diet may be determined directly from stable isotope analyses of human bone (e.g., Richards et al. 2000), or deduced in part from the study of faunal remains in archaeological sites. Mobility and territory size are more difficult to measure, but seem to be reflected to some extent in the long-distance movement of raw materials from their sources (e.g., Roebroeks et al. 1988), while aggregation and dispersal of groups can sometimes be recognized from the size and structure of occupation areas (e.g., Schledermann 1978). Technological complexity can also be assessed from the archaeological record, although much of past and recent hunter-gatherer technology is based on perishable materials (e.g., plant fibers, wood, animal hides) and must be inferred indirectly from various sources of evidence.

Human Ecology and Symbols

The appearance of symbols in the paleoanthropological record roughly 50,000 years ago coincides with evidence for major behavioral changes among anatomically modern humans. Although by themselves symbols seem to have little ecological significance, their apparent relationship to technology and organization has enormous implications for human evolutionary ecology. Symbols are a central concern of social anthropology, and most anthropologists—perhaps understandably—display little interest in efforts to explain modern human social and economic life that do not adequately address the role of symbols (e.g., Sahlins 1976). The appearance of symbols—and the widely inferred acquisition of symbolic language—among modern humans seems to have affected every aspect of their evolutionary ecology discussed earlier. Understanding the significance of symbols therefore becomes essential to explaining the "great leap" to modern humans.

Symbols confer a unique cognitive ability on modern humans to formulate, manipulate, and communicate abstract concepts pertaining to their environment. Parallels between the underlying structure of language and the organization of material culture have been noted (Deetz 1967, 83–101). Indeed, the technology of modern humans after 50,000 years ago (i.e., following the appearance of symbols) exhibits a fundamental difference from all preceding forms of technology. Even the simple tools and devices of the Upper Paleolithic—throwing sticks, small-mammal traps, rotary drills, and so on—reveal a novel ability to manipulate the natural environment in complex ways. The change has been characterized as a shift from "domain-specific" cognition to general intelligence (Mithen 1994, 1996).

The impact of symbol-based technology did not become obvious in the archaeological record until after the end of the Pleistocene. Within a few millennia, new types of technology appeared—including water wheels, metallurgy, monumental architecture, and others—that were without precedent in human history. Preservational bias and the traditional concern of Paleolithic archaeologists with stone tools has helped obscure the transition in technological ability that took place 50,000 years ago. Many elements of the new symbol-based technology of behaviorally modern humans must be inferred or deduced from the relatively impoverished archaeological record of the Pleistocene, and stone tools—which were the special ("domain-specific") forte of the Neanderthals and their archaic human contemporaries—are the least revealing measure of this transition.

Symbol-based technology completely altered human evolutionary ecology, allowing rapid adaptation to an almost unlimited range of environments. Modern humans dispersing into northern Eurasia designed new technologies for heat conservation and protection in cold climates that were probably superior to the evolved morphological adaptations of the Neanderthals. They also developed more complex tools and weapons to increase foraging efficiency in habitats where resource availability was often limited in time and space. Throughout the Old World, technological innovations permitted modern humans to expand their niche breadth far beyond that of their predecessors. Watercraft, nets, and other fishing devices were designed to exploit aquatic prey, and traps and snares were developed to harvest small mammals and birds. The expanding hominid niche eventually began to exclude other species, generating waves of extinctions that continue to the present day. But the new technological capabilities undoubtedly played a central role in the global dispersal of modern humans (and the corresponding disappearance of archaic humans).

Symbols also appear to facilitate unique forms of organization among modern humans. The alliance networks of modern hunter-gatherers are based on shared symbols—linguistic and material—that seem to provide the chief means of maintaining the reciprocal relationships among individuals outside their "blood relatives." The fundamentally manipulative nature of these relationships (as opposed to parent-child or sibling relationships) underlies the complex web of friendship and deception that constitutes modern human social and political life.

The impact of symbol-based organization also did not become obvious in the archaeological record until following the end of the Pleistocene. Within a few millennia, complex forms of organization developed in various parts of the world that—like the new technology that accompanied it—had no precedent in human history. But detecting evidence of an organizational shift that coincides with the appearance of symbols 50,000 years ago is even more difficult than documenting the change in technological abilities. The archaeological record prior to 50,000 years ago provides very little information on the organizational structure of groups, although there is some evidence for larger groups and wider networks after this time (Mellars 1996; Gamble 1999).

Nevertheless, symbol-based organizational structures also probably altered the evolutionary ecology of modern humans, and are believed by some anthropologists to have been instrumental in their global dispersal.

Assuming that the Neanderthals and other archaic humans lacked the ability to create large reciprocal alliance networks, they may have been unable to disperse over wide areas containing scattered and unpredictable resources like modern hunter-gatherers in marginal environments. The creation of tribal networks may have allowed modern humans to colonize very cold and dry environments that were beyond the range of the Neanderthals and their archaic contemporaries. Symbol-based organization may have been critical to the occupation of much of Eastern Europe, Siberia, and Australia after 50,000 years ago (Whallon 1989; Gamble 1994).

Also less apparent in the Pleistocene than during the subsequent postglacial period is the relationship between organization and technology. In the past few thousand years, human organizations have steadily expanded in size and complexity as a consequence of technological innovations—and with the aid of communications technology (e.g., writing, printing press, radio) (Boserup 1981; Pacey 1990). At the same time, these organizations provide the essential social context for technological innovation and application (White 1962; Adams 1996). The dynamic interaction between symbol-based organization and technology is surely the driving force of human history, and the phenomenon that set it on a unique trajectory 50,000 years ago. The rate of technological and organizational change continues to accelerate with no end in sight.

Eastern Europe provides the most suitable setting to address the evolutionary ecology of the transition to modern humans. Although the capacity for using symbols probably evolved elsewhere, Eastern Europe offers an unparalleled opportunity to examine the adaptive responses of archaic and modern humans to a highly marginal environment (Hoffecker 1999a). The contrast in their responses reveals much about the difference between Neanderthals and modern humans, and the importance of symbols in the transition from one to the other. While an equally revealing contrast may be found in Siberia, which also represented a marginal environment for archaic and modern humans in northern Eurasia, the limited settlement record of the Neanderthals provides little basis for comparison here (e.g., Goebel 2000). Eastern Europe contains a much richer record of archaic human occupation, and both the spatial-temporal distribution and contents of the sites provide insights into the differences between the evolutionary ecology of modern humans and their predecessors that are less evident in other parts of the world.

CHAPTER 2

Environmental Setting

Modern Landscape and Climate

Eastern Europe may be defined as that portion of the European continent that lies east of the Carpathian Mountains and the Polish Plain (roughly longitude 25 degrees East). It is thus bounded on the north by the Arctic Ocean, and on the south by both the Black and Caspian Seas and the Caucasus Mountains. It is bounded on the east by the Ural Mountains (longitude 60 degrees East), which mark the western limit of northern Asia. In terms of its rapidly changing political geography, Eastern Europe comprises the European portion of the former Soviet Union (i.e., the European part of the Russian Federation, Ukraine, Belarus, Moldova, western Kazakhstan, and the Baltic states of Estonia, Lithuania, and Latvia) (fig. 2.1).

The predominant geographic feature of Eastern Europe is a vast lowland plain that occupies most of its area and rarely exceeds elevations of a few hundred meters above sea level. The *East European Plain* represents a stable platform of Precambrian igneous and metamorphic rock that crops out only in the ice-scoured northwest and in the south-central Ukraine. The Precambrian basement complex is mantled with thick sedimentary rock that has filled in the troughs and basins to produce a relatively level surface. Landscapes in the north have been influenced by Pleistocene glaciation, which created massive moraines, marshes, and lakes (Lydolph 1977).

The most significant topographic relief on the southern half of the East European Plain is provided by several low plateaus that represent blocks of pre-Quaternary sediment uplifted by horizontal stresses produced by movement of the platform. The westernmost plateau is the *Volyn-Podolian Upland*, which adjoins the northeastern slope of the Carpathian Mountains, and averages elevations of 200–400 meters above sea level. Major streams have cut deeply into the younger sedimentary rock of the Volyn-Podolian Upland—most notably the Dnestr and Prut Rivers. In the center of the plain lies the broad *Central Russian Upland*, which averages only 200 meters above sea level. South of this plateau is the smaller *Donets Ridge* (averaging 200–250 meters in elevation), which is situated north of the Sea of Azov, and farther east is the *Volga Upland* (200–300 meters in elevation).

The plateaus are divided by the major river basins that empty into the seas on the southern margin of Eastern Europe. The *Dnepr Basin*, which

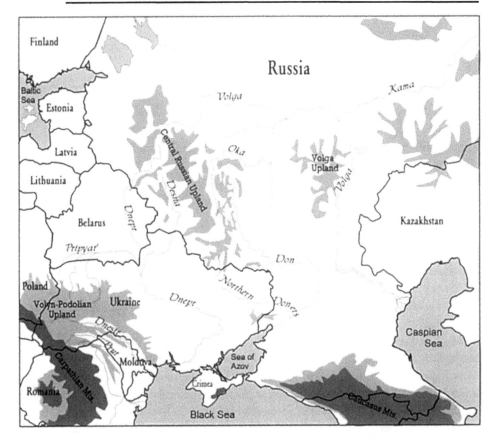

Figure 2.1. Map of Eastern Europe, illustrating modern political boundaries and major geographic features.

also drains the massive Pripyat' Marshes region north of the Volyn-Podolian Upland, lies between the latter and the Central Russian Upland. The *Oka-Don Lowland* separates the Central Russian and the Volga Uplands, and the Volga Basin lies between the Volga Upland and the western slope of the Ural Mountains. Along the rivers, side-valley streams have cut deep ravine systems into the terraces and upland margins.

Pleistocene glaciation has also influenced landscapes in the river basins on the southern half of the plain. At times, glaciers expanded as far south as roughly 50 degrees North in the Dnepr Basin and Oka-Don Lowland, leaving prominent moraines in the valleys. Large quantities of unconsolidated sediment generated by the glaciers were eventually washed into the southward-flowing river systems and subsequently eroded, creating terraces. As many as eight terrace levels have been identified in the canyonlike Dnestr Valley, while only several terraces are visible along the

Dnepr and Don Rivers. During the cold phases, winds deposited a thick blanket of loess across both uplands and basins. In the valleys, much of the loess was washed onto lower terrace slopes, where it often buried the debris of human occupations (Velichko 1961a ; Klein 1973, 18–32; Ivanova 1977).

On the southern margin of the East European Plain lie the Crimean and Caucasus mountain ranges, which are part of the same geosynclinal zone. The former—situated along the south coast of the *Crimean Peninsula*—rarely exceeds 1,500 meters in elevation. By contrast, the Caucasus Mountains contain the highest peak in Europe (Mount Elbrus) at 5,642 meters above sea level. Both ranges possess relatively gentle northern slopes comprising a series of ridges of eroded sedimentary rock. Northward-flowing streams have incised deep valleys and created numerous caves and rockshelters in these ridges (Lydolph 1977, 237–245; Ferring 1998).

Isolated from the moderating influence of the North Atlantic Ocean, Eastern Europe is subject to more continental climates than other parts of Europe. Winter temperatures are especially cold in the northern and central regions of the East European Plain, and precipitation is particularly low in the southern plain. In the central plain at 51 degrees North (Kursk), mean January temperature is –9 degrees C, compared with between 0 and 5 degrees C at similar latitudes in Western Europe; minimum January temperatures at Kursk reach –34 degrees C. In the southern plain at latitude 48 degrees North (Volgograd), mean January temperature is –8 degrees C and annual precipitation is only 373 millimeters, compared with –2 degrees C and 928 millimeters at the same latitude in Central Europe (Munich). Because of the warm continental summers in the southern plain, Volgograd and Munich actually experience the same mean annual temperature of 8 degrees C (Konyukova et al. 1971) (fig. 2.2).

The modern flora and fauna of Eastern Europe exhibit strong latitudinal zonation, which seems to have been less pronounced during cold periods of the Pleistocene. In the far north, a narrow tundra zone extends along the Arctic Ocean coast and is bounded on the south by a broad belt of boreal forest. The west-central plain is occupied by mixed forest composed of spruce and pine with oak and elm trees. Toward the eastern part of the central plain, much of the mixed forest zone is replaced with forest-steppe, where stands of oak, beech, and hornbeam alternate with herbaceous steppe flora. The southern plain is covered with open steppe (although the Donets Ridge supports a small island of forest-steppe), reflecting the arid climate of the region. Open steppe environments are not found in other parts of Europe today (van der Hammen et al. 1971; Lydolph 1977). Climates are warmer and less arid in the Crimean and Caucasus Mountains, and forest communities—including broadleaf

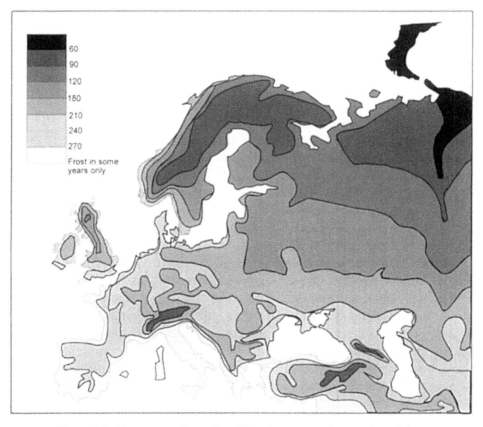

Figure 2.2. Temperature in northern Eurasia expressed as number of days per year with temperatures above 0 degrees C (based on data from the *New Oxford Atlas* 1978, 98).

deciduous forest—are found in altitudinally zoned bands. Zones of coniferous forest and alpine tundra lie at higher elevations in the Northern Caucasus.

Pleistocene Stratigraphic Framework and Paleolithic Sites in Eastern Europe

In a marked departure from the established traditions of geology, the stratigraphic framework of the Pleistocene epoch (1.8–0.01 million years ago) has been built largely on a record of climatic change (*climatostratigraphy*). Formerly based on a highly incomplete succession of glacial till and outwash deposits, the framework is now founded on the stable oxygen-isotope stratigraphy of deep-sea sediments. The latter provides a proxy record of global climate oscillation that reflects changes in terrestrial ice volume, and reveals a lengthy sequence of glacials and interglacials—equated with *oxygen-isotope stages*—over the past two million years. Chronological control is provided by paleomagnetism and radiometric dating (Bowen 1978; Lowe and Walker 1997).

A variety of terrestrial deposits also contain proxy climate records that can be potentially correlated with segments of the oxygen-isotope climatostratigraphy. These include deep pollen cores from lake sediments, ice cores from continental glaciers, and deposits of loess containing buried soils. They also include the often fragmentary record of glacial moraines, stream terraces, and travertine deposits. Correlation with oxygen-isotope stages is often facilitated by fossil fauna and flora in these deposits—which may have both biostratigraphic and paleoclimatic significance—as well as relative and absolute dating (Bradley 1985).

The longest and most complete proxy climate record for the Pleistocene in Eastern Europe is provided by *loess stratigraphy* (Udartsev 1980; Velichko 1990). As in parts of Central Europe and Northern Asia, thick blankets of loess—chiefly silt derived from wind-swept plains of glacial outwash—accumulated during cold periods on the East European Plain. In river valleys, much of this loess was washed onto lower slopes and terraces as loess-derived or *loessic colluvium* (Butzer 1971, 199; Mücher 1986). During warmer intervals, loess deposition slowed or ceased and soils formed on stable land surfaces. Forest soils usually developed during full interglacials, while chernozem and frost-gley soils formed during cooler interstadial episodes. All of the nine glacial-interglacial cycles (oxygen-isotope stages [OIS] 19 through 2) of the last 800,000 years appear to be represented in the Middle and Late Pleistocene loess stratigraphy of the East European Plain (Velichko 1990, 105, table 1) (fig. 2.3).

Most Paleolithic sites on the East European Plain are open-air localities buried in loess and loessic colluvium on river terraces in the major valleys (fig. 2.4). These sites often can be correlated with the loess stratigraphy, and indirectly with the oxygen-isotope framework for the Pleistocene. The most important stratigraphic markers in loess deposits are the buried soil horizons. Several sites in the southern plain contain artifacts associated with soils of the late Middle Pleistocene (OIS 13–6), and a number of occupations are related to interglacial and interstadial soils of the earlier Late Pleistocene (OIS 5). The scarcity of artifacts in the loess bed that overlies the latter suggests that much of the plain was abandoned during 73,000–55,000 years ago (OIS 4). Younger sites are found in interstadial soils and loess units of the final phases of the Pleistocene (OIS 3–2), and most of these fall within the range of radiocarbon dating (Velichko 1961a; Klein 1973; Hoffecker 1987; Praslov 1995; Sinitsyn et al. 1997).

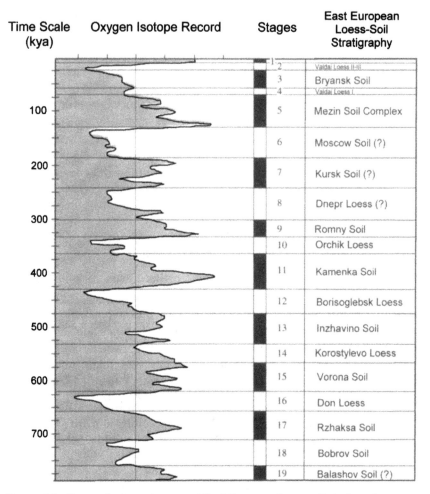

Figure 2.3. Oxygen-isotope stages and East European loess-soil stratigraphy.

A small number of Paleolithic localities on the East European Plain contain artifacts buried in stream deposits. On the basis of their relative stratigraphic position and paleobiotic contents, these deposits can also be correlated with stages in the oxygen-isotope record. In the Dnepr Basin, several sites are found in buried alluvium apparently deposited during the Last Interglacial (OIS 5e). At several other sites, occupation debris was at least partly buried in floodplain sediment during the Last Glacial (OIS 2) (Velichko 1961a, 1969).

Paleolithic sites on the northern slopes of the Crimean and Caucasus Mountains are typically found in caves and rockshelters. A few sites in the southwestern region of the East European Plain are also located in natural shelters. The deposits in these sites are sometimes more difficult to

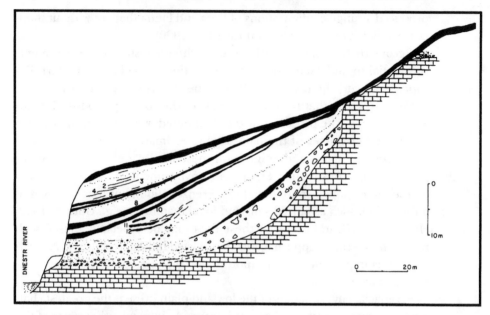

Figure 2.4. Paleolithic open-air site buried in loess-derived colluvium: Korman' IV in the Middle Dnestr Valley (redrawn from Ivanova 1977, 154, fig 12).

correlate with the global framework, although cold periods are often represented by horizons of rockfall. The latter have been used—along with fossil flora and fauna—for dating and cross-correlation of stratigraphic sequences in many rockshelters in the Crimea (Kolosov et al. 1993, 47, fig. 20), and to some extent in the Northern Caucasus (e.g., Nesmeyanov 1999, 312–314). The Northern Caucasus contains sites that date to the early and late Middle Pleistocene (OIS 15–6) and the full span of the Late Pleistocene (OIS 5–2), while occupations in the Crimea appear to postdate the Last Interglacial (i.e., OIS 5d–2).

Reconstructing Pleistocene Environments in Eastern Europe

Throughout the Pleistocene, climates at middle latitudes in the northern hemisphere oscillated between interglacial conditions similar to those of the present day and full glacial conditions similar to the modern Arctic. During interstadial periods, climates occupied an intermediate position between these extremes. The varying climatic conditions created a series of midlatitude environments across the northern continents that differ from those of the present. Interglacial climates were somewhat cooler or warmer than today, while glacial and interstadial conditions often produced strange configurations of flora and fauna that are without modern parallel (Frenzel 1968; Guthrie 1982, 1990).

Reconstructing the salient features of these Pleistocene environments is essential to understanding human evolutionary ecology in Eastern Europe. However, the task is a difficult one that is complicated from the outset by the fact that the environments of the colder periods lack modern analogues. The process is further hampered by the highly incomplete and biased character of the data used to estimate the distribution and abundance of plants and animals on these landscapes. Traditional approaches to reconstructing Pleistocene environments have relied heavily on the use of pollen-spore data (e.g., Frenzel 1968), but other sources of information, such as buried soils and fossil mammals (e.g., Guthrie 1982; Holliday 1987), indicate that these data may be especially biased with respect to the cold steppic habitats that existed in Eastern Europe. Accordingly, it is important to draw upon the broadest range of paleoenvironmental data available.

Periglacial environments of the lowland plain present the greatest challenge to reconstruction in Eastern Europe. As in many other parts of the northern hemisphere, these environments have been variously described as "tundra-steppe" or "periglacial steppe" because of their odd combination of animals that are presently confined to either the tundra or midlatitude grasslands (Hibbert 1982). On the East European Plain, they are most appropriately termed *loess-steppe*, because of the important role of loess deposition and weathering in their creation. Loess-steppe prevailed across most of the plain during full glacial periods, and comparable conditions seem to have existed during the cooler interstadials. Similar environments were present in the lowlands of the Carpathian Basin and in parts of Siberia and Beringia (Lozek 1967; Kukla 1975; Grichuk 1984; Guthrie 1990).

Loess was the product of glaciation. As the massive ice sheets advanced from Scandinavia over the northwest plain of Eastern Europe, they transformed tons of bedrock—including limestone—into unconsolidated till. With the help of glacial meltwater and intensified frost weathering, vast quantities of calcareous silt accumulated on exposed outwash surfaces near the ice. Fierce katabatic winds generated by the glaciers deposited the loess over large areas of the central and southern plain. In the extremely cold and arid climate of the periglacial zone, mineral salts were precipitated at the surface, while other forms of weathering were inhibited (Frenzel 1968, 645; Lozek 1968, 75–78; Kukla 1975).

The unique conditions of periglacial loess deposition and weathering—high solar insolation and evaporation (compared to arctic latitudes), low moisture, strong winds, and a mineral-salt substrate—supported a surpris-

Table 2.1. Late Pleistocene and Modern Climates in Eastern Europe

	Southwest Plain	Central Plain	Crimea	Northern Caucasus
Modern				
January mean temperature (°C)	−5	−9	−1	−1
July mean temperature (°C)	19	18	22	22
Effective temperature (°C)[a]	12.3	11.8	13.1	13.1
Annual precipitation (mm)	600	550	500	700
Last Interglacial (OIS 5e)				
January mean temperature (°C)	−1	−3	0	−2
July mean temperature (°C)	20	19	22	22 (?)
Effective temperature (°C)[a]	12.8	12.4	13.2	13.0
Annual precipitation (mm)	1000	700	1000	1000
Late Pleniglacial (OIS 2)				
January mean temperature (°C)	−21	−30	no data	no data
July mean temperature (°C)	14	16	no data	no data
Effective temperature (°C)[a]	10.7	10.9	no data	no data
Annual precipitation (mm)	350	200	no data	no data

Sources: Konyukova et al. 1971; Velichko 1984.
[a]Calculated according to Bailey 1960.

ingly rich and diverse herbaceous flora. Despite estimated winter temperatures of between −20 and −30 degrees C, warm continental summers probably raised effective temperature above 10.5 degrees C (see table 2.1). Primary productivity must have been significantly higher than that of modern arctic tundra environments, and closer to that of a temperate grassland (perhaps in the range of 500 grams per square meter per year [Leith 1975, 205, table 10.1]). Low moisture conditions may have helped to enrich and warm loessic soils (Guthrie 1990, 200–225).

Pollen profiles for the loess-steppe indicate a high percentage of modern steppe taxa such as *Artemisia*, combined with tundra plants (e.g., *Lycopodium*), and many halophytic species (Frenzel 1968; Grichuk 1984). However, despite the development of special techniques for collecting pollen-spore samples from loess (e.g., Grichuk 1984, 155), it is likely that grasses and other cold steppe flora are underrepresented in these profiles (Guthrie 1990, 229–238). While arboreal pollen is present, much of it was probably introduced from distant sources; some trees such as pine (which is well represented in samples from the East European Plain) produce large quantities of easily transported and preserved pollen (Butzer 1982, 177–181). Outside the Volyn-Podolian Upland, trees were scarce or absent in most areas of the plain during cold periods, and this is supported by the absence of wood fragments in the sediment. Perhaps most significant is the fact that humans living on the central and southern plain during the Last Glacial Maximum (OIS 2) relied almost exclusively on bone fuel (Klein 1973, 45–48).

Lacking buried soil profiles and reliable pollen data, most inferences about the flora of the loess-steppe are derived from the fauna (Guthrie 1982). It is the peculiar associations of animals from widely dispersed modern habitats that originally inspired the concept of the loess-steppe as an environment without modern analogue (Hibbert 1982, 153). Large mammals include steppe grazers such as mammoth (*Mammuthus primigenius*), bison (*Bison priscus*), horse (*Equus caballus*), and saiga (*Saiga tatarica*), as well as typical tundra inhabitants such as reindeer (*Rangifer tarandus*), musk ox (*Ovibos moschatus*), and polar fox (*Alopex lagopus*). In areas like the southwest plain, where some forest vegetation was extant, woodland taxa (e.g., roe deer [*Capreolus capreolus*]) are also present (Vereshchagin and Baryshnikov 1984). A similar pattern is evident among the small mammals, terrestrial molluscs, and insects (Kukla 1975, 106; Markova 1984; Elias 1994).

The extraordinary diversity of the vertebrate and invertebrate fauna suggests a fine-grained mosaic of plant communities. Evidence for the latitudinal zonation of biomes found today in the northern hemisphere is limited (Velichko 1973; Guthrie 1982, 315). During the slightly warmer interval at the end of OIS 3, a frost-gley soil developed across the East European Plain that reveals only minor latitudinal differences (Velichko and Morozova 1982a, 88–89). But at the time of the Last Glacial Maximum (OIS 2), it is apparent that the vegetation of the drier southern plain was less diverse than other areas, because steppe bison dominate the large mammal assemblages (Boriskovskii and Praslov 1964; Leonova 1994).

Secondary productivity and herbivore biomass were almost certainly higher on the periglacial loess-steppe than in modern environments with comparable climates. Grasslands tend to support a high vertebrate biomass relative to primary productivity (Whittaker 1975, 222–230), but loess-steppe conditions must have been especially favorable for large grazing herbivores (Guthrie 1990, 200–225). Low annual precipitation (much of which may have fallen during warmer months) and high winds probably limited snow cover, allowing easier movement and access to food during the winter. The loessic soils—accumulating constantly—supported a flora rich in mineral salts that would have eliminated seasonal mineral deficiencies.

The Productivity Paradox of the Loess-Steppe and Human Ecology

The periglacial loess-steppe was created by the imposition of an arctic climate on middle latitudes combined with the deposition of large quantities of wind-blown loess. Loess-steppe existed in various forms in many parts of the northern hemisphere during Pleistocene glacial periods, and has no analogue among modern environments. The latter is evident from the bizarre mixture of tundra and grassland animals that lived in this habitat (Hibbert 1982). On the East European Plain, loess-steppe may provide important insights to the differences between Neanderthals and modern humans. The former seem to have been largely unable to cope with this cold and treeless environment, while the latter apparently thrived in it during all but the most extreme intervals (see chapters 4–6).

Reconstructing loess-steppe is difficult because of the lack of modern analogues. Various lines of evidence indicate an extremely cold and arid climate comparable to today's arctic (Velichko 1984; see table 2.1), which implies relatively low plant (or primary) productivity (e.g., Whittaker 1975, 201–206). This in turn suggests low secondary productivity and widely scattered herbivore resources, which would have placed high mobility demands on human foragers (see chapter 1). However, the size and diversity of the large mammal community that inhabited the loess-steppe contradict these predictions and suggest much higher plant and animal productivity than are found on the modern tundra (Guthrie 1982).

The productivity paradox is not easy to explain. Received solar energy was higher at middle latitudes on the East European Plain than it is in the modern arctic, but the existence of loess-steppe environments in Beringia indicates that this variable was not critical. On the other hand, waterlogged tundra soils were absent, and continuous loess deposition would have provided a rich source of nutrients for plant growth (Butzer 1971, 287–289). Warm continental summers and reduced snow cover probably generated a protracted and intense growing season. The limited snow cover combined with a wealth of loessic mineral salt would have removed two major limiting factors for large mammal abundance (Guthrie 1990, 205–225). Thus, low winter temperatures and the lack of trees might have been more important factors in human settlement history.

Despite the evidence and arguments for high productivity, loess-steppe on the central East European Plain may still have imposed heavy mobility requirements on human foragers. When modern humans occupied this environment during the Last Glacial Maximum (see chapter 6), they appear to have been moving distances comparable to recent hunter-gatherers in arctic tundra (e.g., Kelly 1995, 128–132).

Another unusual feature of periglacial loess-steppe environments that had considerable significance for their human occupants was the formation of large bone accumulations (Vereshchagin and Baryshnikov 1984, 492–493). Cold temperatures and a favorable geochemical setting (i.e., abundance of calcium carbonate and absence of soil acids) would have preserved mammal bones and teeth on the loess surface for years. Seasonal thaws and slopewash activity probably worked to concentrate them in ravines and stream valleys over time. Large concentrations of mammal bone evidently piled up at stream confluences and ravine mouths in the Beringian loess-steppe (Péwé 1975, 98). In a largely treeless landscape, natural occurrences of bone became an important resource for Paleolithic people.

Reconstruction of interglacial and warm interstadial Pleistocene environments in Eastern Europe presents a lesser challenge. During these periods, diagnostic forest and chernozem soils formed on the loess deposited in the preceding cold interval (Kukla 1975; Velichko et al. 1984). The forest, wooded steppe, and steppe habitats represented by these soils seem to have modern analogues, and the pollen profiles obtained for the forested environments probably provide an accurate picture of the arboreal vegetation. The mammalian and molluscan faunal assemblages of the interglacials are similar to the present day, and lack the unusual associations of the loess-steppe. Although the mammalian faunas are still more diverse than those of today, this is a consequence of the wave of extinctions at the close of the Pleistocene, followed by local extinctions during the Holocene. Reflecting cooler climates, mammal faunas of the interstadials sometimes contain arctic elements (e.g., reindeer), and environments during these phases exhibit reduced zonality similar to the loess-steppe (e.g., Velichko et al. 1984, 113).

Pleistocene environments on the higher mountain slopes that border the East European Plain seem to have retained at least some degree of altitudinal zonation. However, the boundaries of vegetation zones were shifted downward to lower elevations during cold periods. In the Northern Caucasus, for example, pollen data and the distribution of small mammals indicate that alpine meadow flora and fauna were found 600–700 meters below their modern elevation (Levkovskaya 1994, 78).

Middle and Late Pleistocene Stratigraphy and Landscapes

The environmental context for human settlement of Eastern Europe extends back to the earlier Middle Pleistocene with the initial occupation of sites in the Northern Caucasus and southern plain (although widespread colonization did not take place until the Late Pleistocene). During the six or seven glacial-interglacial cycles that took place over this span of time, the longitudinal climatic gradient of Europe played a significant role in shaping this environmental context. The warmest interglacial periods witnessed movement of West-Central European broadleaf forest into the East European Plain, and eastward displacement of the southern steppe zone beyond the Volga River. As was noted earlier, latitudinal zonation was reduced during the glacial phases, as continental climates and steppic environments moved westward into Central and northwestern Europe (Frenzel 1968; Velichko et al. 1984).

Middle Pleistocene (OIS 19–6)

The lower boundary of the Middle Pleistocene has been established on the basis of paleomagnetic stratigraphy and coincides with the beginning of the Brunhes polarity epoch (currently calibrated to 780,000 years ago). The shift from reversed to normal polarity that marks the beginning of the Brunhes epoch has been identified in several loess profiles on the East European Plain (Velichko 1980). Above this level lies a long sequence of loess units and buried soils that represent most, if not all, of the glacial-interglacial cycles recognized in the marine oxygen-isotope record for the Middle Pleistocene (Zubakov 1988; Velichko 1990).

A major glacial event took place during the early Middle Pleistocene (*Don Glaciation* in East European terminology) that may be correlated with OIS 16 and dated to ca. 660,000–620,000 years ago. A massive ice lobe expanded as far south as latitude 50 degrees North in the Oka-Don Lowland. The oxygen-isotope record indicates that this was followed by several warm interglacials (OIS 15 through OIS 9), which date between roughly 620,000 and 300,000 years ago. They are represented by a series of well-developed forest soils, at least some of which may be correlated with the warm-loving *Singil'* fauna (Alekseeva 1990; Velichko 1990). Climates seem to have been warmer than present during most of these interglacial periods, with fir-hornbeam forest covering much of the East European Plain (Zubakov 1988, 20). The oxygen-isotope curve suggests that the intervening cold intervals were brief or of limited severity. The glacial corresponding to OIS 10 (ca. 360,000–340,000 years ago) was especially short, and is correlated with a relatively thin layer of loess (Velichko 1990, 107).

In contrast to the preceding period, the later Middle Pleistocene was characterized by generally cool climates. According to the oxygen-isotope record, OIS 9 was followed by two protracted glacial periods (OIS 8 and OIS 6), separated by a lengthy but rather cool interglacial (OIS 7). During this period, another massive glacial event took place (*Dnepr Glaciation* in East European terminology), and an ice lobe expanded into the Dnepr Basin as far south as 49 degrees North. The Dnepr till is associated with a thick bed of calcareous loess (3–4 meters in depth) that reflects the presence of periglacial loess-steppe environments in unglaciated parts of the plain (Velichko 1990, 109–112). These deposits probably correlate with OIS 8 (e.g., Zubakov 1988, 18), but alternatively might date to OIS 6.

Above the Dnepr loess lie several weakly developed soil horizons (including the *Kursk Soil*) that may represent the milder conditions of OIS 7 (ca. 245,000–185,000 years ago). Mammal remains include steppe and tundra elements (e.g., saiga, reindeer) (Alekseeva 1990, 56–57). The final phase of the Middle Pleistocene (OIS 6) is represented by another relatively extensive glaciation (*Moskva Glaciation*) and interval of loess deposition (Velichko 1990). The Moskva loess directly underlies the Last Interglacial soil and apparently dates to ca. 185,000–130,000 years ago (Velichko et al. 1997). Associated mammalian fauna from localities on the central plain include arctic species (e.g., musk ox [*Ovibos moschatus*]) and steppe forms, and reflect typical periglacial loess-steppe environments (Alekseeva 1990, 57–58).

Last Interglacial and Early Glacial (OIS 5)

The oxygen-isotope record indicates that warm interglacial conditions prevailed at the beginning of the Late Pleistocene (128,000–115,000 years ago), and this interval is recognized as the *Last Interglacial* climatic optimum (OIS 5e). On the East European Plain, OIS 5e is represented by a widespread buried soil horizon (*Salyn* Phase of the Mezin Soil Complex) that may be correlated with Pedokomplex III in Central Europe (Velichko and Morozova 1982, 86; Velichko et al. 1984, 102–105) (fig. 2.5).

By all indications, climates were warmer and wetter than those of the present day, with mean January temperatures on the central plain rising to an estimated –3 to 0 degrees C (i.e., comparable to Central Europe today) (Velichko 1984, 261). The Salyn horizon is characterized by a thick forest soil that extends several hundred kilometers south of the modern forest zone. Pollen spectra suggest that it supported a deciduous forest dominated by white beech, oak, and linden (Grichuk 1984, 167–171). The Salyn soil assumes a more chernozemic form in the southern plain, and pollen profiles indicate that a forest-steppe zone covered most of this region. Open steppe was largely confined to areas east of the Volga River. Mammal remains from deposits dating to the climatic optimum in the central and southern plain lack cold-loving taxa, and include mammoth, red deer, horse, and bison (Vereshchagin and Kolbutov 1957, 84; Zavernyaev 1978, 29; Alekseeva 1990, 59–66) (fig. 2.6).

During the Last Interglacial (OIS 5e), the geography of the southern plain was further altered by changes in the Black and Caspian Seas. As

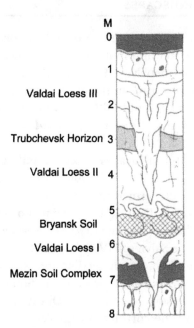

Figure 2.5. Late Pleistocene loess and buried soil profiles for the East European Plain (redrawn from Velichko 1990, 106, fig 2).

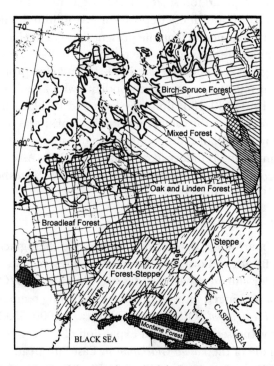

Figure 2.6. Environments of the Last Interglacial climatic optimum (OIS 5e) (redrawn from Grichuk 1984, 170–171, fig 17.10).

global ice volume shrank, sea levels rose, and the Black Sea was flooded with Mediterranean water and molluscan fauna, increasing 8–12 meters above its present level (Chepalyga 1984, 230–231; Zubakov 1988, 11). The northern isthmus of the Crimean peninsula was inundated, and the latter apparently became an island. Although isolated from the oceans, the Caspian Sea also expanded due to increased moisture in the region (Chepalyga 1984, 232).

Following the Last Interglacial were several mild interstadial periods, which date to between 115,000 and 70,000 years ago (OIS 5c and 5a), and are sometimes collectively termed the Early Glacial. They were preceded by brief cold oscillations (OIS 5b and 5d), the first of which correlates with a zone of frost disturbance (*Smolensk Cryogenic Horizon A*) and a thin layer of loess that overlies the Salyn soil (Velichko et al. 1984, 98). Across much of the East European Plain, the succeeding interstadials are represented by a single buried soil (*Krutitsa* Phase of Mezin Soil Complex), but some areas contain multiple soils (e.g., Middle Dnestr Valley [Ivanova 1981, 128]). These soils may be correlated with Pedokomplex II in Central Europe.

The Krutitsa soil is a chernozem that exhibits little latitudinal zonation and reflects a widespread forest-steppe environment on the central plain. Although glaciation and loess deposition was minimal, climates were colder and more continental than the present day. On the basis of the soils data, mean January temperatures are estimated to have ranged between –20 and –4 degrees C (Velichko et al. 1984, 113). Pollen profiles from the southwest plain reflect an open pine woodland with a few broadleaf taxa (Levkovskaya 1981, 130–133; Bolikhovskaya and Pashkevich 1982, 125–129), while vegetation in the central and southern plain was probably more steppic. The mammalian fauna contains many steppe taxa, and some arctic forms, including reindeer (e.g., Tatarinov 1977). Assemblages from the Crimea contain both reindeer and arctic fox (Vereshchagin and Baryshnikov 1980; Kolosov et al. 1993).

Pleniglacial (OIS 4–2)

According to marine oxygen-isotope curves, the Pleniglacial (70,000–10,000 years BP [or radiocarbon years before present]) was a sustained period of colder climates that includes two glacial events (OIS 4 and 2) separated by a prolonged interval of milder oscillating temperatures (OIS 3). The *Early Pleniglacial* (OIS 4) is dated to ca. 70,000–55,000 years BP and evidently witnessed a significant expansion of the Fennoscandian ice sheet (early stage of the *Valdai Glaciation* in East European terminology), although till deposits are scarce (Faustova 1984, 3–4). On the East European Plain, OIS 4 probably correlates with a layer of frost deformation that overlies the Krutitsa soil (*Smolensk Cryogenic Horizon B*) and a loess bed (*Valdai Loess I*) that averages only 0.5–1.0 meters in thickness (Velichko et al. 1984, 114).

During the Early Pleniglacial, there is little evidence for normal soil formation, and loess-steppe vegetation apparently covered most of the East European Plain. Paleobiotic data for this interval are limited, and the existence of loess-steppe environments is primarily based on the loess deposits. Pollen profiles from the central plain contain a high proportion dwarf birch and other arboreal taxa (some of it possibly introduced from distant sources), but reflect increasing dominance of herbaceous flora in the upper part of the loess (Grichuk 1972, 21–26).

The *Middle Pleniglacial* (OIS 3) is currently dated to ca. 55,000–25,000 years BP, and the later part of this period falls within the effective range of radiocarbon dating. Glaciation and loess deposition seem to have been very limited during the Middle Pleniglacial, which is chiefly represented by a widespread buried soil (*Bryansk Soil*) dating to the terminal phase of OIS 3 (30,000–25,000 years BP). The latter is a frost-gley soil with a carbonate illuvial horizon that exhibits little latitudinal variation, and has been severely disturbed by subsequent frost action. It is apparently without modern analogue (Velichko and Morozova 1982a, 88–89; Velichko et al. 1984, 107–110). The Bryansk soil may be correlated with Pedokomplex I in Central Europe, and the *Denekamp Interstadial* in the North European sequence (Lowe and Walker 1997, 336–337).

Pollen samples from the Bryansk soil reveal a high proportion of arboreal taxa, including some broadleaf forms (oak, hornbeam, and hazel), and some localities have yielded macrofossils of these species (Grichuk 1972, 26–31). This suggests relatively warm conditions and expanded forest vegetation. But the mammalian fauna does not differ significantly from the succeeding glacial period, and contains most of the loess-steppe taxa (e.g., mammoth, horse, bison, reindeer, arctic fox) (Markova 1984, 215). Estimated mean January temperatures range from –21 to –19 degrees C, and total annual precipitation from 450 to 350 millimeters (Velichko et al. 1984, 114).

In addition to the terminal OIS 3 soil, at least one earlier Middle Pleniglacial warm oscillation appears to be represented at some localities on the East European Plain. This horizon probably correlates with another brief interstadial dating to ca. 39,000–36,000 years BP (*Hengelo Interstadial* in North European terminology) (Lowe and Walker 1997, 336–337). A partly redeposited soil in the Middle Dnestr Valley, and a layer of organic colluvium (Lower Humic Bed) at Kostenki on the Middle Don River may be assigned to this period (Ivanova 1981, 128, fig. 38; Anikovich 1993, 6). Climates were probably similar to those of the younger interstadial.

The Late Pleniglacial or *Last Glacial Maximum* (OIS 2) is dated to 25,000–13,000 years BP, and is fully within the range of radiocarbon dating (although suitable media for dating are often scarce on the central and southern plain). Expansion of the Fennoscandian ice sheet reached a maximum at ca. 20,000–18,000 years BP on the East European Plain (Late Valdai Glaciation); the terminal moraine is only a few kilometers north of Smolensk (latitude 55 degrees North). The glaciers created an extensive system of proglacial lakes in the unglaciated northeastern region (Grosswald 1980, 15–20).

The initial phase of the Last Glacial Maximum is represented by a layer of severe frost disturbance (*Vladimir Cryogenic Horizon*), followed by the deposition of a thick bed of loess that averages 3–4 meters in depth on the central plain (*Valdai Loess II*). The loess is homogeneous, unweathered, and rich in calcium carbonate. A brief warm oscillation dated to ca. 17,000 years BP is represented in places by a thin gley horizon (*Trubchevsk Horizon*), which was followed by a wave of deep ice wedge formation (*Yaroslavl' Cryogenic Horizon A*) and more loess deposition (*Valdai Loess III*). The uppermost loess bed averages 3 meters in depth and is similar to the underlying unit; it constitutes the parent material of the modern soils that have developed since the end of the Pleistocene (Velichko et al. 1984, 110–111).

Throughout the Last Glacial Maximum, periglacial loess-steppe environments prevailed across the central East European Plain. Pollen profiles are dominated by nonarboreal pollen, including grasses, sedges, Chenopodiaceae, and *Artemisia*. Most of the arboreal pollen, which contains many dwarf forms (dwarf birch and alder), were probably derived from distant sources (Grichuk 1984, 171–178). As noted earlier, little wood fuel was available to the people who inhabited the region at this time. Mammals of the modern high arctic zone, such as musk ox and collared lemming (*Dicrostonyx torquatus*), are found in the faunal assemblages, along with other tundra and steppe forms (e.g., mammoth, rhinoceros, horse, bison, reindeer, and arctic fox) (Vereshchagin and Baryshnikov 1982; Markova 1984, 215–216) (fig. 2.7).

Climates were extremely cold and dry on the East European Plain during OIS 2, and the estimated mean January temperatures for the central region—based on the flora and fauna—are between –40 and –30 degrees C (Velichko 1984, 273–279). On the other hand, estimated mean July temperatures are only 3–5 degrees below the present day (roughly +14 degrees C). Total annual precipitation in the central plain is thought to have been below 200 millimeters (i.e., lower than the modern steppe zone)

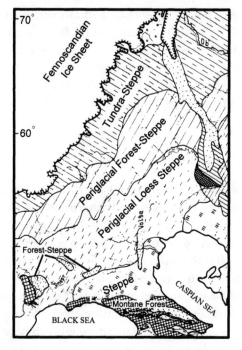

Figure 2.7. Environments of the Last Glacial Maximum (OIS 2) (redrawn from Grichuk 1984, 176–177, fig 17.14).

(Velichko et al. 1984, 115). Climates were less dry in the southwest part of the plain, where some forest vegetation (pine, alder, and birch) survived throughout the Last Glacial Maximum (e.g., Borziyak et al. 1997, 292). Permafrost extended 400–600 kilometers south of the ice margin, covering the entire central plain (Velichko and Nechayev 1984).

As during the Last Interglacial, changes in sea level had significant effects on the geography of the southern plain. With the increase in global ice volume, world ocean levels fell sharply. The Black Sea dropped to more than 90 meters below the present shoreline, severing its link to the Mediterranean and becoming a semifreshwater basin (Chepalyga 1984, 235–236). By contrast, the Caspian Sea—swollen with glacial meltwater from the northern plain—expanded to 150 percent of its modern area.

After ca. 13,000 years BP, climates in the northern hemisphere warmed substantially and the process of deglaciation accelerated. The period dating to 13,000–10,000 years BP is widely designated as the *Late Glacial*. On the East European Plain, loess deposition continued and a final episode of frost wedge formation (*Yaroslavl' Cryogenic Horizon B*) may be correlated with a cold oscillation at 11,000–10,500 years BP (*Younger Dryas* event in

North Atlantic terminology) (Velichko and Morozova 1982b, 120; Lowe and Walker 1997, 342–343). Pollen profiles for this interval indicate a pine-birch woodland interspersed with steppe in the deglaciated northern plain, birch-pine forest with some broadleaf trees in the central region, and steppe in the south (Grichuk 1982, 108–109). This pattern reflects the reemergence of a marked zonation of biomes with the decline of glacial conditions.

Landscape and Paleoanthropology in Eastern Europe

The strikingly different landscapes of Eastern Europe produced a paleoanthropological record unlike that of Western Europe—the birthplace of Paleolithic archaeology. The harsh continental climates and open habitat of the East European Plain discouraged human colonization until relatively late in prehistory, and the temporal depth of the record is shallow (Klein 1973, 11–18). In any case, extensive glaciation during the late Middle Pleistocene would have destroyed any traces of early occupation in much of the northern and central plain (Praslov 1995). Throughout the Late Pleistocene, the lowlands remained marginal environments and were subject to fluctuating settlement.

The lack of natural shelters on the East European Plain has played a major role in shaping the Paleolithic record in this part of the world. Most Middle and Upper Paleolithic sites in Western Europe—as well as many in Central Europe—are found in caves and rockshelters. These sites are highly visible today, and often contain lengthy sequences of occupation and well-preserved faunal remains. But with the exception of some small caves in the southwest region, natural shelters in Eastern Europe are confined to the Crimea and Northern Caucasus.

The majority of East European sites are open-air occupations and few contain more than one or two occupation levels. Without the protective bedrock casing of a natural shelter, these open-air sites are more susceptible to erosion and loss. Their locations are less obvious than caves, and they are typically recognized by the discovery of associated large mammal remains—especially the bones and tusks of mammoths. Despite a favorable geochemical medium for preservation (i.e., loess or loess-derived loam), most of these remains eventually weather away in open-air settings. The East European record is therefore more strongly biased against older and smaller sites (Hoffecker 1999a, 130).

Human skeletal remains are especially vulnerable to weathering and destruction in open-air sites, and the problem is further exacerbated by the climatic conditions on the East European Plain. Frozen ground prevented deep burials, and frost-heaving probably destroyed many graves, as in modern Arctic environments (Zimmerman 1985). Predictably, most Neanderthal remains are found in the caves of the Crimea and Northern Caucasus, and the majority of modern human skeletons are found in sites dating to interstadial periods.

CHAPTER 3

Middle Pleistocene Settlement

Hominids began to colonize latitudes above 40 degrees North during the earlier Middle Pleistocene, but their occupation of northern environments was largely confined to Western Europe, where climates are exceptionally mild and moist due to the influence of warm ocean currents. The remaining colder and drier areas of northern Eurasia (i.e., Eastern Europe and Siberia) were mostly unoccupied until the Late Pleistocene. Moreover—prior to the late Middle Pleistocene—settlement in Western Europe seems to have occurred primarily during the warmer interglacial periods (Roebroeks and van Kolfschoten 1995). This early settlement of northern latitudes may not have entailed significant new morphological or behavioral adaptations, and simply reflected an extension of the *Homo* geographic range permitted by climatic change and related factors (e.g., local carnivore extinctions [Turner 1992]).

The earliest known sites in Eastern Europe date to the Middle Pleistocene, but they are relatively rare and confined to the southernmost regions. The most securely dated early and late Middle Pleistocene sites are caves on the northern slope of the Caucasus Mountains. However, several open-air localities are found in the southern part of the East European Plain (below latitude 48 degrees North). Assessing the character and distribution of settlement in Eastern Europe during the Middle Pleistocene is complicated by several factors, including the effects of past glaciation and sea level fluctuations. Nevertheless, there is sufficient (negative) evidence that occupation of Eastern Europe and Siberia (i.e., northern Eurasia east of the Carpathian Mountains) was limited and sporadic prior to the Late Pleistocene.

It is still not clear precisely why Middle Pleistocene hominids were unable to settle widely across the cooler and drier regions of northern Eurasia, although low winter temperatures and the reduced availability of digestible plant foods seem likely to have been critical factors. Prior to the appearance of fully developed Neanderthals in the late Middle Pleistocene, there is little evidence for the adaptations to colder environments discussed in chapter 1. The pre-Neanderthal occupants of Europe were probably highly dependent on plant foods, and lacked both the morphological features and technological skills necessary to cope with most East European and North Asian environments (even under warm interglacial conditions). By contrast, the Neanderthals—who successfully colonized at least some of these environments—seem to have evolved many cold adaptations, although comparison is difficult because of the fragmentary character of the earlier Middle Pleistocene hominid record.

Hominid Colonization of Northern Latitudes

Like most higher primates, the genus *Homo* evolved in equatorial environments, and its first major range expansion was confined to lower and middle latitudes of the Old World. The colonization of regions above roughly 40 degrees North—where mean annual and January temperatures begin to decline substantially—did not take place for another million years. Approximately 1.8 to 1.7 million years ago, the first hominid sites appear outside sub-Saharan Africa. These sites, which contain *Homo ergaster* fossils and/or Lower Paleolithic stone tools, are found in the Near East and in the lower and middle latitudes of East Asia. Although the dating of the earliest Asian localities (e.g., Riwat, Longgupo) remains controversial (Swisher et al. 1994; Wanpo et al. 1995), the Dmanisi site in Georgia (at 41 degrees North) indicates a minimum date of 1.7 million years for *Homo* in Eurasia (Gabunia et al. 2000). Artifacts in the Nihewan Basin of China (also 41 degrees North), which are dated to ca. 1.0 million years ago, represent the oldest reliably dated East Asian localities (Schick and Dong 1993).

The initial expansion out of Africa coincides with the appearance of *Homo ergaster* and is probably related to changes in morphology and behavior that are evident at this time. The former include increased brain volume, smaller cheek teeth, reduced forelimb length, decreased sexual dimorphism, and increased overall body size (Klein 1999, 287–295). The broadly contemporaneous spread of drought-resistant grasses in sub-Saharan Africa indicates a significant expansion of open habitat, and *Homo ergaster* probably represents adaptation to the latter. New foraging tactics that entailed better planning (larger brain), increased mobility (fully modern bipedalism and body size), and provisioning of young by male-female pairs (reduced sexual dimorphism) may have been critical to occupation of more arid and seasonally variable environments (Gamble 1994; Cachel and Harris 1998).

Adaptation to more open environments in sub-Saharan Africa seems to have simultaneously permitted *Homo ergaster* expansion into similar habitats across midlatitude Eurasia. Associated changes in tool technology seem to have been minimal, because the earliest sites in Eurasia contain tools comparable to the Oldowan pebble tool industry (Gabunia et al. 2000). However, large bifaces characteristic of the Acheulean industry subsequently appeared in Africa and western Eurasia, and may reflect some new technological developments in these areas. By contrast, the East Asian population seems to have become isolated from African and western Eurasian *Homo*, and a regional species (*H. erectus*) persisted without major morphological (or technological) change until the end of the Middle Pleistocene (Schick and Dong 1993; Klein 1999).

The oldest sites in Europe apparently date to the end of the Early Pleistocene and beginning of the Middle Pleistocene. They include Atapuerca Gran Dolina in northern Spain and possibly Ceprano in central Italy (both at 42 degrees North), which are dated to ca. 0.8–0.7 million years ago (Carbonell et al. 1995; Ascenzi et al. 1996). These sites contain human skeletal remains that do not exhibit close similarities to either earlier African or later European representatives of *Homo*, and simple flake and pebble tools (i.e., no bifaces). The remaining European localities antedating 0.5 million years lack human fossils and contain only fractured stone and animal bone fragments that may not represent human occupations (Gamble 1994; Roebroeks and van Kolfschoten 1995). Alternatively, at least some of them may reflect sporadic episodes of colonization that preceded sustained settlement of these latitudes.

After the beginning of OIS 13 (540,000 years ago), there is more substantial evidence of northern settlement in the form of well-dated human fossils and undisputed artifacts in Western and Central Europe as far north as latitude 52 degrees North (Roebroeks and van Kolfschoten 1995). These include *Homo* fossils from Mauer, Arago, Boxgrove, Bilzingsleben, and Vértesszöllös, and tool assemblages from Torralba-Ambrona, Hoxne, Kärlich, Korolevo, and many other localities dating to 500,000–300,000 years ago. Few caves survive from this time range, and most sites are open-air locations along the margins of ancient streams, lakes, and springs. The tools include handaxes (absent in Central Europe), choppers, and a variety of flake implements and unmodified flakes. The occupation floors lack convincing traces of artificial shelters and constructed hearths, although there is evidence for the use of fire in the form of burned bone and charcoal fragments at some localities (e.g., Kretzoi and Dobosi 1990).

The period between 500,000 and 300,000 years ago, which spans OIS 13–9, was generally mild in northern Eurasia. The interglacials that correspond to OIS 11 and 9 were especially lengthy and warm. This shift to predominantly warm climates, after a sustained interval of generally cool conditions, may be one reason for the movement into higher latitudes roughly half a million years ago (Hublin 1998). Turner (1992) also suggested that the extinction of several large northern carnivores at this time reduced competition for scavengers and provided new opportunities for obtaining meat in these environments; the European carnivore guild became similar to the African one. These circumstances may have temporarily created environmental conditions in Western and Central Europe comparable to those found at more southern latitudes.

Evidence for new morphological adaptations in the West and Central European sites during 500,000–300,000 years ago is relatively limited. Mitochondrial DNA analyses of European Neanderthal specimens indicates that the split between European and African representatives of *Homo* may have occurred between 690,000 and 550,000 years ago (Krings et al. 1997; Ovchinnikov et al. 2000). However, a pronounced pattern of morphological cold adaptation is not apparent until the emergence of the Neanderthals after 300,000 years ago, and the changes that appear prior to this may simply reflect genetic drift. Most European hominid fossils dating to 500,000–300,00 years ago may be classified as *Homo heidelbergensis* (Rightmire 1998), although it is possible that a separate group inhabited parts of Central Europe (Svoboda et al. 1996).

Evidence of changes in diet and foraging strategy is also limited and rather ambiguous. A number of tool assemblages in northern Eurasia dating to 500,000–300,000 years ago are associated with the remains of various large mammals such as elephant, rhinoceros, horse, bison, giant deer, and others. At some localities, stone tool cut marks have been identified on bone fragments (e.g., Ambrona, Vértesszöllös, Hoxne, and Boxgrove [Shipman and Rose 1983; Kretzoi and Dobosi 1990; Singer et al. 1993; Parfitt and Roberts 1999]). The presence of cut marks indicates that the bones probably represent food debris, and thus reflect some meat consumption. But it is not clear whether the bones were derived from human hunting or simply collected from carnivore kills or other cases of natural mortality. The pattern does not seem to differ significantly from that of earlier *Homo* sites in southern latitudes, which also contain bone fragments with cut marks (Shipman 1986; Potts 1988).

No major changes in stone tools are associated with the colonization of the north, but there is some evidence of technological change. While the sites lack indisputable traces of artificial shelters and well-constructed hearths (Gamble 1994, 137–139), they do yield the first reliable evidence for the use of controlled fire, notably at Vértesszöllös in Hungary (Kretzoi and Dobosi 1990). It is not known precisely how the occupants of these sites were acquiring and using fire (the apparent absence of hearths may reflect a pattern of use unlike that of modern people), but it may have been essential to the successful colonization of latitudes above 40 degrees North. Gamble (1986, 386–390) suggested that wooden probes and controlled fire might have been used to locate and thaw frozen mammal carcasses buried beneath the snow during winter months.

One reason for the limited evidence of adaptive change associated with the move north may be its low visibility in the paleoanthropological record.

For example, morphological changes might have included a thicker coat of body hair. Also, human remains from European sites older than 300,000 years are especially scarce, and the existing data may simply not provide sufficient basis for determining the presence or absence of skeletal features that reflect adaptation to cooler temperatures or to other aspects of northern environments. The occasional scatters of stone and bone fragments that represent occupation floors preserved from this time period may fail to reveal important changes in foraging tactics and nonlithic technology that differentiate the northern and southern sites.

But another and perhaps more important reason for the lack of evidence for change may be that the pattern of northern colonization during this interval required few novel adaptations beyond the control of fire. To begin with, the range of environmental tolerance in *Homo* was already a broad one, as indicated by the geographic distribution of fossils and sites across southern Eurasia prior to 500,000 years ago. Second, as noted earlier, populations of *Homo* seem to have ridden into Europe and other parts of northern Eurasia on a wave of warm climates at the beginning of OIS 13. Although there are traces of occupation during the colder phases of the interval between 500,000 and 300,000 years ago (Roebroeks et al. 1992; Gamble 1995, 281), most sites of this period are associated with the warmer phases, and there appears to be a general bias against glacial environments (Roebroeks and van Kolfschoten 1995). If humans abandoned many of the regions above 40 degrees North during the glacial periods, this may be the most significant contrast between the northern and southern records during 500,000–300,000 years ago, because occupation of southern latitudes was continuous throughout this interval.

The Problem of East European Colonization

Recognizing and documenting a prehistoric episode of colonization presents special problems to the archaeologist, and some of these problems are further compounded by the geography of Eastern Europe. It is unclear what sort of archaeological traces a colonizing population would leave in its wake. Such traces might have been highly ephemeral and almost impossible to find in a landscape like the East European Plain.

Archaeologists have no ethnographic or ethnohistoric examples of modern hunter-gatherer colonization to guide the collection and interpretation of their data. The process of hunter-gatherer expansion into previously uninhabited environments is unknown. Did it occur incrementally, as the expanding population reached carrying capacity in each newly occupied portion of the environment? Or did it entail broad and rapid dispersal at initial low density? What types of sites were occupied by the colonizing hunter-gatherer groups, and how did they differ from those of noncolonizing groups? These questions are especially difficult to answer in the case of earlier forms of *Homo*, who almost certainly pursued land-use strategies unlike those of any recent hunter-gatherers.

The inherently stressful nature of a newly colonized environment (unless it is shielded from previous settlement only by some natural barrier such as a body of water) suggests that a colonizing hunter-gatherer population might have lived at very low density. The overall distribution of sites could have been highly dispersed, and individual occupations might have been of limited duration and intensity. Such a colonizing population would probably generate an archaeological record of low visibility. Their sites might lack well-developed hearths and other features, and their tools might be confined to crude, expedient forms indistinguishable from naturally fractured bone and rock. Sites would be especially susceptible to destruction by natural processes, and the few surviving examples could be difficult to recognize or discover. The exception to the pattern would be caves or rockshelters, which would protect the remains from natural destruction and provide a marker for survey archaeologists. The scarcity of natural shelters across much of Eastern Europe renders it a particularly difficult place to search for evidence of early settlement.

Some archaeological examples of prehistoric colonization may exist, but they are predictably controversial. The oldest purported sites in Western and Central Europe include problematic localities such as Le Vallonet, Kärlich A, Červený kopec, and others (Roebroeks and van Kolfschoten 1995). Most of the uncertainty attending these sites concerns the interpretation of the artifacts, which may represent naturally fractured stone. However, the fact that these sites would have been occupied by an early form of *Homo* employing a relatively primitive lithic technology encourages cautious appraisal of these remains. Similar doubts have been raised about the artifacts at Diring Yuriakh in central Siberia, which represents the earliest credible evidence of settlement above 60 degrees North (more than 260,000 years old) and also would have been occupied by an archaic form of *Homo* (Waters et al. 1997). The best example of initial colonization at low density may lie in Alaska and the Yukon, where the oldest dated occupations (more than 11,500 years BP) comprise small scatters of lithic debris (West 1996).

The Middle Pleistocene archaeological record of Eastern Europe differs significantly from that of Western Europe (and bears more similarity to that of northern Asia). Between the Carpathian and the Ural Mountains, there are no more than a handful of sites that may be firmly dated to between 780,000 and 130,000 years ago. The comparative lack of these sites may be plausibly ascribed to environmental barriers to hominid settlement (and evidence that the *Homo* population of Western Europe avoided cold environments prior to 300,000 years ago is especially important to this question), but also could reflect factors that influence the preservation and discovery of archaeological remains. These factors include the reduced preservability of open-air occupations relative to caves, and some special factors that apply to Middle Pleistocene landscape history in Eastern Europe (see chapter 2).

Time eventually acts as to erase the differences between regions containing natural shelters and those lacking such features. Caves and rockshelters normally weather away after several hundred thousand years, and most sites in Western Europe dating to 500,000–300,000 years ago are open-air localities (Roebroeks and van Kolfschoten 1995). Nevertheless, some caves survive for longer periods, and the presence of important sites such as Atapuerca, Arago, and Petralona in Western and Central Europe, indicates that the reduced preservability of open-air sites must still be accounted for—although to a lesser degree—in interpreting the East European archaeological record of the earlier Middle Pleistocene.

But more significant are the effects of Middle Pleistocene glaciations and sea level fluctuations in Eastern Europe (Praslov 1995). Perhaps the most widespread glaciation of the East European Plain (Don Glaciation) took place during OIS 16 (660,000–620,000 years ago), when massive ice lobes extended down the Don Valley as far as 50 degrees North. The Dnepr Glaciation, which occurred during OIS 8 (303,000–245,000 years ago) or possibly later, was almost equally extensive (Velichko 1990). During these and other glacial advances, meltwater spilled over from the Caspian into the Black Sea, inundating large areas of land north of the Caucasus Mountains (Zubakov 1988). Only a small portion of the East European Plain remained uncovered by ice or water during these major cold events, which would have swept away most traces of prior human occupation. Praslov (1995, 61) suggested that investigator bias has also worked against the discovery of Middle Pleistocene sites on the East European Plain; skeptical archaeologists have disdained surveys for such sites and hotly contested any purported finds.

Middle Pleistocene Sites in Eastern Europe

The most firmly dated Middle Pleistocene sites in Eastern Europe (as defined in chapter 2) lie on its southernmost margin in the Northern Caucasus. Occupations dating to the earlier Middle Pleistocene (probably OIS 15–11) are found in Treugol'naya Cave, and later settlement is documented at Myshtulagty lagat (Weasel Cave) (OIS 7–6) (Doronichev 1992;

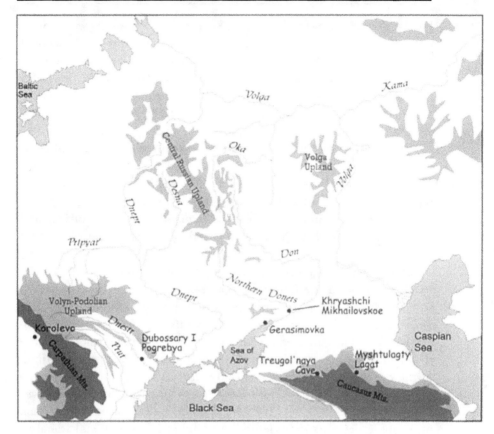

Figure 3.1. Map of Middle Pleistocene sites in Eastern Europe.

Hidjrati et al. 1997). A small number of open-air localities are reported from the southern parts of the East European Plain, but at least some of these are problematic. The best evidence of Middle Pleistocene occupation on the southern plain is found at the sites of Khryashchi and Mikhailovskoe on the Northern Donets River, which contain indisputable artifacts in buried soil horizons that probably date to OIS 9–7 (Praslov 1968, 1995). None of these sites has yielded human skeletal remains, and the morphology of their occupants is unknown (fig. 3.1).

Early Middle Pleistocene Sites (OIS 19–9)

The earliest evidence of settlement in Eastern Europe has been recovered from the open-air locality of *Gerasimovka* on the coast of the Sea of Azov. The site, which was discovered and investigated by N. D. Praslov in 1959–1964, yielded a small quantity of chert and quartzite pieces of suspected human manufacture (Praslov 1968, 18–21; 1995). The pieces were apparently deposited in a coarse gravel layer that underlies marine sediment of the Baku and Chaudian transgressions dating to the Early/Middle Pleistocene boundary (Zubakov 1988). But while the dating of the finds (discovered in association with remains of *Mammuthus trogontherii* and rodents of the Tiraspolian faunal complex) seems to be firmly established, their status as artifacts is ambiguous. Approximately a dozen items were selected from a much larger sample of naturally fractured quartzite and chert pieces in a stream deposit containing many pebbles of the same raw material; all of the selected pieces were heavily weathered and rolled. None of the pieces described and illustrated by Praslov (1968, 21, fig. 2) represents an artifact of unquestionable human manufacture (e.g., biface), and the possibility that all of them were created by natural processes is strong (Klein 1966, 32).

More convincing traces of East European habitation are found on the northern slope of the Caucasus Mountains at *Treugol'naya Cave*. This site, which was discovered and excavated during 1986–1991 by V. B. Doronichev and L. V. Golovanova, represents a rare example of a natural shelter that has survived the forces of weathering and erosion for more than half a million years (Doronichev 1992). It lies at the uppermost limit of the present-day tree line at 1,500 meters above sea level (although during the early Middle Pleistocene it was roughly 500 meters lower [Nesmeyanov 1999, 309]). The cave contains a deep succession of loam and rubble deposits that yielded over 3,800 large mammal remains, including early Middle Pleistocene representatives of bison (*Bison schoetensacki*), horse (*Equus altidens*), and rhinoceros (*Stephanorhinus hundsheimensis*) (Baryshnikov 1993; Hoffecker 1999b). The lowest artifact-bearing levels produced an interglacial fauna and six ESR dates—obtained on terrestrial mollusc shell—averaging 583,000 years ago, indicating possible correlation with OIS 15. Only a handful of simple flake tools and unretouched flakes—often of imported chert—were found in these levels. The large mammal remains appear to have been collected by carnivores or washed into the cave by a small stream (Hoffecker 1999b). Overlying levels yielded more interglacial fauna and ESR dates of 360,000 and 420,000 years ago (possibly OIS 11) (fig. 3.2).

Between roughly 425,000 and 300,000 years ago, two major interglacial periods took place. These periods, which correlate with OIS 11 and OIS 9, were intervals of pronounced and sustained warmth. In the traditional West European chronology, one or both of these stages correspond to the Holsteinian or Hoxnian (formerly referred to as the "Great Interglacial" in Britain), when climates were warmer and more oceanic—especially in Central and Eastern Europe—than the present day (Frenzel 1973, 104–

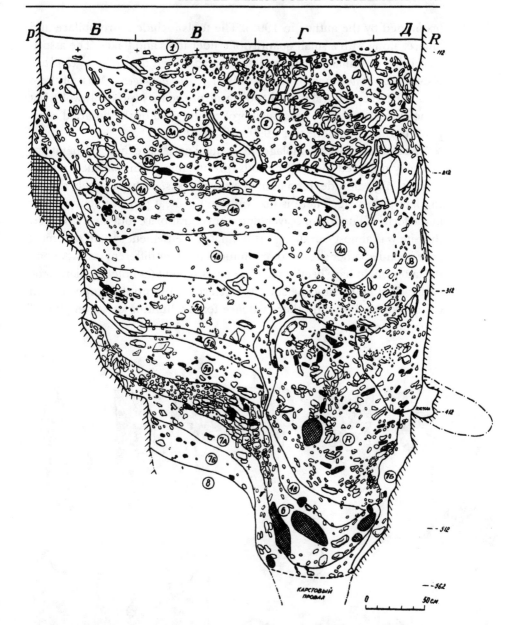

Figure 3.2. Stratigraphy of Treugol'naya Cave (after Doronichev 1992, 127, fig. 3).

111). The final Middle Pleistocene interglacial (OIS 7) was significantly cooler, and conditions of comparable warmth did not reoccur until the beginning of the Late Pleistocene.

Treugol'naya Cave contains layers that most probably date to OIS 11 and/or OIS 9 and yield stone artifacts of indisputable human manufacture (examined by the author in 1998). The tools include a protobiface and limace, along with a small quantity of cruder tools and flakes. The associated large mammal remains are dominated by red deer (*Cervus elaphus*), bison (*Bison schoetensacki*), and cave bear (*Ursus deningeri*). Most of them accumulated by natural mortality or carnivore activity (probably wolf), but some of the bones may have been broken and used by the hominid occupants of the cave (Hoffecker 1999b) (fig. 3.3).

The small artifact assemblages in Treugol'naya Cave indicate that it was probably occupied for brief periods (presumably during warmer months given its high altitude) for a limited array of tasks, perhaps largely related to plant gathering. There is no evidence for meat procurement—either in the form of hunting or scavenging—or central-place foraging. However, this site was almost certainly part of a larger settlement system that must have included other sites at lower altitudes used for different tasks. Among the latter might have been some meat procurement (as noted earlier, some earlier Middle Pleistocene sites in Western and Central Europe contain cut-marked bones and other evidence for the hunting or scavenging of large mammals).

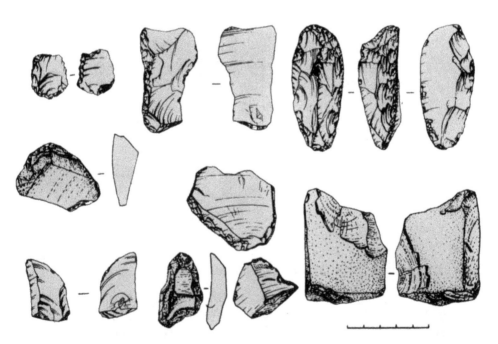

Figure 3.3. Stone artifacts recovered from Layer 4 at Treugol'naya Cave in the Northern Caucasus (modified from Doronichev 1992, 130, fig. 7).

Several localities on the southern margin of the East European Plain that most probably date to OIS 11 contain possible traces of human occupation. These open-air sites are found along major rivers below latitude 48 degrees North, and have yielded stone artifacts (some of which probably represent naturally fractured rock) and isolated large mammal remains. Their interpretation presents archaeologists with a classic problem in the recognition of human colonization of new habitat in this time range. As noted earlier, humans might have occupied areas north of these localities that were subsequently destroyed by glacial advances (Praslov 1995).

On the southwestern edge of the East European Plain, stone artifacts were recovered from ancient high terraces along the Lower Dnestr River (Anisyutkin 1987; Anisyutkin and Levkovskaya 1992). At *Pogrebya*, artifacts were found in loess and a buried soil assigned to the Likhvin Interglacial (broadly correlated with OIS 11), and in association with the tooth of an early Middle Pleistocene mammoth (*Mammuthus trongontherii*), on the sixth terrace. At *Dubossary I*, artifacts were also found in the loess and possibly associated with the same buried soil on the seventh terrace. At each site, roughly 95 percent of the finds were collected from the surface and a more recent group of artifacts was distinguished from the older assemblage (Anisyutkin and Levkovskaya 1992, 84–88). Thus, there is potential for mixture with undated and possibly much younger material.

The assemblages include a variety of cores and simple flake and pebble tools (examined by the author in 1988). Only a small percentage represent indisputable stone artifacts (e.g., side-scrapers, limace fragments), and it is unclear whether any of these were found in buried context. Most items exhibit the appearance of randomly fractured rock. However, the buried materials lie in windblown (and not fluvial) sediment, and it is unclear what natural processes would have deposited and fractured the rock fragments and pebbles. These localities may contain evidence of human occupation in the form of very primitive stone implements and flaking debris.

A pair of important sites on the Northern Donets River was discovered by the geologist G. I. Goretskii in 1950, and investigated by Praslov during 1964–1966. The sites of *Khryashchi* and *Mikhailovskoe* are found several kilometers upstream from the Don River confluence on the third terrace level. Alluvium of this terrace grades into the Mariin Terrace of the Don, which is assigned to the Likhvin Interglacial [i.e., probably OIS 11 (Zubakov 1988)]. The alluvium is composed of cobbles and gravels in a sand matrix and contains a Middle Pleistocene freshwater molluscan fauna of interglacial character. Above the alluvium lies a sequence of three buried soils capped with a thick bed of loams and sandy loams. The uppermost soil is tentatively dated to the Last Interglacial (OIS 5e) at the beginning of the Late Pleistocene. The underlying buried soils—on the basis of stratigraphic position, pedology, and palynology—appear to represent warm periods (interglacials or major interstadials) of the late Middle Pleistocene (OIS 9 and/or, possibly, 7) (Praslov 1984b, 97) (fig. 3.4).

At Khryashchi, artifacts were recovered from the alluvium as well as from one of the overlying soils (see later discussion). Only a few artifacts were actually recovered *in situ* from the ancient alluvial sediments; most pieces were collected from the surface of the underlying bank on to which they had been redeposited. A total of 60 artifacts is assigned to what is called the "early alluvial complex," and they are almost evenly divided between items of quartzite and chert. The artifacts are heavily rolled and deeply patinated, and many of them may be naturally fractured pieces (examined by the author in 1998). They include cores, retouched flakes, and pebble tools (Praslov 1968, 28–33). Although one of the pebble tools exhibits a bifacially worked edge and almost certainly represents a humanly manufactured artifact, it was apparently recovered from the bank and could have been derived from the younger buried soils that overlie the alluvium. At Mikailovskoe, *in situ* artifacts were found only in one of the overlying soils (see later discussion). Thus, as in the case of the Lower Dnestr River localities, the early materials from Khryashchi must be regarded as problematic.

Late Middle Pleistocene Sites (OIS 8–6)

During the late Middle Pleistocene (300,000–130,000 years ago), climates were cooler on average than in the preceding interval. A glacial episode (OIS 8) was followed by a relatively cool interglacial (OIS 7), and a prolonged and very cold glacial (OIS 6). During this period, a fully developed set of "classic" Neanderthal features appears in the human fossil record of Western Europe (Hublin 1998; Rightmire 1998; Klein 1999), where there are many well documented sites and some evidence of large mammal hunting (Gamble 1986; Mellars 1996). In Eastern Europe, although there was continued occupation of the southern upland margin (Northern Caucasus) and the first incontestable traces of human settlement on the southern edge of the East European Plain, the impoverished archaeological record contrasts sharply with that of Western and Central Europe.

At the localities of Khryashchi and Mikhailovskoe on the Northern Donets, artifacts were recovered from the buried soils that overlie the alluvium and underlie the soil assigned to the early Late Pleistocene. As noted earlier, these soils probably date to one or more interstadials or interglacials of the late Mid Pleistocene (OIS 7 or 9). Each locality yielded

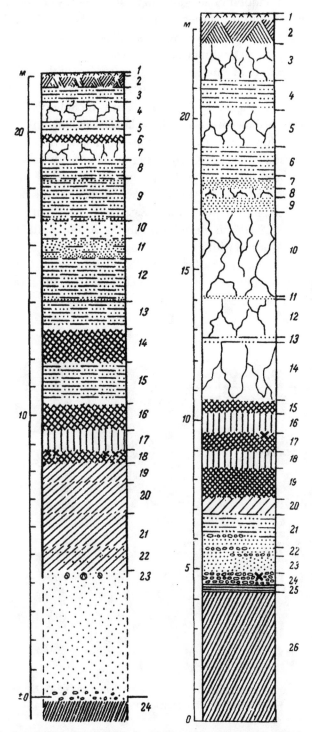

Figure 3.4. Stratigraphy of Mikhailovskoe (left) and Khryashchi localities on the Northern Donets River (modified from Praslov 1968, 25, fig. 4 and 41, fig. 13).

stone artifacts of unquestionable human manufacture, some of which were recovered *in situ* from the buried soils (Praslov 1968, 1984b, 1995).

At Mikhailovskoe, chert flakes were found in the lowermost (dark gray) soil, which is up to 50 centimeters in thickness, and approximately 200 artifacts were collected from the embankment. Most of the artifacts are of locally available chert, but some are quartzite; small flakes and fragments (less than 8 cm in length) predominate. Cores include discoidal, protoprismatic, and Levallois forms, and the flakes also exhibit much variability. Among the roughly 20 retouched tools are points, denticulates, scrapers, and a limace. Isolated large mammal fragments were also recovered, and include bones and teeth of red deer, horse, and probably bison (N. D. Praslov, pers. comm., 1998). Mikhailovskoe represents the oldest incontestable evidence of settlement on the East European Plain (fig. 3.5).

At Khryashchi, artifacts were found in the second soil, which is described as "black-brown with a light brownish hue" and measures up to 65 centimeters in thickness (Praslov 1968, 23). Fifteen items (thirteen originally reported) of chert and quartzite, including four side-scrapers, were found at the top of the soil horizon. No faunal remains are reported.

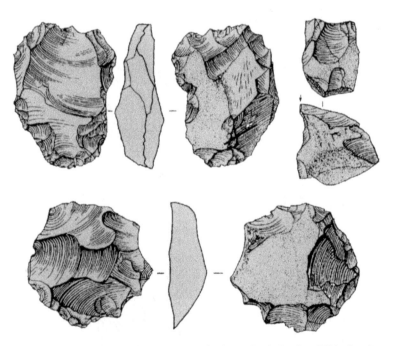

Figure 3.5. Flake cores recovered from the lower buried soil at Mikhailovskoe on the Northern Donets River (modified from Praslov 1968, 45, fig. 14).

Khryashchi also contains indisputable evidence of human occupation in stratigraphic context.

Continued occupation of the southern uplands is also evident at *Myshtulagty lagat* (Weasel Cave), located in the north-central Caucasus at an elevation of 1,125 meters above sea level. This site contains a deep sequence of Late Pleistocene and late Middle Pleistocene occupation layers (Hidjrati et al. 1997). Layers 15–18 are thought to date to a cool period prior to the Last Interglacial (possibly OIS 6), and contain remains of the Middle Pleistocene vole *Arvicola chosaricus*. A volcanic ash unit in Layer 18 has yielded preliminary single-crystal-laser-fusion dates of not less than 200,000 years. On the basis of palynology, underlying Layer 19 was deposited during a warm period, and may correlate with the final Middle Pleistocene interglacial (OIS 7). All of these layers, as well as more deeply buried units (Layers 20–25) contain stone artifacts and large mammal remains. Detailed descriptions are not yet available, but the tools are reportedly similar to those in the upper layers (Layers 5–14), which are characterized as Denticulate Mousterian of Levallois Type, and the large mammals include deer, goat, and cave bear (Hidjrati et al. 1997).

Conclusions: Middle Pleistocene Settlement in Eastern Europe

Despite inherent biases against the preservation and discovery of Middle Pleistocene sites in much of Eastern Europe, there is sufficient basis for the conclusion that hominid settlement of the latter was very limited prior to the Late Pleistocene. Human occupations were chiefly confined to the southernmost margin, especially the Northern Caucasus. Even accounting for the reduced visibility of open-air sites, as well as the destructive effects of glaciation and sea level fluctuation, traces of occupation on the East European Plain are few in comparison to those in Western and Central Europe during this period.

The extreme scarcity of sites is especially significant in the southwest region of the plain—along the Dnestr and Prut Rivers and their tributaries—during the final Middle Pleistocene (OIS 7–6). These areas are rich in promising geomorphic contexts for archaeological remains in this time interval, including stream terraces, loess deposits with buried soils, and some natural shelters (Ivanova 1977; Velichko 1990). They were unaffected by Middle Pleistocene ice sheets or marine transgressions, and they have been subject to substantial archaeological investigation, particularly since 1945 (e.g., Chernysh 1959, 1973; Ketraru 1973). Although these valleys contain the highest concentration of Late Pleistocene sites on the

East European Plain, they are virtually devoid of earlier remains (Praslov 1984b, 1995).

The lack of evidence for Middle Pleistocene settlement in Eastern Europe seems consistent with the broader pattern of hominid distribution and adaptation prior to the appearance of the Neanderthals. The pre-Neanderthal inhabitants of northern Eurasia were largely confined to relatively warm and humid settings, and exhibit few if any of the adaptations to cooler and drier environments described in chapter 1 (summarized in fig. 1.6). There is no evidence for cold-climate morphological features such as shortened distal limb segments among the (admittedly scarce) sample of human skeletal remains in Europe that antedate OIS 8 (Hublin 1998, 298–300; Klein 1999, 295–312). Nor is there evidence for a diet high in meat, or an economy based primarily on the hunting of large mammals; the heavy wear observed on the teeth of the *Homo heidelbergensis* jaw from Mauer suggests a diet high in plant foods (Puech et al. 1980). Foraging range and territory size appear to have been small, and—with the possible exception of controlled fire—no significant advances in technology are evident (Gamble 1994, 134–143).

The combination of low winter temperatures and reduced abundance of digestible plant foods probably accounts for the scarcity of East European sites prior to OIS 5 (as well as the scarcity of West and Central European sites occupied during glacial periods prior to OIS 8). The *Homo* populations that inhabited latitudes above 40 degrees North prior to the later Middle Pleistocene (OIS 8–6) may have occupied a niche very similar to that of the *H. ergaster* and *H. erectus* populations that had colonized southern Eurasia during the Early Pleistocene. Very favorable climatic conditions in Western Europe during OIS 13–9 (especially the interglacial intervals) may have permitted expansion into this part of northern Eurasia without major adaptive change. Widespread colonization of Eastern Europe—and subsequent occupation of parts of Siberia—seem to have become possible only after the evolution of a specialized northern variant of *Homo* after 300,000 years ago.

The Neanderthals occupied much of Eastern Europe, and the apparent contrasts in morphology and behavioral ecology with their predecessors lend further credence to these conclusions (see chapter 4). Many Neanderthal skeletal features appear to represent cold adaptations (Coon 1962; Trinkaus 1981; Hublin 1998), and their morphology has been characterized as "hyperpolar" relative to modern humans (Holliday 1997, 254). Stable isotope analyses of their bones indicate a high meat diet (e.g., Richards et al. 2000), and taphonomic studies of large mammal remains in their sites suggest that they were capable hunters of big game (e.g., Chase 1986; Gaudzinski 1996). There is evidence of wider movements—

especially in the context of cooler and drier environments (Roebroeks et al. 1988; Mellars 1996, 161–165), and some important advances in technology, such as hafting of stone implements (e.g., Anderson-Gerfaud 1990).

Interpretation of the Middle Pleistocene record in Eastern Europe is further complicated by evidence for differences in morphology and behavior between Western and Central European hominids prior to the appearance of the Neanderthals. Acheulean bifaces are absent in Central Europe during this period, and lithic industries are characterized by pebble and flake tools (Gladilin and Sitlivyi 1990). Although skeletal remains are very scarce, isolated fragments from Bilzingsleben (eastern Germany) and Vértesszöllös (Hungary) exhibit some *H. erectus* features (Svoboda et al. 1996, 38–46). Both the artifacts and morphology suggest links with East Asia, and it has been hypothesized that the Central Europeans represent a separate dispersal into Europe from Asia (Rolland 1992; Klein 1999, 339–341).

No hominid fossils have been recovered to date from a Middle Pleistocene context in Eastern Europe, and the relationship between its peripheral inhabitants and humans in other parts of Eurasia is inferred from the stone artifact assemblages. The latter present a confusing picture that may be largely a function of the very small size of the sample. But it also probably reflects the fact that Eastern Europe seems to have been a settlement vacuum during the Middle Pleistocene, alternately drawing in—and perhaps periodically expelling—colonizing populations from the fringes of Central Europe and the Near East.

The early Middle Pleistocene pebble and flake tool assemblages from Treugol'naya Cave (probably OIS 13) suggest a connection with Central Europe (Doronichev 1992, 121–122). The problematic assemblages from Pogrebya and Dubossary I in the Lower Dnestr Valley (possibly OIS 11) also lack bifaces and exhibit similarities to the Central European industries (Anisyutkin and Levkovskaya 1992, 91–92). Although redeposited bifaces have been recovered from several localities in the Northern Caucasus, they appear to be much younger (terminal Middle Pleistocene or later) (Lyubin 1998, 156–168).

The later Middle Pleistocene artifacts from Mikhailovskoe and Khryashchi on the Northern Donets River (possibly OIS 9–7), and Myshtulagty lagat (Weasel Cave) in the Northern Caucasus (OIS 7–6) are somewhat different. The Northern Donets assemblages contain some Levallois prepared cores (Mikhailovskoe) and rather crude-looking Middle Paleolithic flake tool types (Praslov 1984b, 96–98), and may represent the initial appearance of the Mousterian in Eastern Europe. Such an early Mousterian industry might be derived from Central Europe, where

Mousterian assemblages (containing Levallois cores) are present as early as OIS 7 at Korolevo in the Carpathian Basin (Gladilin and Sitlivyi 1990, 47–55). Alternatively, it might be related to industries in the Near East, where a Levallois Mousterian may also be present by OIS 7 (Bar-Yosef 1998, 44–50). The Myshtulagty lagat assemblages, which are described as Denticulate Mousterian of the Levallois type with numerous blades, have been compared to some contemporaneous industries in the Near East (Hidjrati et al. 1997).

CHAPTER 4

Neanderthal Adaptations

The Neanderthals were a northern form of *Homo* that evolved gradually in Western Europe, and later spread into other parts of northern Eurasia. The hominid fossil record—including fossil DNA—indicates that they were the product of a lineage that had diverged from the southern *Homo* population about half a million years ago. Their characteristic features and way of life seem to have developed by the late Middle Pleistocene (roughly 200,000 years ago). They were probably the first hominids to adapt to cold environments, and this explains much about them.

The Neanderthals remain an enigma in many respects, and they continue to exercise a fascination on the public that has no equal in paleoanthropology. They have been the subject of popular films and novels, as well as heated academic debates. Much of the disagreement reflects the fact that they defy simple characterization; they seem to have been very similar to us in some ways and very different from us in other ways. Now that their evolutionary history and phylogenetic relationship to modern humans are better understood, they may be seen not so much as a primitive version of ourselves, but more as an alternative form of *Homo* (Trinkaus and Shipman 1992; Stringer and Gamble 1993).

Eastern Europe has much to tell us about the Neanderthals. They were the first hominids to settle widely across this part of the world, which was almost certainly a consequence of their unprecedented ability to cope with cold and dry environments. Both their morphology and hunting skills were probably critical to the occupation of the East European Plain. But the distribution of Neanderthal sites in space and time suggests that they were unable to inhabit the periglacial loess steppe environments that existed across much of the latter during intervals of extreme cold. The contrasting record of modern humans, who subsequently colonized these environments, suggests that the Neanderthals may have been fatally constrained by a lack of ability to develop innovative technology and to forage across large areas.

Neanderthal Evolution and Adaptation

Neanderthal Origins

The East European Neanderthals did not evolve locally, but arrived with a fully developed suite of morphological and behavioral traits from Western

Europe. Despite the belief that evolutionary change will often occur on the geographic margin of the species range—in response to the differing adaptive demands of marginal environments—there is no compelling evidence that the Neanderthals were the product of "geographic speciation" (Mayr 1970, 278–295). Such an event would presumably have taken place in the colder and drier regions of Central and Eastern Europe. Instead, it is apparent that the characteristic features of the Neanderthals evolved over a period of several hundred thousand years in Western Europe. These features, which are accompanied by evidence for behavioral changes, seem to have developed in response to temporal rather than spatial variations in environment. In other words, they represent an adaptive shift to the periods of glacial climate in Western Europe. The paleoclimatic record reveals that in the late Middle Pleistocene (between roughly 300,000 and 130,000 years ago), northern Eurasia was subject to prolonged intervals of extreme cold (Hublin 1998, 303–305).

F. Clark Howell first suggested in a series of influential papers authored during the 1950s that the European Neanderthals were the product of isolation in glacial Europe. He envisioned the Neanderthals as a geographic race, cut off from the larger archaic *Homo* population of western Eurasia by expanding glaciers and uninhabitable periglacial areas. They were isolated from the larger gene pool at the close of the Middle Pleistocene, and their specialized features evolved rapidly through the combined effects of genetic drift and selection (Howell 1952, 1957).

It now appears that the Neanderthal lineage was isolated much earlier from other *Homo* populations (or at least the *Homo* population that was ancestral to *H. sapiens*). Fossil DNA evidence points to an early Middle Pleistocene split (Krings et al. 1997; Ovchinnikov et al. 2000). Moreover, the barriers to gene flow from outside Europe, which seem to have been maintained over multiple interglacials and glacials, were probably more conventional and not a product of cold climates. The principal barriers were likely to have been large bodies of water and mountain ranges (and they were probably not absolute barriers but simply limited potential gene flow). Another isolating factor was probably the East European Plain. As noted in the previous chapter, it seems to have been largely or wholly uninhabited—during interglacials as well as glacials—until the appearance of the Neanderthals.

At least some of the skeletal features considered diagnostic of the Neanderthals are visible in west and south European hominid fossils dating to 500,000–300,000 years ago (corresponding roughly to OIS 13–OIS 9). Examples include the midfacial projection expressed in skulls from the caves of Petralona (Greece) and Arago (France), and the incipient suprainiac fossa on occipitals from Swanscombe (England) and Reilingen (Germany) (Hublin 1998, 299–300; Rightmire 1998, 223–224; Klein 1999, 296–306). Many Neanderthal cranial traits are recognized among the large sample of remains from the Sima de los Huesos at Atapuerca in northern Spain, which is currently dated to approximately 300,000 years ago. These specimens exhibit both midfacial projection and a suprainiac fossa, as well as large anterior teeth and a retromolar space (Stringer 1993; Arsuaga et al. 1997).

The full complement of specialized Neanderthal traits had emerged by the final interglacial/glacial cycle of the Middle Pleistocene (i.e., after roughly 250,000 years ago) (Hublin 1998, 300). The "classic Neanderthal" pattern is evident in fossils like those from Ehringsdorf (Germany) and Biache (France), which are dated to OIS 7 and OIS 6, respectively. Significantly, this development is associated with changes in the archaeological record, including the transition to a Middle Paleolithic tool industry (e.g., Maastricht-Belvedere [Netherlands] dated to OIS 7 [Roebroeks 1986]). Most of the European Neanderthal sample is derived from Late Pleistocene deposits that may be assigned to OIS 5–OIS 3 and dated to between 127,000 and 30,000 years ago (Stringer and Gamble 1993).

Neanderthal Morphology

The skeletal morphology of the Neanderthals exhibits some temporal and geographic variability, but represents a generally consistent complex of traits (e.g., Howell 1957; Trinkaus and Howells 1979; Stringer and Gamble 1993). The differences between the Neanderthals and modern humans embodied in this trait complex appear sufficient to justify classification of the former as a separate species (*Homo neanderthalensis*) (Tattersall 1986), although a number of anthropologists prefer a subspecies classification (*Homo sapiens neanderthalensis*) (e.g., Smith 1982; Wolpoff 1999). Salient features of the cranium include a long and low vault, high brain volume (averaging slightly over 1,500 cubic centimeters), low and receding frontal, and a well-developed supraorbital torus. The occipital bone extends outward to form what is called an occipital bun, and possesses a transverse torus that borders a pronounced depression (suprainiac fossa). The face projects forward, and is characterized by inflated cheeks and a large nasal cavity. The mandible typically lacks a chin and exhibits a gap between the third molar and the ascending ramus (retromolar space). The front teeth are large relative to the cheek teeth, and they usually display heavy wear, especially among adult specimens.

The postcranial skeleton is very robust with well-developed areas for muscle and ligament attachments. The vertebrae possess large transverse and spinous processes. The ribs are thick, forming a broad and deep chest. The clavicle is long and curved with large articular ends. The pubic bone is long and thin relative to modern humans. The scapula is broad and displays a deep groove along the margin of the dorsal surface. The limb bones are robust with thick shaft walls, and the distal limb segments are short relative to the proximal segments (i.e., low brachial and crural indices). The bones of the hands and feet are generally large, and the phalanges (fingers and toes) are short.

When F. Clark Howell proposed that the Neanderthals were the result of isolation in glacial Europe, he attributed their specialized morphology to a combination of genetic drift and selection "in response to the imposed rigors of a harsh subarctic climate" (Howell 1957, 337). He declined to speculate on the adaptive significance of particular skeletal features (e.g., Howell 1952, 401–403), and it was left to Carleton Coon to explain the morphology of the Neanderthals. In his 1962 synthesis *The Origin of Races*, Coon argued that many of their characteristic features were cold adaptations, and drew comparisons with the morphology of modern humans of the circumpolar regions. He particularly emphasized features of the face, such as the large nose (for warming inhaled air like a "radiator") and large infraorbital foramina (for generous blood supply to the face). But he also alluded to the adaptive significance of postcranial features, such as the shape of the chest and the shortened distal limb segments, which conformed to the predictions of Bergmann's and Allen's "ecological rules" of cold climate adaptation (see chapter 1). Coon noted the lack of "archaeological evidence of cultural improvements that would help mitigate the severity of the climate," and concluded that Neanderthal adaptation "was probably anatomical and physiological" (1962, 534).

Unfortunately, Coon had become embroiled in an ugly controversy concerning his purported views on race at the time that his book was published. The man and his work fell into disfavor among anthropologists, and his insights regarding the morphology of the Neanderthals were hotly disputed or largely ignored in the ensuing years (Trinkaus and Shipman 1992, 318–324). However, renewed research on this issue during recent decades has provided much support for Coon's original argument, especially with respect to the postcranial skeleton. For measurements that show a significant correlation with latitude or mean annual temperature among modern humans, Neanderthals fall at or beyond the end of the spectrum, exhibiting a "hyperpolar" morphology (Trinkaus 1981; Holliday 1997; Hublin 1998). These include brachial and crural indices (relative length of distal limb segments), length of limb segments relative to trunk height, index of femoral head diameter to femoral length (relative body linearity), and length of the clavicle (reflecting the breadth of the chest). The overall large size of the hands and feet, combined with short fingers and toes, also appears to fit a pattern of cold adaptation.

Coon's "radiator nose theory" proved to be the most controversial of his proposed Neanderthal cold adaptations (Trinkaus and Shipman 1992, 321). Although some anthropologists find Coon's argument compelling (e.g., Wolpoff 1999, 668–669), it remains problematic (e.g., Laitman et al. 1996; Franciscus 1999). But other cranial traits, such as the large infraorbital foramina and high endocranial volume, are found among modern human populations that have occupied high latitudes for many generations (Coon 1962, 534; Holloway 1985, 320–321). The latter suggests that the impressive size of the Neanderthal brain may be as plausibly ascribed to climate as cognitive ability.

The disputed explanation of the Neanderthal nose underscores the fact that most interpretations regarding the adaptive significance of specific traits in both fossil and extant species cannot be falsified. Such interpretations always run the risk of becoming untestable accommodative arguments that cannot exclude the many other forces that can influence the size, shape, function, or behavior of an organism (Gould and Lewontin 1979). Many Neanderthal skeletal traits appear to have been inherited from the earlier *Homo* population in Europe (e.g., low receding frontal, large supraorbital torus) and perhaps had no adaptive function. Some characteristic features probably reflect the effects of other components of the skeleton (e.g., the retromolar space as a function of the forward placement of the teeth). And other traits may have been the result of genetic isolation and random drift (Hublin 1998).

Nevertheless, the presence of multiple features that correlate so strongly with high latitude and low temperature among modern humans provides for a powerful argument that Neanderthals were morphologically adapted to cold climate. Furthermore, the Neanderthals also display some latitudinal variation in these features. The Near Eastern group, which appears to represent a relatively late expansion into slightly lower latitudes (roughly 32–40 degrees North), exhibits less extreme values for these features than the European sample (Trinkaus 1981, 215). By contrast, although the West and Central/East European Neanderthal groups manifest some differences, features related to cold adaptation are well expressed in both groups. Most important, the temporal and spatial distribution of the European Neanderthals shows a broad correspondence with cold environments, despite the fact that some remains are dated to the Last Interglacial warm peak (OIS 5e), when temperatures were higher than the present. In fact, Gamble (1986, 369–370) notes the limited evidence for sites dating to 127,000–115,000 BP, and suggests that much of Europe was unoccupied or sparsely inhabited during this interval.

Some Neanderthal features seem to have had an adaptive function, but one not related—at least directly—to cold climate. The large size of the front teeth is widely thought to reflect their regular use for gripping and chewing hides and other tough material (Coon 1962, 541–542; Brace 1964; Trinkaus 1983). In this case, the functional inference is supported by patterns of extremely heavy wear on the teeth commonly observed on all but the youngest individuals. The large and well-developed muscle attachments (including the deep groove on the margin of the dorsal scapula) indicate especially powerful limbs and hands, which presumably reflect strenuous activity requirements. More problematic is Trinkaus' (1983) suggestion that the long pubic bone indicated a wide birth canal in females, and might correspond to a gestation period of more than nine months (apparently invalidated by the pelvis recovered from Kebara, Israel [Stringer and Gamble 1993, 86–88]).

Neanderthal Behavior and Ecology

Coon's provocative observation that Neanderthal adaptation was based primarily on morphology and physiology rather than "cultural improvements" represents a fundamental theme in Neanderthal studies. It was restated by C. Loring Brace (1964) and by other anthropologists in subsequent years. One of the more important implications of this thesis is that Neanderthal adaptations can be properly understood only by combining the analysis of skeletal morphology with archaeology. Each aspect of their adaptation must be balanced with the other. In more recent years, it has become apparent that the Neanderthals also may have developed some significant behavioral adaptations to cold environments.

To begin with, the dating of fully developed Neanderthals to the late Middle Pleistocene (roughly 250,000 years ago) shows that their emergence is broadly coincident with some improvements in stone tool technology. These include the Levallois prepared core technique and other production methods that allowed greater control over the size and shape of tool blanks (Van Peer 1992; Boeda 1993; Mellars 1996). The analysis of microscopic wear patterns on the finished tools indicates that some of them were hafted (Beyries 1988; Anderson-Gerfaud 1990, 406–410), and this may explain the need for increased standardization in the size and shape of tools. Although the design of the handles (which were almost certainly fashioned from wood) is unknown, their use could have dramatically enhanced tool performance and efficiency.

The Neanderthals also may have developed a wooden tool technology significantly more sophisticated than that of their predecessors. Microwear studies now reveal that a high proportion of their stone tools were used for working wood. The microwear patterns include traces of scraping, whittling, planing, chopping, and sawing (Anderson-Gerfaud 1990, 401–404). Because only isolated specimens of wood artifacts have been preserved and recovered from Lower and Middle Paleolithic sites in Europe (e.g., Lehringen [Movius 1950], Schöningen [Thieme 1997]), it is currently impossible to assess the similarities and differences between the wooden tools of the Neanderthals and their predecessors.

Another realm of possible "cultural improvements" overlooked by Coon is foraging strategy and diet. At the time that Coon published his book in 1962, the bones and teeth of mammals found in Lower and Middle Paleolithic sites were assumed to represent the hunted prey of their occupants. Not until the early 1980s was this assumption widely challenged (e.g., Binford 1981), and replaced by critical analyses of faunal remains from these sites. Taphonomic studies of Lower and Middle Paleolithic large mammal assemblages from Europe indicate that there may have been some important differences between the ecology of the Neanderthals and that of their predecessors. While sites associated with the Neanderthals contain substantial evidence for hunting large mammals and central-place foraging (e.g., Chase 1986; Gaudzinski 1996), earlier sites (i.e., antedating 250,000 years ago) are ambiguous (see chapter 3). The latter often yield bones with traces of stone tool cut marks indicating that they were stripped of meat, but it is not clear that they represent the remains of hunted prey (Gamble 1994, 136–137). Because few caves or rockshelters are preserved from this time period, the pre-Neanderthal archaeological record is biased against the types of sites that are typically used to infer hunting and central-place foraging, and the issue is unresolved.

Nevertheless, the Neanderthals may have developed new foraging tactics involving the regular hunting of large mammals, and substantially increased the meat component of their diet. Such changes in European hominid ecology could have been critical to the colonization of colder and drier environments, where caloric requirements were greater and available plant foods were less abundant. Changes in foraging strategy might be related to increases in technological complexity and efficiency (e.g., hafted implements).

As they spread eastward at the close of the Middle Pleistocene (Hublin 1998, 306), Neanderthal foraging adaptations—perhaps including more complex and efficient tools—may have been as much a prerequisite for occupation of areas like the East European Plain as their cold-climate morphology. However, Coon's insight into the importance of morphology and physiology in their adaptations retains much value. In contrast to the first modern human occupants of Eastern Europe, the Neanderthals apparently failed to develop any of the advanced technologies for cold protection (e.g., tailored clothing, heated shelters) found among recent hunter-gatherers in northern latitudes (Hoffecker 1999a).

Neanderthal Sites in Eastern Europe

The spatial and temporal distribution of Neanderthal sites in Eastern Europe is an issue of critical importance to understanding their adaptation and the subsequent transition to modern humans. Both the presence and absence of their sites in specific spatial and temporal settings provides a measure of their ability to cope with varying environmental conditions. Comparison with the distribution of modern human sites offers a revealing contrast between the adaptive strategies of the two forms of *Homo* in a marginal environment (fig. 4.1).

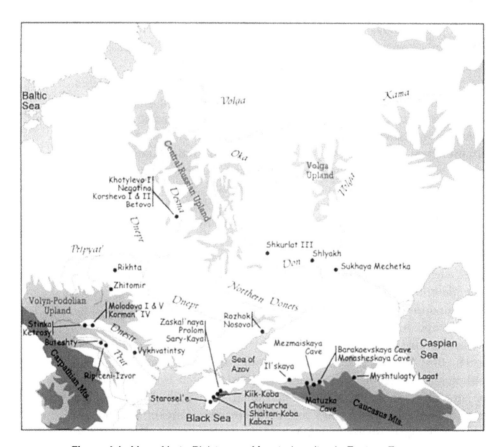

Figure 4.1. Map of Late Pleistocene Mousterian sites in Eastern Europe.

Unfortunately, mapping Neanderthal site distribution in Eastern Europe is fraught with problems and uncertainties. While the widespread occurrence of caves and rockshelters in Western and Central Europe ensures both the preservation and recognition of sites, many areas of Eastern Europe are largely devoid of natural shelters, and sites are more vulnerable to destruction and more difficult to find (see chapter 2). Accordingly, the known distribution of sites is especially susceptible to the influence of local geologic conditions, recent human settlement, and archaeological survey.

A more serious problem is the dating of Neanderthal sites in Eastern Europe. Although some radiocarbon dates in the 45,000–30,000 years BP range are available, many of these dates are suspect because they lie at or near the effective temporal limits of the method and have a high potential for contamination from young carbon (Sinitsyn et al. 1997, 23–24). The majority of occupations are significantly older, and while absolute dating methods with greater range (e.g., OSL, ESR) have been applied to a small number of them (e.g., Rink et al. 1998), most are currently dated by other means. The latter generally include dating with reference to paleobiological remains, buried soil horizons, and/or overall stratigraphic context; they often provide very low temporal resolution and retain many uncertainties. Occupation levels are typically assigned to broad temporal or stratigraphic units (e.g., OIS 3), and even these are often problematic.

Another concern is the identification of Neanderthal sites in the archaeological record. Less than a dozen sites in Eastern Europe contain Neanderthal skeletal remains (see later discussion), and most of their sites are identified as such by the presence of stone artifacts characteristic of the *Mousterian* industry. This approach entails some risk, because Mousterian artifacts have been found associated with both Neanderthal and modern human skeletal remains in the Near East, and some perceive a similar pattern in Eastern Europe (e.g., Praslov 1984b, 109–110). Many archaeologists have avoided the issue by decoupling the Mousterian from any specific set of skeletal remains, but this precludes the critically important analysis of the archaeological record in conjunction with the morphology of its creators.

At present, it appears that all Mousterian artifacts in Western and Central Europe—and almost all Mousterian artifacts in Eastern Europe—were produced by Neanderthals. In Eastern Europe, Neanderthal remains are firmly linked to the Mousterian industries of the Crimea (modern human remains originally assigned to the Mousterian site of Starosel'e have now been redated to the Holocene [Marks et al. 1997]). Most Mousterian industries in the Northern Caucasus are also associated with Neanderthal remains with the possible exception of artifacts in the lower levels at Myshtulagty lagat (Weasel Cave), which lack associated skeletal remains and bear similarities to Near Eastern industries (Hidjrati et al. 1997). Because some of the latter were made by early modern humans (*Homo sapiens*), Myshtulagty lagat could reflect their presence on the southeastern margin of Eastern Europe. Skeletal remains from Mousterian sites on the East European Plain are rare and ambiguous. These sites may be tentatively attributed to the Neanderthals on the basis of the similarities of their artifacts to industries produced by Neanderthals in adjoining regions (Central Europe and the Crimea).

In Western and Central Europe, it is apparent that Neanderthals also manufactured some industries traditionally classified as Upper Paleolithic, including the Chatelperronian (Franco-Cantabrian region) and probably the Szeletian (Central Europe) during OIS 3 (e.g., Allsworth-Jones 1986; Mellars 1996). Similar industries have been found in Eastern Europe, but lacking any associated skeletal materials (e.g., "Brynzeny Culture"), and the possibility remains that at least some of these were also produced by local Neanderthals (discussed in chapter 5).

General Patterns of Settlement

The Neanderthals settled widely across the southern half of Eastern Europe, although the number of sites varies according to area and time period. The absence of sites above latitude 53 degrees North may be explained in part by glacial advances across the northwestern portion of the East European Plain, which would have destroyed traces of habitation. However, large areas in the northeastern part of the plain remained unglaciated, and the lack of known sites in this region seems more likely to reflect the absence of settlement. Even under the full interglacial conditions of the present day, these areas experience January mean temperatures of between −10 and −15 degrees C (Konyukova et al. 1971).

The localities of Mikhailovskoe and Khryashchi on the Northern Donets River (see chapter 3) may indicate a Neanderthal presence on the southern plain during the final warm intervals of the Middle Pleistocene (possibly OIS 9–7). However, most evidence of Neanderthal settlement in Eastern Europe is confined to the Late Pleistocene, and the earliest group of sites is dated to the Last Interglacial (OIS 5e), which is calibrated to between 127,000 and 115,000 years ago. Climatic conditions were warmer than present during much of this brief interval, and it is represented by a widespread and characteristic buried soil horizon (i.e., Salyn phase of the Mezin Soil [see chapter 2]).

Curiously, there are few traces of Last Interglacial occupation in the southwestern region of the East European Plain, which contains the richest concentrations of Neanderthal sites from other periods (Praslov 1984a). Last Interglacial sites are also absent in the Crimea, which may reflect the fact that it had apparently become an island at this time (Soffer 1992, 247). However, they are found on the northern and eastern fringes of the distribution of known sites, including Khotylevo and other localities on the Middle Desna River, Shkurlat on the Middle Don River, and Sukhaya Mechetka on the Lower Volga River (Zamyatnin 1961; Zavernyaev 1978; Shevyrev and Khrisanfova 1984). Caves on the northern slopes of the Caucasus Mountains also contain occupations dating to OIS 5e (e.g., Myshtulagty lagat [Hidjrati et al. 1997]). Sites of the OIS 5e thermal maximum are not common in Western and Central Europe, and their scarcity is thought to indicate unfavorable conditions created by dense forests and marshes (Gamble 1986, 367–370). It may be significant that Last Interglacial occupations in Eastern Europe are found in the cooler and drier lowland regions and at higher elevations.

Sites dating to the Early Glacial period (OIS 5a–5d), which is calibrated to between 115,000 and 71,000 years ago, are relatively common. Most of these occupations are associated with several prolonged intervals of intermediate warmth that are represented by well-developed buried soils. Sites are found on the western and southern margins of the East European Plain, especially in the Dnestr and Prut Valleys (e.g., Molodova V) and near the coast of the Sea of Azov (e.g., Rozhok I), and in the southern upland areas of the Crimea (e.g., Zaskal'naya V) and Northern Caucasus (e.g., Matuzka) (Praslov 1984a; Kolosov et al. 1993). Some occupations may be present in the central East European Plain along the Middle Desna River (e.g., Korshevo I and II [Tarasov 1989]). Sites of the Early Glacial are also common in Western and Central Europe, and conditions throughout much of the continent seem to have been especially favorable for the Neanderthals (Gamble 1986, 160–177).

The Early Pleniglacial cold phase (OIS 4), which is currently dated to 71,000–55,000 years BP, apparently saw conditions similar to those of the Last Glacial Maximum. During this period, loess accumulated on the East European Plain, and rockfall horizons were deposited in caves of the southern uplands. At least a few occupations in the Crimean and Northern Caucasus caves are tentatively dated to OIS 4 (e.g., Barakaevskaya Cave [Lyubin 1994]). However, much of the East European Plain may have been unoccupied. The widespread loess unit assigned to the Early Pleniglacial in the central part of the plain (i.e., Valdai Loess I [Velichko 1990]) is reportedly devoid of artifacts. In the southwestern region, there is a gap in the succession of occupations at major stratified sites in both the Dnestr and Prut Valleys that appears to date to this interval (Carciumaru 1980; Ivanova 1987). Abandonment of "polar desert" areas during OIS 4 is also thought to occur in Central Europe (Gamble 1986, 374–375).

The final period of Neanderthal settlement coincides with the Middle Pleniglacial (OIS 3), which began at roughly 55,000 years BP and lasted until the beginning of the Last Glacial Maximum or Late Pleniglacial at 25,000 years BP. This was a prolonged interval of cool climate interrupted by several slightly warmer episodes represented by loess and one or more weakly developed soil horizons. The youngest Neanderthal occupations are dated to roughly 30,000 years BP (i.e., shortly before the close of this period) and appear to overlap with the earliest modern human settlement (dated to more than 35,000 years BP). Middle Pleniglacial Neanderthal sites are firmly documented in the Crimea (e.g., Starosel'e, Kabazi II [Marks and Chabai 1998]) and the Northern Caucasus (e.g., Mezmaiskaya Cave [Golovanova et al. 1999]). Traces of occupation are also present in the southwest region of the East European Plain at Korman' IV and Molodova V in the Dnestr Valley (Chernysh 1977, 1987). However, evidence of occupation on the central plain is ambiguous. The Mousterian artifacts at Betovo on the Desna River are associated with evidence for cold conditions and could date to the Middle Pleniglacial, although they may be much older (Tarasov 1977, 1989). Also, Neanderthals might have produced some of the earliest Upper Paleolithic assemblages at Kostenki on the Middle Don River, but this remains to be demonstrated (see chapter 5).

In sum, during the Middle Pleniglacial, East European Neanderthal occupations are common in the southern uplands, but seem to have been rare or absent on the central plain, where climatic conditions were almost certainly harsh. The apparent reoccupation of the southwestern region—where conditions were probably milder—reflects a similar pattern in northern areas of Central Europe (Gamble 1986, 374–375). The recent dating of occupation levels in the Crimea and Northern Caucasus to approximately 35,000–30,000 years BP suggests that the Neanderthals were present—at least in these parts of Eastern Europe—as late as they were in Western and Central Europe (Marks and Chabai 1998; Ovchinnikov *et al.* 2000).

East European Plain: Southwest Region

The most important Mousterian sites in this part of Eastern Europe are located along the Middle Dnestr River roughly 100 kilometers northeast of the city of Chernovtsy in the Ukraine. The valley along this segment of the river forms a canyon more than 200 meters deep, and contains up to eight terrace levels (now partially flooded by a hydroelectric dam reservoir). Although Paleolithic remains have been found on most terrace levels and both sides of the river, major Mousterian sites are concentrated along the south bank on the second terrace (25–40 meters above the prereservoir level of the river). These include *Molodova I*, *Molodova V*, and *Korman' IV*, all of which were investigated by the late A. P. Chernysh over a period of several decades (Chernysh 1959, 1977, 1982, 1987) (fig. 4.2).

The base of the second terrace is composed of alluvial sand and gravels deposited during the Last Interglacial. At Molodova I, the alluvium grades upslope into a thick buried soil layer containing terrestrial molluscs typical of warm interglacial conditions (e.g., *Helix pomatia*) that also correlates to OIS 5e. The layers containing Mousterian artifacts overlie these deposits and represent a sequence of colluvial loams and buried soils that date to various phases of the Early Glacial (OIS 5a–5d) and Middle Pleniglacial (OIS 3). Much of the loam was originally composed of loess that accumulated on surfaces throughout the valley, and almost continuously washed downslope by gravity and water. Many of the soil horizons were also eroded or reworked by slope action (Klein 1973, 18–26; Ivanova 1977, 1982) (fig. 4.3).

These locations were occupied repeatedly—presumably by Neanderthals—who often accumulated large quantities of lithic debris, mammal bones, and traces of hearths. The lithic assemblages reflect use of Levallois prepared core techniques, production of blades, and manufacture of unifacial knives, scrapers, and points (Chernysh 1982, 1987). Among the associated faunal remains, mammoth and horse are most common, but this is probably influenced by weathering (i.e., differential preservation of larger bones and teeth) (Alekseeva 1987, 154–156). At the time of their occupation, the sites were on the lowest terrace level—approximately 20 meters above the river floodplain—adjacent to side-valley ravines. The locations were consistently attractive because they probably represented reasonably well drained and sheltered encampments (near the valley floor) in close proximity to freshwater and multiple resources. The latter included raw material for stone tools in the form of flint cobbles and fragments in the alluvium, and perhaps local concentrations of specific plants and animals (although there is no evidence for exploitation of aquatic animals in the river).

Another important site in the Middle Dnestr Valley is located roughly 50 kilometers upstream from Molodova. *Ketrosy* also is situated on the modern second terrace of the south bank adjacent to a large tributary stream (Anisyutkin 1981a, 7–8). A biface fragment recovered from the alluvium near this site represents one of the few possible traces of Last Interglacial occupation in the region (Anisyutkin 1981b, 57). Most artifacts and associated debris were found in the overlying loams and are dated to an Early Glacial interstadial (possibly OIS 5a) (Ivanova et al. 1981). Mammoth predominates heavily among the faunal remains (David 1981, 135–137). The small lithic assemblage is similar to those from the Molodova

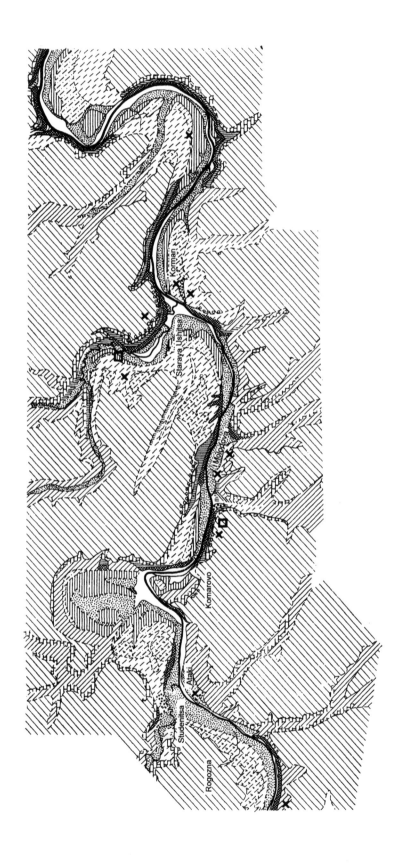

sites (i.e., Levallois cores, scrapers), but unlike the latter, Ketrosy lacked a consistent attraction over a period of tens of thousands of years. Neanderthals apparently visited the site on multiple occasions, but over a restricted period of time. It was subsequently flooded, later gradually buried in slope deposits, and never reoccupied.

Neanderthals also camped on the higher terrace levels in the Dnestr Valley. The site of *Stinka* is located near Ketrosy on a fourth terrace promontory roughly 70 meters above the modern river. A small assemblage of artifacts was found at a depth of less than a meter in loam unconformably overlying the ancient alluvium of the fourth terrace. The sediment containing the artifacts yielded pollen and spores indicating a forest-steppe and relatively cool climates, and the occupation might date to the Early Glacial or the Middle Pleniglacial (Anisyutkin 1969; Ivanova 1969). Faunal remains were not preserved. The technology and tool types at Stinka are different from those at the sites previously mentioned; there is little evidence of Levallois techniques and the tools include many denticulates and some bifaces. Both the artifacts and topographic setting of the site contribute to a complex picture of Neanderthal adaptations in this region.

Roughly 250 kilometers downstream from Stinka on the Dnestr third terrace level is the small rockshelter of *Vykhvatintsy*, which was formerly a cave (Chernysh 1965, 103–105). A large quantity of mammal remains (carnivores, giant deer, red deer, bison, and horse) and roughly 200 artifacts were recovered from a clay layer near the base of the deposits filling the shelter. The fauna is apparently lacking cold-loving taxa, and may date to the Last Interglacial; the associated artifacts include cores and relatively crude flake tools (Ketraru 1973, 21–24). A smaller collection of artifacts, including bifaces, and mammal remains (e.g., mammoth, reindeer) were found in an overlying clay unit.

Some major sites are also present in the neighboring Prut Valley. Roughly 100 kilometers south of Molodova, along the west bank of the Prut in Rumania, lies *Ripiceni-Izvor*. This site rests on a low bedrock terrace and contains a sequence of Mousterian (Neanderthal) occupation levels buried in primary loess (Paunescu 1965; Paunescu et al. 1976). The three lower levels are associated with mixed forest vegetation (based on the analysis of pollen and spores) that seems to reflect rather mild interstadial conditions, and are dated to the Early Glacial (Carciumaru 1980). An overlying loess unit is associated with cold steppe vegetation and may correlate with the Early Pleniglacial cold phase; it is devoid of artifacts. The three upper Mousterian levels were deposited during slightly milder

Figure 4.2. *(opposite)* Topographic setting of Mousterian sites in the Middle Dnestr Valley (after Ivanova 1959, 229, fig. 5).

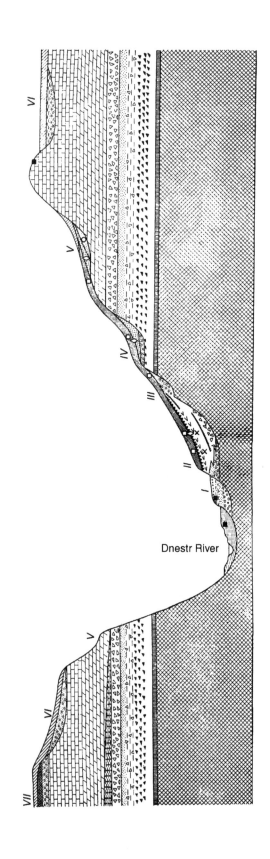

conditions, and are thought to date—with supporting radiocarbon estimates—to the Middle Pleniglacial (Mertens 1994, 516–517).

Ripiceni-Izvor exhibits many parallels with the Molodova sites on the Dnestr River. Both seem to have been occupied repeatedly over a period of time spanning at least parts of the Early Glacial (OIS 5a–5d) and Middle Pleniglacial (OIS 3) with a hiatus during the Early Pleniglacial (OIS 4). The occupations contain large quantities of lithic debris and mammal bones, and apparently were the loci of a variety of activities and perhaps extended habitation. However—as in the case of Stinka—there are significant contrasts in lithic production techniques and tools, especially during the later periods of occupation. Although Levallois flaking technology was employed throughout the sequence, the tools from the upper levels include many bifacial forms.

Rockshelters located within the Prut Valley and its tributaries were also apparently inhabited by Neanderthals. The remains of a rockshelter—now completely destroyed—were found at *Buteshty* on a small tributary of the main river (Kamenkutsa River) in Moldova. Several hundred artifacts were recovered from this locality in association with mammal bones and teeth that may be tentatively dated to the earlier Late Pleistocene (OIS 5). The cores and tools are similar to those from the Molodova sites, but the heavy predominance of cave-dwelling carnivores among the mammal fauna (more than 80 percent) suggests only intermittent use by Neanderthals (Ketraru 1969, 1973, 43–46). The small rockshelter of *Buzduzhany I* is also located along a tributary valley (Rakovets River). As at Buteshty, several hundred Mousterian artifacts were recovered in association with mammal remains that included a high percentage of carnivores. However, horse remains were also relatively common at this site. The age of the occupations is unclear, although the presence of reindeer among the fauna indicates that they postdate the Last Interglacial (Ketraru 1973, 48–51). A similar occupation is present in the lowest layer of the *Trinka I* rockshelter, which is located on another small Prut Valley tributary (Dragishte River) in this region (Anisyutkin et al. 1986, 33–56).

East European Plain: Central Region

Mousterian sites are less common in the central East European Plain than they are in the western and southern parts of Eastern Europe. This may reflect a pattern of settlement restricted to periods of very mild climate (firmly dated sites are confined to the Last Interglacial). But it also may be influenced by processes that affect the preservation and discovery of

Figure 4.3. *(opposite)* Geomorphic context of Mousterian sites in the Middle Dnestr Valley (modified from Chernysh 1965, 10–11, fig. 2).

archaeological remains. Many sites in this region are found in stream deposits, and represent occupation debris that has been disturbed or redeposited. At the end of the Last Interglacial, the landscape seems to have been subject to widespread erosion and mass-wasting (Tarasov 1989, 170).

Two important localities have been discovered in the west Dnepr Lowland near the northern margin of the Volyn-Podolian Upland. This area (known as the *Poles'e*) is a poorly drained basin that was almost entirely covered by the massive Dnepr ice sheet during the late Middle Pleistocene (OIS 8). *Zhitomir* is located long a small river (Svinoluzhka) near the terminal moraine of this glacier (and east of the city of Zhitomir in the Ukraine). At this locality, several thousand artifacts were collected from the surface, and from shallow depths in bedded sandy loam and sand overlying eroded Dnepr moraine deposits. None of the material appears to be in primary context, and artifacts from various periods—including the Upper Paleolithic and Neolithic—are mixed together with naturally fractured rock (Mesyats 1962). Several groups of early artifacts have been distinguished on the basis of weathering, and they comprise forms that are typical of the East European Mousterian, including small handaxes and other bifacial tools (Kukharchuk 1995). Although the site remains undated (faunal remains are absent), it may be tentatively assigned to the Neanderthals.

Rikhta, which is located north of Zhitomir on another small river (Rikhta River), is similar to the latter. The site occupies a low promontory only a few meters above the modern river level. Over 12,000 artifacts were found at shallow depths in a sand layer containing inclusions of redeposited Dnepr till (Smirnov 1979, 9–10). No faunal remains were encountered. Some artifacts typical of the Upper Paleolithic are present, and this assemblage may also include materials from more than one time period. Nevertheless, most of the artifacts are similar to the weathered finds from Zhitomir, and they are generally attributed to the Mousterian (Praslov 1984b, 108). Both of these sites may have been located near the Dnepr moraine in order to exploit a rich source of lithic raw material in the form of flint cobbles and nodules.

The primary concentration of sites in the central plain is found along a segment of the Middle Desna River northwest of the city of Bryansk (Russian Federation). The most remarkable of these is *Khotylevo I*, which was discovered and investigated by F. M. Zavernyaev during 1958–1964. This site extends for a full kilometer along a terrace on the west bank of the river roughly 20 meters above the floodplain. Large-scale excavations exposed approximately 500 square meters, and recovered an assemblage of roughly 18,000 artifacts and associated faunal remains from alluvial sand and gravels. The most common vertebrates are mammoth, red deer, horse, and bison. Because the mammalian fauna is characteristic of the Late Pleistocene and the molluscan fauna indicates climatic conditions warmer than those of the present day (Motuz 1967), the artifact layer may be firmly assigned to the Last Interglacial thermal maximum (OIS 5e). The overlying layers comprise buried soils and loessic loams dating to OIS 5–2 (Zavernyaev 1978) (fig. 4.4).

As at the Poles'e sites, Khotylevo I may contain artifacts from several contexts, although presumably none of them postdates the Last Interglacial. Many of the artifacts seem to be redeposited, but they exhibit sharp edges and lack traces of rolling, and apparently were carried only a short distance from their original location (Zavernyaev 1978, 32–33). The site probably represents a group of loci situated on or near the Last Interglacial floodplain that were occupied at different times over an extended

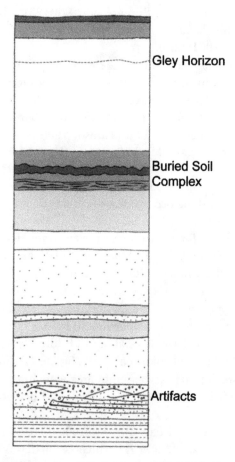

Figure 4.4. Stratigraphic profile of Khotylevo I on the Middle Desna River (redrawn from Zavernyaev 1978, 11, fig. 3b).

period. The lithic assemblage contains a diverse array of cores—including numerous Levallois cores—small handaxes, bifacial points, and other tools. The high ratio of cores to tools indicates that much primary flaking occurred at the site (and the alluvial gravels are a raw material source), but the battered edges of many tools suggest that other activities were performed at Khotylevo as well (Zavernyaev 1978; Praslov 1984b, 108).

Roughly 30 kilometers upstream from Khotylevo lie two localities that yielded artifacts in a similar setting. At *Negotino* and *Negotino-Rudnyanka*, which are found on opposite banks of the Desna, small quantities of redeposited flakes and Mousterian tools were recovered from alluvial gravels (Tarasov 1989, 167–169). Lightly rolled cores, tools, and flakes were also found in sandy alluvium many kilometers downstream near the town of Novgorod-Severskii at *Chulatovo III* (Zarovskaya Krucha) in the Ukraine. As at Khotylevo I, they were associated with the remains of Late Pleistocene mammals (Voedvodskii 1947; Beregovaya 1960, 23). All of these finds occur in Last Interglacial alluvium with inclusions of eroded Dnepr moraine (Velichko 1961b, 59).

The Bryansk area also contains some sites that probably represent later Mousterian occupations under cooler climatic conditions. The twin sites of *Korshevo I* and *II*, located upstream from Khotylevo on the west bank of the Desna River, contain *in situ* occupation layers buried in a humic bed at the base of several meters of loess (Tarasov 1986, 1989, 170–172). The humic bed appears to be the Mezin Soil Complex, which may be broadly correlated with OIS 5 (Last Interglacial and Early Glacial). Pollen and spores from this unit indicate boreal and mixed forest found today at more northern latitudes, and the occupations appear to date to an Early Glacial interstadial (OIS 5a or 5c).

At the nearby site of *Betovo*, artifacts were excavated from an analogous context (Tarasov 1977) (fig. 4.5). However, pollen and spores from the occupation layer reflect much colder conditions with few trees, and associated vertebrates include modern tundra species such as collared lemming (*Dicrostonyx torquatus*) and arctic fox (*Alopex lagopus*). The dating of the Betovo occupation layer is problematic. It has been variously assigned to the Last Interglacial, Early Glacial, and Middle Pleniglacial (Tarasov 1977, 29–30; 1989, 174; Velichko et al. 1997, fig. 2). All three of these sites contained Mousterian artifact assemblages with high percentages of denticulate and notched tools, and Korshevo I and Betovo also produced several bifaces.

Settlement on the central East European Plain is also documented on the Middle Don River at *Shkurlat III*, which is situated 9 kilometers east of the main river along a tributary (Gavrilo River) near the town of

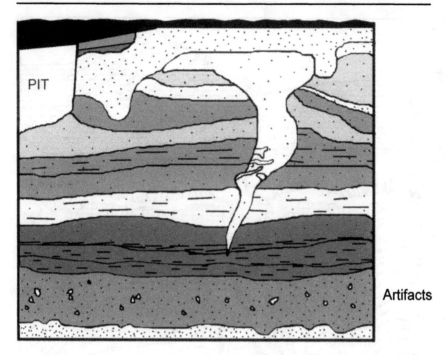

Figure 4.5. Stratigraphic profile of Betovo on the Middle Desna River (redrawn from Tarasov 1977, 21, fig. 2).

Pavlovsk. At this locality, alluvial sands and gravels of the third terrace yielded an assemblage of warm-loving mammals and molluscs that are dated to the thermal maximum of the Last Interglacial (Shevyrev et al. 1979; Alekseeva 1990, 59–63). The remains included a human scapula fragment that exhibits features characteristic of both the Neanderthals and archaic modern humans (e.g., bisulcate lateral border); no artifacts or other traces of human occupation are reported (Shevyrev and Khrisanfova 1984).

East European Plain: Southern Region

The most important Mousterian site in the southern plain is *Sukhaya Mechetka* (Volgograd), located along a large ravine on the west bank of the Volga River. It was excavated in 1952 and 1954 by S. N. Zamyatnin, who exposed a total of approximately 650 square meters and mapped a Mousterian—presumably Neanderthal—occupation floor of unusual size and complexity (fig. 4.6). Artifacts and features were found on the former surface of a dark-reddish-brown buried soil that underlies a thick sequence of loams, sands, and clays (Zamyatnin 1961, 9–12). The sands and clays

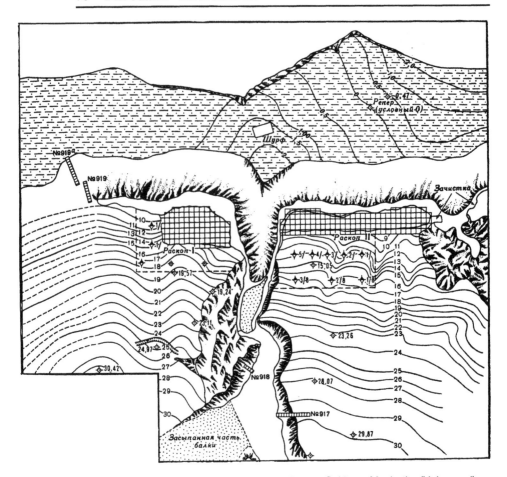

Figure 4.6. Topographic setting and excavations at Sukhaya Mechetka (Volgograd) on the Lower Volga River (after Zamyatnin 1961, 9, fig. 2).

contain marine molluscs and were deposited during the Khvalyn Transgression of the Caspian Sea, which is correlated with the Pleniglacial (OIS 4–2). They overlie loams containing at least one buried soil that may date to the Early Glacial (OIS 5a–5c), which in turn overlie the occupation layer—dated to the Last Interglacial (OIS 5e). Small samples of pollen and spores from the associated reddish-brown soil suggest steppe conditions similar to those of the present (Chiguryaeva and Khvalina 1961).

Roughly 8,000 stone artifacts were recovered from the occupation layer, including a variety of non-Levallois cores, small handaxes, bifacial foliates, scrapers, and other tools. Raw material was evidently obtained from nearby deposits. Many of the tools were found with concentrations of flaking debris or mammal bone fragments (primarily steppe bison [*Bison priscus*]). Some concentrations are associated with traces of former hearths containing wood charcoal (Zamyatnin 1961, 12–15). The charcoal is derived from coniferous trees that probably grew on the slopes of ravines in the area (Chiguryaeva and Khvalina 1961, 39).

Sukhaya Mechetka represents a location probably occupied repeatedly for a variety of activities, but only during the period when the reddish-brown soil was forming. At this time the surface was stable, and the artifacts and other remains seem to have accumulated across a large area as separate and occasionally overlapping scatters of debris and ash. The site was not reoccupied in later periods, although this may be due largely to the fact that it was underwater during much of the Pleniglacial.

A different type of site is found at *Rozhok I* on the coast of the Gulf of Taganrog. Discovered and excavated by N. D. Praslov in 1961–1962, this locality contains a sequence of six occupation horizons buried in colluvial loams that washed into an ancient ravine. They lie more than a meter below a buried soil horizon thought to date to the Middle Pleniglacial. Pollen and spores from the loams containing the artifacts indicate relatively moist conditions with the presence of some deciduous and coniferous trees. Another buried soil horizon underlies the colluvial loams. The Rozhok I occupations are widely correlated with OIS 5 (possibly the OIS 5a or 5c interstadial) although they could conceivably postdate OIS 5 (Praslov 1968, 64–71, 1984a, 31, fig. 5; Klein 1969a, 90–91).

The occupation horizons are strikingly different from Sukhaya Mechetka. On all levels, artifacts are less numerous than mammal remains, which primarily consist of steppe bison, but also include wild ass, giant deer, and others. This is at least partly the result of unusually good preservation and limited fragmentation of bone and teeth. The lithic assemblages are small and generally contain a high proportion of tools—chiefly scrapers (Praslov 1968, 72–93). Rozhok I seems to represent a limited-activity site occupied periodically for relatively brief intervals. A similar occupation seems to exist at the nearby site of *Nosovo I*, although only one occupation layer is present, no faunal remains were preserved, and the tools include several bifaces (Praslov 1972).

The recently discovered site of *Shlyakh* is found along the Lower Don River near the town of Frolovo. The site is located on a large ravine roughly 14 kilometers east of the river. Artifacts are buried in a sequence of loams and sandy loams that appear to be at least partly of colluvial origin. The main occupation layer lies beneath a buried soil of unknown age, and yielded two radiocarbon dates of 45,700 and 46,300 years BP. The artifacts from this layer include Levallois blade cores and typical Mousterian tools (points, side-scrapers, and knives) (Nekhoroshev 1999). Although this assemblage is thought to represent a very late Mousterian industry that anticipates Upper Paleolithic blade technology, it should be noted that Levallois blades are found in other East European Mousterian sites (e.g., Molodova, Myshtulagty lagat), and the dates must be treated as minimum estimates.

Crimea

The southern Crimea contains the richest concentration of Neanderthal sites in Eastern Europe, which almost certainly reflects not only the mild climate but the increased preservation and visibility of sites in caves and rockshelters (although many of these have since eroded and collapsed). The first discovery of Mousterian artifacts in Eastern Europe was reported in 1880 by K. S. Merezhkovskii at the cave of Volchii Grot (near the city of Simferopol'), and the first Neanderthal skeletal remains from this part of the world were found at Kiik-Koba in 1924 by G. A. Bonch-Osmolovskii (Vekilova 1979) (fig. 4.7).

Several major rockshelters and open-air sites are located at the eastern end of the Crimean Mountains near the city of Belogorsk. These sites have been discovered and/or investigated by the Ukrainian archaeologist

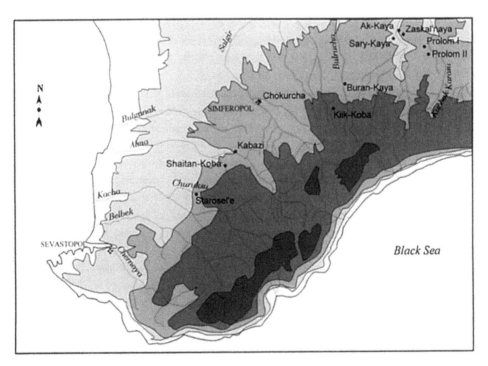

Figure 4.7. Map of Mousterian sites in the Crimea (redrawn from Ferring 1998, 21, fig. 2–4).

Yu. G. Kolosov over the past three decades. Roughly 7 kilometers north of the city, on the east side of the Biyuk-Karasu River, lie *Zaskal'naya V* and *VI*, which are most widely known for discoveries of Neanderthal skeletal remains. Both localities represent former caves that have now collapsed onto the upper slope of a large ravine. At each site, seven occupation layers are buried in a thick sequence of sand and rubble, and the lower layers are thought to date to the Early Glacial. A buried soil horizon associated with the third occupation layer at Zaskal'naya V (fourth layer at Zaskal'naya VI) is tentatively assigned to a late interstadial during this interval (possibly OIS 5a). Rockfall units within the succession of occupation layers are correlated with various cold episodes during the Early Glacial and Pleniglacial. The faunal remains include some cold-loving forms (reindeer and arctic fox) that indicate climates significantly cooler than those of the present day (Kolosov 1983, 20–29; Kolosov et al. 1993, 68–89) (fig. 4.8).

The occupation layers at Zaskal'naya V and VI contain large quantities of artifacts, ash, bone, and other habitation debris. The lithic assemblages include a modest number of cores (mostly non-Levallois) but numerous flakes and blades. A high proportion of the tools are bifaces of various types, along with scrapers. Both sites appear to represent localities close to water and raw material sources that were repeatedly and intensively occupied for a wide range of activities.

On the opposite bank of the Biyuk-Karasu River lie two open-air sites. At *Sary-Kaya I*, roughly 450 square meters were exposed with the aid of a bulldozer. A single horizon of artifacts was found buried in loam with some poorly preserved remains of mammoth and horse. The artifacts apparently are associated with a buried soil layer, but the age of the occupation is unknown. Only a few hundred artifacts—the majority of which are tools such as scrapers and bifaces—were recovered from the large excavation, and Sary-Kaya I is interpreted as a short-term and limited-activity occupation (Kolosov et al. 1993, 100–107). The nearby site of *Krasnaya Balka*, which was also broadly exposed with a bulldozer, revealed a similar pattern (Kolosov et al. 1993, 89–92). One of the likely attractions of the location was an immediate source of high-quality chert, which was heavily used at both sites.

Approximately 10 kilometers southeast of these sites, along a tributary of the Biyuk-Karasu River, are a pair of rockshelters. *Prolom I* still retains a rock overhang covering an area of roughly 30 square meters, and faces southeast, but *Prolom II* has collapsed onto the upper valley slope 22 meters above the modern river level. At Prolom I, two occupation layers were found in a shallow sequence of sand and rubble with some mammal remains (chiefly saiga [*Saiga tatarica*]). These layers yielded over 10,000

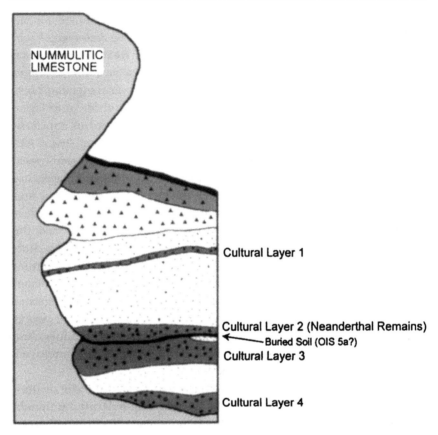

Figure 4.8. Stratigraphy of Zaskal'naya V (redrawn from Kolosov 1983, 23, fig. 8).

artifacts, including some radial cores, as well as points, scrapers, and bifaces (Kolosov et al. 1993, 124–133). At Prolom II, four occupation layers were identified in a somewhat deeper sequence of loam and rubble. A massive rockfall horizon that separates the upper two layers may correlate with OIS 4. Numerous vertebrate remains were recovered from these layers, especially saiga, horse, fox, and hyena; the presence of reindeer from all levels indicates cool climates. The lithic assemblages, which range from a few artifacts to about 2,000 items, contain scrapers, denticulates, and bifaces (Kolosov et al. 1993, 92–100).

Farther west, along the Zuya River near Simferopol', is the famous rockshelter of *Kiik-Koba*. The site was discovered by Bonch-Osmolovskii, who excavated virtually the entire occupation area (measuring roughly 50 square meters) in 1924–1926 (Klein 1965, 36–45). It faces southeast and lies 120 meters above the modern river level. Two relatively thick occupation levels (termed "complexes") were defined in a shallow sequence of loam and rubble. Most of the associated vertebrate remains, which include giant deer, saiga, horse, mammoth, and others, were recovered from the upper complex. Reindeer is absent, although cool climates seem to be indicated by the presence of alpine chough (*Pyrrhocorax graculus*) and possibly arctic fox. Associated charcoal fragments (also primarily derived from the upper complex) are almost entirely from juniper, while pollen/spore analysis indicated a general predominance of nonarboreal vegetation (Bonch-Osmolovskii 1940; Kolosov et al. 1993). Roughly 13,000 artifacts were reported from the lower complex, including over 2,000 retouched pieces, but it appears that a significant percentage of them represent naturally fractured chert (resulting from frost action and/or trampling effects) (Klein 1969a, 94–96). Nevertheless, the assemblage contains some bifacial tools and scrapers, and hominid occupation is confirmed by traces of a burial. Largely on the basis of the primitive character of the cores and tools, the lower complex is often thought to be one of the oldest in the Crimea (e.g., Kolosov et al. 1993, 29), but this assessment is undoubtedly influenced by the effects of natural processes on the artifacts. The upper complex yielded a few cores (typically radial) and several hundred tools, including points, scrapers, and bifaces. It seems to reflect an occupation similar to that at the Zaskal'naya rockshelters—an intensively used multiple-activity site.

The cave of *Chokurcha I* (now reduced by erosion to the size of a rockshelter) is located on a tributary of the Salgir River at the eastern edge of Simferopol'. As at Kiik-Koba, most of the occupation area (roughly 120 square meters) was exposed during excavations, primarily undertaken in 1928–1931 by N. L. Ernst (Klein 1965, 56–57). The shelter opens to the south, and currently lies 25–30 meters above the river level. Three occupation layers were defined in a shallow sequence of loam and rubble. Over 5,000 large mammal remains were identified (not subdivided by occupation level), including mammoth, bison, wild ass, and others; mammoth is especially abundant and a large concentration of bones and tusks is associated with the lowest occupation level (Vereshchagin and Baryshnikov 1980). As at many other Crimean Neanderthal sites, the presence of reindeer indicates cool conditions and suggests that the occupations postdate the Last Interglacial.

Most of the artifacts collected from Chokurcha I were lost during World War II, but they apparently comprised about 1,000 retouched items. Some discoidal cores are recorded, and tools reportedly included numerous bifaces and scrapers. Kolosov et al. (1993, 116) suggested that only the lowest occupation represents a long-term habitation, and that the upper levels reflect short-term visits during a period when the cave was being used primarily as a hyena den. Hyena remains—including coprolites—are common in the uppermost level.

Another group of major sites is found in the foothills of the western part of the Crimean Mountains. These sites are located along rivers that flow north and west out of the mountains and empty into the Black Sea along the western coast. The deeply stratified open-air locality of *Kabazi II*, which is situated on the north bank of the Alma River, is the most important site among this group and perhaps in the Crimea. Discovered by Kolosov in 1985, Kabazi II was initially thought to represent another collapsed rockshelter. However, subsequent investigations revealed that the site had been buried in slope deposits comprising loam and limestone rubble. A test pit (or *sondage*) excavated to a depth of 13 meters in 1987 uncovered over 30 occupation horizons and levels containing Mousterian artifacts. The upper units (geologic strata 11–2) have been dated by radiocarbon, ESR, and U-series to approximately 70,000–27,000 years ago, and thus span the Middle and Early Pleniglacial (OIS 3–4) (McKinney 1998; Rink et al. 1998). More deeply buried artifacts in stratum 13 are associated with soil formation and correlated with the Last Interglacial (or more broadly with OIS 5); six more occupation levels lie beneath this level (Chabai and Ferring 1998).

The lithic assemblages at Kabazi II exhibit substantial variability over time. Stone tools recovered from some of the middle horizons include numerous bifaces, along with scrapers. But some of the upper horizons contain assemblages with Levallois cores, numerous blades, and a variety of scrapers but no bifaces. Associated mammal remains are chiefly composed of wild ass (*Equus hydruntinus*). Occupations throughout the lengthy sequence appear to reflect short-term encampments or "ephemeral stations" (Chabai and Marks 1998, 362).

Several kilometers southwest of Kabazi II, along the north bank of an Alma River tributary, is the site of *Shaitan-Koba*. This small rockshelter is situated 21 meters above the modern river level, and faces southwest. Despite the limited occupation area (less than 30 square meters) and very shallow deposits, excavations conducted by Bonch-Osmolovskii in 1929–1930 yielded over 25,000 artifacts. As in some of the upper levels at Kabazi II, Levallois cores are present, and the tools include many scrapers but few bifaces. The age of the site is unclear, but the presence of cold-loving mammals indicates that it postdates the Last Interglacial (Kolosov 1972).

Farther south lies the well-known site of *Starosel'e*, which is found on the outskirts of the city of Bakhchisarai on a tributary of the Kacha River. The site is located near the mouth of a deep canyon on a bedrock bench about 12 meters above the canyon floor. Starosel'e was originally excavated in 1952–1953 by A. A. Formozov, who interpreted it as a partially collapsed rockshelter containing a single thick occupation level and the burial of a modern human child (Formozov 1958; Klein 1965, 45–51). In 1993–1995, the site was investigated by a Ukrainian-American team that drew very different conclusions about its history (Marks and Chabai 1998). The new excavations revealed a sequence of four occupation levels buried in alluvial and colluvial deposits containing limestone slabs exfoliated from the canyon walls. The burial appears to be intrusive, and probably dates to the late Holocene (Marks et al. 1997, 1998). Radiocarbon, ESR, and U-series dates on the occupation levels indicate that most of them are probably of Middle Pleniglacial age (OIS 3), but the lowest level may be older (OIS 4 or OIS 5a–5d) (McKinney 1998; Rink et al. 1998).

Unlike in the cases of Kabazi II and Shaitan-Koba, there is no evidence for the use of Levallois techniques at Starosel'e. Most of the lithic assemblages contain a number of bifacial tools in addition to scrapers, although the third occupation level yielded an anomalous assemblage comprising scrapers on thick flakes and numerous denticulates (Marks and Monigal 1998). Large quantities of wild ass remains—apparently representing food debris of the occupants—were recovered from the site (Burke 1999a, 2000). The ratio of finished tools to production waste is generally high at Starosel'e, and all of the occupation levels seem to represent short-term camps at a location that clearly possessed consistent attraction(s) (Marks and Chabai 1998).

Northern Caucasus

Despite the profusion of natural shelters in this region, the first Mousterian artifacts were discovered in 1898 at the open-air site of *Il'skaya I* on a tributary of the Kuban' River in the lower foothills near Krasnodar. Large-scale excavations were conducted in 1926–1928 by Zamyatnin (1929) and in 1936–1937 by V. A. Gorodtsov (1941), and some additional work was undertaken by Praslov (1964) during the 1960s. Over 700 square meters have now been exposed. In 1979, another locality (*Il'skaya II*) was discovered adjacent to the first site, and excavations were conducted here during 1981–1991 (Shchelinskii 1985) (fig. 4.9).

Although the Il'skaya sites currently rest on a terrace 15 meters above the modern Il' River, they were situated on the high floodplain at the time of occupation (Nesmeyanov 1989, 256). Unfortunately, descriptions of the stratigraphy manifest many inconsistencies, and the interpretation of the geomorphology and dating remains confused. Isolated artifacts recovered from alluvial gravels at the base of the sequence at Il'skaya II are thought to date to the Last Interglacial. Most of the overlying occupation levels (which may number as many as 14 at Il'skaya I) seem to have been buried in colluvial loams during OIS 5d–5a or later, and some of these may be associated with episodes of soil formation (Nesmeyanov 1999, 176–182). The large body size of some of the carnivores associated with the Il'skaya I occupation levels indicates cold climates (possibly OIS 4–3) (Hoffecker et al. 1991, 141).

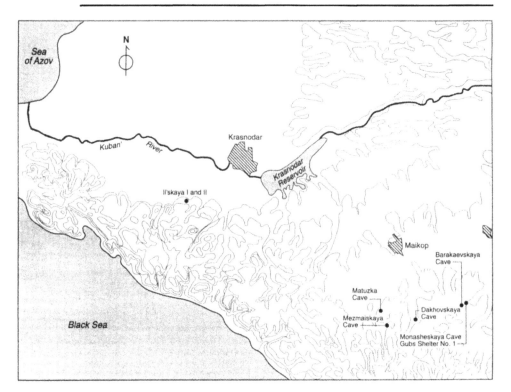

Figure 4.9. Map of Mousterian sites in the northwestern Caucasus.

The lithic assemblages contain large numbers of cores (chiefly radial and subprismatic with occasional examples of Levallois technique) and a high ratio of production waste to finished tools. The latter are dominated by scrapers and points, but include some bifacial forms. More than 85 percent of the highly fragmented large mammal remains are steppe bison (*Bison priscus*). Although the high floodplain location is unusual, the overall pattern at Il'skaya seems similar to that of localities like Molodova and Zaskal'naya—a repeatedly occupied long-term habitation and multiple-activity site ("base camp") near sources of fresh water and lithic raw material (Hoffecker and Cleghorn 2000, 371–372).

Matuzka Cave is located at an elevation of over 700 meters above sea level in the northwestern foothills, on a small tributary of the Pshekha River. It represents a large southwest-facing karst cavity containing loam and rubble deposits over 4 meters deep (Golovanova et al. 1990). Matuzka seems to possess one of the longest occupation sequences in Eastern Europe. The lowest layer yielded remains of warm-loving fauna and flora, and probably dates to the Last Interglacial thermal optimum (OIS 5e).

Overlying levels reveal evidence of cooler climates and appear to span much of the remaining Late Pleistocene (OIS 5d–3). All layers contain small assemblages of Mousterian artifacts with high percentages of tools, as well as abundant remains of cave bears (*Ursus deningeri*) that probably died during winter hibernation (Baryshnikov and Hoffecker 1994, 11). Matuzka Cave presents a sharp contrast to Il'skaya, and may be unique among East European sites—a natural shelter occupied on a recurrent basis, apparently for a very limited range of activities, throughout much of the Neanderthal epoch.

Farther east at slightly higher elevations (800–900 meters above sea level) lie a group of small natural shelters in Borisovskoe Gorge. They include *Monasheskaya Cave* and the adjoining site of *Gubs Shelter No. 1*, and the nearby site of *Barakaevskaya Cave*, which were investigated by V. P. Lyubin (1977, 1994), P. U. Autlev (1964), and E. V. Belyaeva (1992) between 1961 and 1991. All of these sites contain shallow deposits of loam and rubble that appear to date (on the basis of associated faunal remains and pollen-spore samples) to the Pleniglacial (OIS 4–3). Neanderthal skeletal remains were found at Barakaevskaya and Monasheskaya Caves. The Borisovskoe Gorge sites also yielded large quantities of lithic waste and tools, reflecting some use of Levallois techniques and production of scrapers and points, and a diverse array of large mammal taxa (including some alpine species). They appear to represent intensive multiple-activity occupations or "base camps" used to exploit resources at middle altitude (Hoffecker and Baryshnikov 1998, 205–206).

At an even higher elevation (1,300 meters above sea level) overlooking a small tributary valley of the Kurdzhips River is *Mezmaiskaya Cave*. This site was discovered and investigated during 1987–1997 by L. V. Golovanova, who exposed over 40 square meters (Golovanova et al. 1999). Four occupation layers containing Mousterian artifacts were identified in a sequence of clay, loam, and rubble deposits that date to the Middle Pleniglacial (OIS 3) and possibly earlier, and contain Neanderthal skeletal remains (fig. 4.10). The associated large mammal bones and teeth are predominantly of bison, goat, and sheep, and many exhibit the characteristics of hunted prey remains (Baryshnikov et al. 1996). The lithic assemblages include a limited number of cores and a relatively high percentage of tools; bifaces are common in the lower occupation layers but almost completely absent in the upper layers. Mezmaiskaya Cave seems to represent a short-term camp—probably occupied during the warmer months—for exploitation of high-altitude resources (Hoffecker and Cleghorn 2000, 374–375).

A major site is also found in the central north Caucasus near the city of Ordzhonikidze. *Myshtulagty lagat* (Weasel Cave) is located on the east

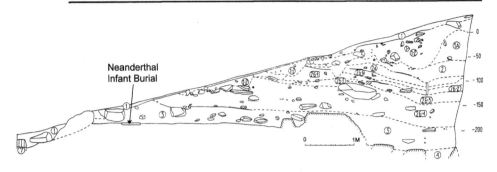

Figure 4.10. Stratigraphy of Mezmaiskaya Cave in the Northern Caucasus (modified from Golovanova et al. 1999, 79, fig. 2).

side of the Gizel'don River (tributary of the Terek River) at 1,125 meters above sea level (see chapter 3). The cave, which is relatively large with a west-facing entrance, was discovered and investigated by N. I. Hidjrati and colleagues during the 1980s and 1990s (Hidjrati 1987; Hidjrati et al. 1997). It contains deep deposits of loam and rubble that date as early as the late Middle Pleistocene. A volcanic ash horizon (Layer 18) in the lower part of the sequence is dated to at least 200,000 years ago, and the overlying units (Layers 15–17) are correlated with OIS 6. The analysis of pollen-spore samples from Layer 14 indicates warm climates, and this unit is dated to the Last Interglacial (OIS 5e). Layers in the upper part of the sequence (Layers 5–12) yield evidence for cooler climates and may be correlated with the Early and Middle Pleniglacial (OIS 4–3).

Paleolithic artifacts have been found in most of these units, as well as in deeper levels underlying the volcanic ash horizon (Layer 18). The upper occupation layers (Layers 5–7) contain Mousterian assemblages with non-Levallois cores and scrapers made on flakes. The lower layers (Layers 12–14) contain a different type of industry characterized by Levallois technique, blade production, and a high percentage of denticulate tools. The late Middle Pleistocene units apparently contain an industry similar to that of the lower layers (Hidjrati et al. 1997). Most of the assemblages are small, but the Last Interglacial occupation level yielded a dense concentration of tools and lithic waste (up to 2,000 items per square meter). Associated large mammal remains are predominantly of goat in the upper layers, and more commonly of deer and cave bear in the lower layers. Use of the site seems to have altered over time, and may have been limited to brief summer visits during cold periods (fig. 4.11).

Figure 4.11. *(opposite)* Chronology of Mousterian sites in Eastern Europe.

TIME SCALE (kya)	OXYGEN-ISOTOPE STRATIGRAPHY	STAGES	EAST EUROPEAN PLAIN			CRIMEA	NORTHERN CAUCASUS
			SOUTHWEST	CENTRAL	SOUTHERN		
10							
20	COLD	2					
30	COOL	3	Molodova V (?) Korman' IV (?)	Betovo (?)	Shlyakh (?)	Kabazi II Starosele	Mezmaiskaya Monasheskaya
40							
50							
60	COLD	4	(?)	(?)	(?)	Kabazi II (?)	Myshtulagty Lagat Barakaevskaya (?)
70							
80	MILD	5a–5d	Ketrosy Molodova I–V Korman' IV Ripiceni-Izvor	Korshevo I & II	Rozhok I Nosovo I	Prolom I & II Zaskal'naya V & VI Chokurcha I	Matuzka Il'skaya II
90							
100							
110							
120	WARM	5e	Vykhvatintsy (?)	Khotylevo I Negotino	Sukhaya Mechetka	Kabazi II (?)	Myshtulagty Lagat (14)

East European Neanderthals: Skeletal Morphology

Until recently, the skeletal morphology of the East European Neanderthals was not well understood. This was due to the small sample of remains, and to confusion regarding the stratigraphic context and archaeological association of specific fossils. Prior to 1970, the morphology of the people who occupied Mousterian sites in Eastern Europe was based almost entirely on the finds from Kiik-Koba and Starosel'e. These included skeletal remains assigned to the Neanderthals and to modern humans, although the provenience of the modern human burial at Starosel'e was questioned by some researchers (e.g., Klein 1969a, 99). The discovery of an isolated tooth lacking any Neanderthal features at Rozhok I appeared to lend further support to the thesis that modern humans had manufactured some of the Mousterian assemblages in Eastern Europe (Praslov 1968, 83–84).

During the past three decades, human skeletal remains have been discovered at a number of Mousterian sites in the Crimea and Northern Caucasus, and the size of the sample has increased significantly, although it is still small in comparison with Western Europe and the Near East. The new finds include cranial and postcranial fragments from the former rockshelters of Zaskal'naya V and VI (1970–1972), a mandible and isolated teeth from Barakaevskaya Cave (1979–1980), an infant burial at Mezmaiskaya Cave (1994), and others (Kolosov et al. 1975; Lyubin et al. 1986; Golovanova et al. 1999). All of these remains have been classified as Neanderthal. Moreover, new field research at Starosel'e indicates that the modern human remains at this site are almost certainly intrusive and of recent age (Marks et al. 1997).

Skeletal materials from Mousterian or early Late Pleistocene contexts on the East European Plain remain scarce and of ambiguous affiliation. An isolated tooth recovered from Rozhok I on the southern plain cannot be firmly assigned to either Neanderthals or modern humans (Praslov 1968, 84; Klein 1969a, 90). Another isolated find (femur) was encountered in 1957 at Romankovo on the Dnepr River that might be associated with Mousterian artifacts (Khrisanfova 1965; Nakel'skii and Karlov 1965). However, this bone (which exhibits primarily modern human characteristics) was found in a redeposited context with artifacts and faunal remains from many different periods—including the Upper Paleolithic—and appears more likely to be associated with the latter (Boriskovskii and Praslov 1964, 23; Alekseev 1978, 172). A scapula was recovered from a Don River locality (Shkurlat III) of Last Interglacial age in 1980–1981 that exhibits both Neanderthal and early modern human features (Shevyrev and Khrisanfova 1984). Because the presence of the Levallois blade industry in late Middle Pleistocene and Last Interglacial levels at Myshtulagty lagat

(Weasel Cave) in the Northern Caucasus indicates a possible intrusion from the Near East at that time (Hidjrati et al. 1997), it is conceivable that archaic modern humans may have occupied some portions of Eastern Europe at the beginning of the Late Pleistocene. This might account for the morphology of the scapula from Shkurlat III.

However, the existing sample of skeletal remains indicates that the occupants of most—if not all—Mousterian sites in the Crimea and Northern Caucasus were Neanderthals similar to those of Western and Central Europe. Despite the lack of associated skeletal material, most Mousterian sites of the East European Plain also appear likely to have been occupied by Neanderthals, because of the similarities of the tool assemblages to Neanderthal industries in adjoining areas (Crimea and Central Europe). This conclusion holds important implications for the Paleolithic settlement of Eastern Europe. It suggests that the first widespread colonization of European lands lying east of the Carpathian Mountains was undertaken by hominids who evolved in Western Europe during the Middle Pleistocene (Hublin 1998, 305–306). Analysis of fossil DNA from the Mezmaiskaya Cave infant (dated to ca. 30,000 years BP) yielded an estimated divergence from the West European Neanderthals (based on DNA extracted from the Neander Valley specimen) of between 151,000 and 352,000 years (Ovchinnikov et al. 2000). This is consistent with a late Middle Pleistocene (OIS 9–6) origin for the Central and East European Neanderthals. It also suggests that widespread settlement of Eastern Europe prior to the appearance of modern humans was tied to the cold adaptations that are so characteristic of the European Neanderthals.

The spatial and temporal distribution of Neanderthal skeletal remains in Eastern Europe exhibits some well defined patterns (see table 4.1). The overwhelming majority of finds are derived from the Crimea and Northern Caucasus. This undoubtedly reflects the superior preservation of bones and teeth in the caves and rockshelters on the southern uplands, but also might be influenced by more sustained and intensive occupation of these areas. Most major periods of the Late Pleistocene prior to 30,000 years BP appear to be represented. Although none of the Crimean or Northern Caucasus finds can be assigned to the Last Interglacial, remains from several sites (e.g., Kiik-Koba) appear likely to date to the Early Glacial (OIS 5a–5d). The Early Pleniglacial cold phase (OIS 4) may be represented by the Barakaevskaya Cave mandible, and the Middle Pleniglacial (OIS 3) is represented at Mezmaiskaya and Monasheskaya Caves. Mezmaiskaya Cave may contain the youngest Neanderthal remains in Eastern Europe, dating to the late phases of OIS 3 (Ovchinnikov et al. 2000).

Collectively, the sample of East European Neanderthal remains includes much of the cranial and postcranial skeleton. Diagnostic features

Table 4.1. Neanderthal Skeletal Remains from Eastern Europe

Locality	Layer	Date	Age/Sex	Material
Crimea				
Kiik-Koba	Layer 6?	OIS 5d–5a	adult	77 postcranial bones
	Layer 6	OIS 5d–5a	child (5–7 months)	vertebrae, ribs, humerus, ulna, femur, tibia, fibula
Zaskal'naya V	1.6 meters	OIS 5d–5a	female adult (25 years)	occipital, first metacarpal
Zaskal'naya VI	Layer 3	OIS 5d–5a	female child (10–12 years)	mandible, teeth, metacarpal, phalanges
Northern Caucasus				
Barakaevskaya Cave	Layer 2	OIS 4?	infant (2–3 years)	mandible, 10 isolated teeth
Monasheskaya Cave	Layer 2	OIS 3	adult	10 isolated teeth, 2 phalanges, others
Matuzka Cave	Layer 5b	OIS 5	adult	1 isolated tooth
Mezmaiskaya Cave	Layer 2	OIS 3	infant (1–2 years)	frontal, parietal fragments
	Layer 2?	29,000 yrs BP	infant (≤2 months)	partial skeleton
East European Plain				
Shkurlat-III (?)	Layer	OIS 5e	adult	scapula
Rozhok I (?)	Layer	OIS 5a/5c	adult	tooth

SOURCES: Bonch-Osmolovskii 1941, 1954; Praslov 1968, 83–84; Kolosov et al. 1975; Danilova 1979a, 1979b; Shevyrev and Khrisanfova 1984; Lyubin et al. 1986; Golovanova et al. 1999.

of the West and Central European Neanderthals are well represented among the sample, while traits used to distinguish modern humans from Neanderthals are correspondingly rare or absent. An adult occipital bone from Zaskal'naya V exhibits lambdoidal flattening and a suprainiac depression above the occipital torus (Danilova 1979a, 77–78). The infant occipital from Mezmaiskaya Cave possesses a well-expressed paramastoid process and foramen magnum of oval form (Golovanova et al. 1999, 81). The child mandible from Zaskal'naya VI exhibits a deep sigmoid notch and outward-projecting condyloid processes (Kolosov et al. 1975, 422–425). Isolated teeth—including front teeth—from the caves of Barakaevskaya, Monasheskaya, and Kiik-Koba display extremely heavy wear (Coon 1962, 555; Lyubin 1994, 154). Unfortunately, no bones of the face are preserved.

Postcranial bones from all localities are generally robust. Rib fragments from the Kiik-Koba child are thicker and more rounded than in a modern human of comparable age. The infant scapula from Kiik-Koba possesses an oval-shaped glenoid fossa and an incipient ridge along the thick outer edge of dorsal surface (Vlček 1975, 415–416). The distal extremities are shortened in both the Kiik-Koba and Mezmaiskaya Cave infants relative to modern humans (Vlček 1975, 413; Golovanova et al. 1999, 82). The Kiik-Koba specimen yields a tibia:femur ratio (crural index) of 79.6 (which is close to the mean for West European Neanderthals [Holliday 1997, 249]), and the Mezmaiskaya remains have low crural and brachial indices (figures not currently available) (fig. 4.12). Bones of the hands and feet from adults at Kiik-Koba and Zaskal'naya V exhibit the same increased breadth and expanded articular surfaces as in the West European sample (Bonch-Osmolovskii 1941, 1954; Coon 1962, 556–557; Danilova 1979b, 85).

Lithic Technology and Tools

The Neanderthals were highly skilled lithic technicians. They were descendents of an early *Homo* lineage that had already developed sophisticated techniques of biface manufacture in the Acheulean, and they had further improved their tool production methods with the acquisition of prepared core techniques. These techniques and the relatively standardized tools that they produced are characteristic of the Mousterian, which is the Paleolithic industry most closely linked to the Neanderthals. Comparisons of lithic technology between the Neanderthals and modern humans reveal few if any significant differences, and probably tell us less about the contrast between the two forms than any other measure of skill (Klein 1999, 409–414).

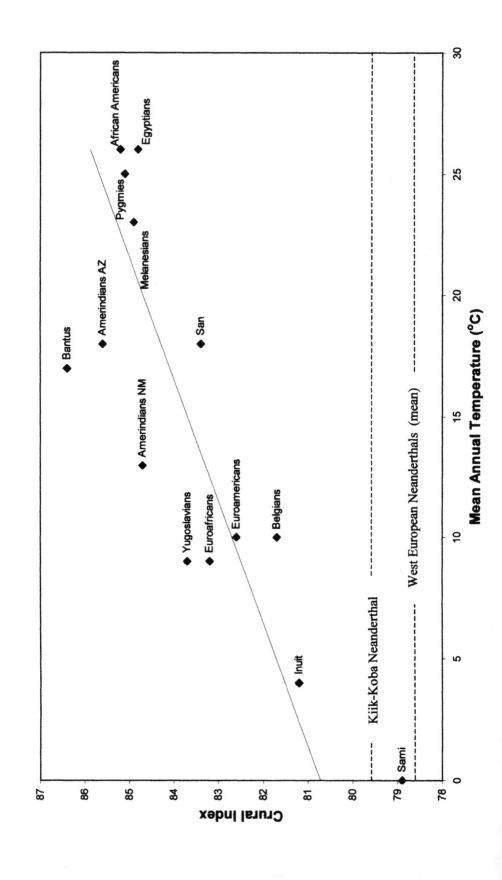

Our understanding of Neanderthal lithic technology is based primarily on studies in Western Europe. The *Levallois technique* was first recognized in France at the beginning of the twentieth century as a method for the controlled production of flakes (Commont 1909). The technique entails careful preparation of the surface and striking platform of a core, and yields blanks of predetermined size and shape. Refitting studies and replication experiments have provided much insight to the Levallois method, and demonstrated that it was applied in a variety of forms to produce flakes, blades, and points (Van Peer 1992; Boeda 1993). In addition to Levallois techniques, the Neanderthals also employed a variety of non-Levallois flaking methods—at least some of which may also have been used to yield blanks of specific size and shape (Turq 1992). However, explaining why the Neanderthals sometimes chose one technique over another is often difficult (Mellars 1996, 88–89), and much remains to be understood about the role of lithic technology in their overall adaptation.

Although many stone blanks were apparently used without modification, the Neanderthals often retouched flakes, blades, and points into tools. These tools included a variety of scrapers, points, denticulates and notches, bifaces, and other less common forms. Like the classification of the cores, their classification is based chiefly on the manufacturing process (Bordes 1961). Many tools were resharpened after use, sometimes altering both their size and shape (Rolland and Dibble 1990). Explaining why Neanderthals in a given time and place chose to make certain types of tools over others is also a difficult challenge, and has been a major focus of controversy in Paleolithic archaeology for many decades (Mellars 1996, 315–355).

The Mousterian of Eastern Europe is fundamentally the same industry that was manufactured by the Neanderthals of Western and Central Europe. The same basic repertoire of flaking techniques and tool types is present in all regions (Klein 1969a; Praslov 1984b; Hoffecker 1999a) (see tables 4.2 through 4.5). Like their skeletal morphology, East European Neanderthals seem to have derived their lithic technology from the western half of the continent, and their eastward expansion at the close of the Middle Pleistocene seems to have spread a uniform tool industry across Europe for the first time. Prior to this event, Central and Eastern Europe was occupied by people who manufactured a crude pebble chopper and flake industry more similar to the Lower Paleolithic of East Asia than the Acheulean of Western Europe. These people may have represented a

Figure 4.12. *(opposite)* Crural index for modern human populations and mean annual temperature showing crural indices for Neanderthals (based on data from Kolosov et al. 1975; Trinkaus 1981).

Table 4.2. Stone Tools in Mousterian Sites of the East European Plain: Southwest Region

Tool Types[a]	Molodova I Layer 4	Molodova V Layer 11	Korman' IV Layer 12	Ketrosy	Stinka lower layer
Points (3-7)	53	11	0	6	0
Side-scrapers (9-29)	47	9	3	50	15
single straight (9)	(x)	(2)	(1)	(11)	(4)
single convex (10)	(x)	(2)	(1)	(19)	(6)
single concave (11)	(x)	(?)	(?)	(5)	(2)
End-scrapers (30-31)	4	1	0	6	7
Burins (32-33)	12	4	0	2	6
Borers (34-35)	6	0	0	6	5
Knives (36-37)	657	53	5	0	0
Backed knife (38)	16	0	3	14	15
Notch (42)	0	0	0	16	39
Denticulate (43)	52	0	2	21	41
Miscellaneous (62)	0	0	0	12	36
Bifacial leaf point (63)	0	0	0	0	4
Total tools:	964	82	18	234	347

Sources: Anisyutkin 1969, 8-13, 1981a, 48, table 7; Chernysh 1977, 67, table 2; 1982, 89, table 5; 1987, 84-85.
[a]Numbers in parentheses refer to standard tool types (Bordes 1961).

Table 4.3. Stone Tools in Mousterian Sites of the East European Plain: Central and Southern Regions

Tool Types[a]	Sites and level				
	Zhitomir Complex 2	Kyotylevo I	Betovo	Sukhaya Mechetka	Shylyakh Layer 8
Points (3–7)	10	119	3	5	2
Side-scrapers (9–29)	18	113	28	202	7
single straight (9)	(?)	(5)	(?)	(19)	(4)
single convex (10)	(x)	(26)	(?)	(63)	(0)
single concave (11)	(?)	(2)	(x)	(17)	(0)
canted (21)	(?)	(8)	(?)	(41)	(1)
transverse (22)	(?)	(20)	(?)	(4)	(0)
End-scrapers (30–31)	59	0	7	0	8
Burins (32–33)	0	6	9	0	10
Borers (34–35)	23	3	9	1	0
Knives (36–37)	243	138	7	3	2
Backed knife (38)	0	0	0	0	4
Notch (42)	0	54	45	15	0
Denticulate (43)	66	30	47	29	3
Tayac point (51)	0	25	0	2	0
Miscellaneous (62)	x	x	0	3	7
Bifacial leaf point (63)	12	11	2	x	0
Handaxe	16	25	2	x	0
Total tools:	938	883	213	301	57

Sources: Klein 1969a, 81, table 3; Tarasov 1977, 27; Zavernyaev 1978, 50–51, tables 4–5; Kukharchuk 1995, 54–57; Nekhoroshev 1999, 52–55.
[a] Numbers in parentheses refer to standard tool types (Bordes 1961).

Table 4.4. Stone Tools in Mousterian Sites of the Crimea

	Sites and level				
Tool Types[a]	Zaskal'naya V Layer 2	Sary-Kaya I	Kiik-Koba upper level	Shaitan-Koba Layer 1	Starosel'e Layer 1
Points (3–7)	50	14	12	53	10
Side-scrapers (9–29)	412	62	268	88	75
single straight (9)	(29)	(18)	(36)	(9)	(8)
single convex (10)	(50)	(10)	(26)	(17)	(10)
single concave (11)	(32)	(2)	(9)	(19)	(4)
canted (21)	(73)	(3)	(124)	(1)	(16)
transverse convex (23)	(34)	(3)	(9)	(5)	(0)
End-scrapers (30–31)	5	3	1	8	1
Burins (32–33)	2	0	3	1	0
Borers (34–35)	0	0	8	0	0
Knives (36–37)	107	36	4	0	0
Backed knife (38)	16	3	0	0	0
Notch (42)	1	7	37	3	11
Denticulate (43)	24	7	59	20	5
Tayac point (51)	0	0	9	0	0
Miscellaneous (62)	x	x	4	0	36
Bifacial leaf point (63)	9	3	x	0	13
Handaxe	4	1	5	0	0
Total tools:	856	238	575	607	159

SOURCES: Klein 1969a, 81, table 3; Kolosov 1972, 24, table 1; 1983, 38–42, table 6; Marks and Monigal 1998, 144–145, tables 7–9.
[a] Numbers in parentheses refer to standard tool types (Bordes 1961).

Table 4.5. Stone Tools in Mousterian Sites of the Northern Caucasus

Tool types[a]	Sites and level			
	Monasheskaya Cave	Gubs Shelter No. 1	Barakaevskaya Cave	Mezmaiskaya Cave, Layer 3
Points (3–7)	7	2	29	30
Side-scrapers (9–29)	57	30	277	259
single straight (9)	(12)	(3)	(157)	(153)
single convex (10)	(16)	(11)	(0)	(0)
single concave (11)	(3)	(1)	(0)	(0)
convergent (18–20)	(5)	(2)	(23)	(26)
canted (21)	(14)	(4)	(51)	(29)
End-scrapers (30–31)	18	5	32	2
Burins (32–33)	2	0	11	0
Borers (34–35)	0	0	32	0
Knives (36–37)	1	0	0	0
Backed knife (38)	14	8	0	0
Notch (42)	14	7	244	0
Denticulate (43)	11	12	52	24
Miscellaneous (62)	3	0	10	0
Bifacial leaf point (63)	0	1	7	10
Handaxe	0	0	0	25
Total tools:	140	80	795	422

Sources: Lyubin 1977, 171, table 27; 183, table 29; Lyubin and Autlev 1994, 137–138, table 22; Golovanova et al. 1999, 81, table 2.
[a]Numbers in parentheses refer to standard tool types (Bordes 1961).

separate *Homo* population—similar to Asian *Homo erectus*—that was effectively replaced by the Neanderthals by the end of the Middle Pleistocene (Rolland 1992).

While the East European Mousterian is essentially similar to that of Western and Central Europe, some differences are evident—especially with Western Europe. Levallois techniques are well documented in all major regions of Eastern Europe, but they are generally less common than in Western Europe. However, one of the most widely used non-Levallois techniques—employing discoidal or radial cores—is quite similar to the Levallois method (Bordes 1961). The same basic array of Mousterian tool types is present, but bifacial tools are more abundant in East European assemblages, especially in the Crimea and on the central East European Plain. They include small handaxes, foliate points, and various other bifacial forms (Praslov 1984b). Bifacial tools are also relatively common in Central Europe, but less so in Western Europe. These two most obvious differences—high percentage of bifaces and low incidence of Levallois technique—may be related to each other.

The challenge of understanding Mousterian lithic technology and its role in East European Neanderthal adaptation is twofold. As in other parts of Europe, archaeologists must struggle to explain the choice of specific flaking techniques and tool types in each assemblage or occupation. For several decades, East European archaeologists have engaged in a lively debate over the nature and meaning of variability in Mousterian assemblages, and a number of local variants or "Mousterian cultures" have been proposed (e.g., Gladilin 1976; Lyubin 1977). Some assemblage contrasts have been attributed to site function (e.g., Shchelinskii 1981; Kolosov et al. 1993). The other challenge lies in explaining the differences between the East European assemblages as a group, and the Mousterian of Western and Central Europe, and the extent to which they might reflect adaptation to local environmental conditions. Although the differences have been noted by some (e.g., Praslov 1984b), there have been few attempts to explain them.

Lithic Technology

The most commonly used cores in the East European Neanderthal sites fall within the broad category of *protoprismatic* (or "parallel") forms (Gladilin 1976, 40–45; Chabai and Demidenko 1998). Many variants have been identified, including unidirectional, bidirectional, orthogonal, convergent, and others; these variants reflect the location of the striking platform(s) and direction(s) of the flaking. On the whole, these cores seem to represent a relatively casual and unsystematic approach to the production of flakes. Despite their protoprismatic shape, they were rarely used to produce blades (i.e., blanks that are more than twice as long as they are wide), which typically comprise fewer than 10 percent of blanks. Proto-prismatic cores are found in most Mousterian assemblages in Eastern Europe (Gladilin 1976, 97–106), and they are also common in other parts of the world (fig. 4.13).

Another common type of core is round or oval in form, and either flaked on one surface (*radial*) or both surfaces (*discoidal*). Flakes were struck from the around the perimeter of the core (i.e., centripetal flaking). The

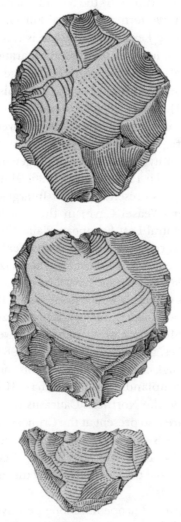

Figure 4.13. Mousterian cores, including discoidal (upper), Levallois (center), and proto-prismatic (lower), from Khotylevo I on the Middle Desna River (redrawn from Zavernyaev 1978, figs. 3, 5, and 12).

analysis of such cores in Western Europe reveals that they required preparation similar to that of Levallois cores, entailing initial trimming around the periphery and successive removal of flakes from one or both surfaces (Bordes 1961; Bordaz 1970, 38–39). Some archaeologists believe that discoidal and radial cores are part of a continuum that includes Levallois forms (Mellars 1996, 73), and it is clear that they reflect a more systematic approach to flake production. They also represent a highly efficient technique that minimizes raw material waste. Discoidal and radial cores are found in most East European assemblages, and are present in all regions and time periods occupied by the Neanderthals (Gladilin 1976); they are also well known in Western and Central Europe.

The use of the classic Levallois technique is less common in Eastern Europe. The technique involves the same preparation steps as discoidal and radial cores, but also includes removal of a large flake, blade, or point from one surface. Like other core types, the Levallois form contains many variants (Bordes 1961). The chief value of the method is a high degree of control over the size and shape of the blank removed from the prepared surface. However, it requires relatively large nodules or cobbles of stone, and yields a small quantity of Levallois blanks (generally less than 20 percent of flakes [Shchelinskii 1974; Mellars 1996, 91–92]).

Levallois technology is present in a Last Interglacial context (OIS 5e) at Myshtulagty lagat (Weasel Cave) in the Northern Caucasus, and Khotylevo I on the central East European Plain. Khotylevo contains several hundred Levallois cores, and many of them exhibit a rare variant of the method entailing removal of a short flake blank (Zavernyaev 1978, 37–38; Praslov 1984b, 108). During the Early Glacial interval (OIS 5d–5a), the Levallois technique is especially well represented in the southwest portion of the plain at Molodova I and V, Ketrosy, Ripiceni-Izvor, and the Buteshty rockshelter. At the Molodova sites, many cores were used to produce blades, bladelike flakes, and points (Chernysh 1982, 1987). During the Early and Middle Pleniglacial (OIS 4–3), Levallois cores are found in some sites of the southern uplands, notably Kabazi II in the Crimea, and Monasheskaya Cave in the Northern Caucasus (Lyubin 1977, 149–158; Chabai 1998). They are also present at the Crimean cave of Shaitan-Koba, which might date to OIS 4–3 or earlier (Kolosov et al. 1993, 176–177). Isolated examples of Levallois cores and blanks often occur in other East European assemblages, but they are most likely the incidental by-products of other flaking methods.

Researchers in Western Europe have long perceived a relationship between Levallois techniques and the availability of sufficiently large nodules and cobbles of suitable raw material (Mellars 1996, 89). This is a plausible explanation for the lack of Levallois cores and blanks at a few

East European sites where large pieces of high-quality stone seem to have been rare or absent (e.g., Rozhok I on the southern plain [Praslov 1984b, 108]). However, as in Western Europe, the occupants of some sites opted for non-Levallois production methods despite the local abundance of large chert nodules and/or cobbles. Examples include Zhitomir and Sukhaya Mechetka on the East European Plain, and Kabazi V in the Crimea (Zamyatnin 1961; Kukharchuk 1995; Yevtushenko 1998). It is difficult to explain why Neanderthals chose other core reduction techniques in these settings.

Levallois techniques might have been linked to the practice of hafting stone tools to wooden handles. The appearance of prepared core methods and evidence for hafting are broadly coincident (Schick and Toth 1993, 292; Klein 1999, 328), and traces of microscopic wear on Mousterian tools indicates that at least some of them were hafted (Beyries 1988, 219–220; Anderson-Gerfaud 1990, 406–410). The high degree of control over blank size and shape provided by Levallois cores would have facilitated production of pieces designed to fit into wooden handles. This could explain why the technique is generally less common in Eastern Europe, where sources of wood—especially hardwood—were sometimes scarce.

Stone Tools and Their Uses

As is true elsewhere in Europe and the Near East, one of the most common stone tools in Mousterian assemblages of Eastern Europe is the *side-scraper (racloir)*. These tools were made by retouching a suitably large flake or blade to create a moderately steep working edge along one or more margins. Side-scrapers include a wide array of types defined on the basis of the number, shape, and position (relative to striking platform) of retouched edges (Bordes 1961). All of the 21 types originally identified in Western Europe are present in the East European sites, including single straight, single convex, double convex, straight convergent, convex transverse, canted, and others (Klein 1969a, 81, table 3). Some additional forms also have been identified in Eastern Europe (e.g., Gladilin 1976, 66–76; Chabai and Demidenko 1998, 43–46). Side-scrapers were often resharpened for continued use, and many of the more heavily retouched forms (e.g., double biconcave) probably represent the same tool after repeated use and resharpening (Rolland and Dibble 1990) (fig. 4.14).

Side-scrapers are present in most East European Mousterian assemblages, but they vary substantially in terms of type frequencies and overall percentage of tools. They represent an especially high proportion of tools in the Crimea and Northern Caucasus (typically 30 to 50 percent), where they usually constitute the most common tool form and often the majority of retouched items (Kolosov et al. 1993; Lyubin 1977, 1994).

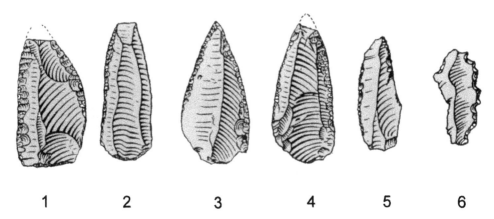

Figure 4.14. Mousterian unifacial tools, including side-scrapers (1 & 2), points (3 & 4), knife (5), and denticulate (6) (1–5 modified from Chernysh 1961, 22, fig. 9; 6 modified from Lyubin and Autlev 1994, 221, fig. 40).

They are less predominant in sites of the East European Plain, where lithic assemblages often contain high percentages of knives, points, and/or retouched flakes (Praslov 1984b). The most common side-scraper types in East European sites appear to be single lateral and single convex, but canted side-scrapers (*racloir déjeté*) are especially well represented in a number of sites (e.g., Kiik-Koba, Starosel'e, Sukhaya Mechetka [Klein 1969a, 105]).

Microscopic analysis of edge-wear patterns on side-scrapers from West European sites indicates that—like many tools manufactured by Neanderthals—they were used primarily for working wood (Beyries 1988; Anderson-Gerfaud 1990). Microwear analysis of stone tools was first developed in the Soviet Union (e.g., Semenov 1940), and has been applied to Mousterian side-scrapers from various parts of Eastern Europe. Although the sample remains small, published results suggest that a higher proportion of side-scrapers in East European sites may have been used on hide and meat (Semenov 1964, 83–84; Shchelinskii 1981, 57; Anderson-Gerfaud 1990, 405). The use of side-scrapers for working hides and other soft materials may have been especially pronounced on the East European Plain at sites such as Rozhok I, Nosovo I, and Sukhaya Mechetka (Semenov 1964, 83–84; Praslov 1968, 81; 1972, 81), where sources of wood were probably scarce. By contrast, side-scrapers from sites in the Northern Caucasus, where forest vegetation was more widespread, appear to have been more heavily used for wood-working (e.g., Monasheskaya Cave [Shchelinskii 1992, 202]). Such apparent differences in side-scraper use between Western and Eastern Europe underscore the fact that the same types of Mousterian tools were often employed for different functions, and that microwear analysis may reveal more about East European adaptations than typological comparisons.

Points are also widely represented in Mousterian industries of Europe and the Near East, and, as in the case of side-scrapers, all of the retouched and unretouched point types recognized by Bordes (1961), as well as some additional forms, are found in Eastern Europe (Klein 1969a, 81; Gladilin 1976, 64–66; Chabai and Demidenko 1998, 41–43). Because retouched points are often difficult to distinguish from convergent and canted side-scrapers (Mellars 1996, 11–112), some of their variation in assemblages simply may reflect differences in artifact classification. In any case, no major spatial or temporal patterns are apparent in the East European sites (although retouched points seem to be inversely correlated with bifacial foliates [Praslov 1984b, 108]). While Dibble (1987) suggested that many retouched points and convergent side-scrapers represented simple scrapers in heavily reduced form, microwear analysis on West European specimens indicates that both tools were often hafted and supports the notion that they were designed to be used as pointed implements (Mellars 1996, 114–115). There is some evidence from microwear studies in Eastern Europe that the pointed tips of these tools were used to penetrate soft material (meat or hide) (Praslov 1968, 87; Shchelinskii 1992, 202).

Another common stone tool type in Eastern Europe is the *knife*. These tools are characterized by a sharp unretouched edge, and a blunt opposing margin formed either by the natural cortex of the stone or by retouching ("backing") (Bordes 1961). Knives are particularly well represented in some assemblages, such as those from the Middle Dnestr Valley sites (28–83 percent), Poles'e sites (Zhitomir and Rikhta [26–54 percent]), and some Crimean caves (Zaskal'naya V and VI [25–28 percent]) (Chernysh 1977, 1982, 1987; Kolosov 1983; Kukharchuk 1995). The reason for their abundance at these sites is unclear. Ordinary unretouched and retouched flakes—many of which exhibit microwear traces of use—are similar to knives in design and function, and the relative proportions of these items may be influenced by artifact classification.

Notches and *denticulates* are present in most East European sites, but rarely in large numbers. These artifacts sometimes appear to be the product of trampling and other natural processes, and their relative abundance is probably influenced by site disturbance (Praslov 1984b, 102–103). This almost certainly accounts for their high percentage in the lower occupation level at Kiik-Koba in the Crimea (Klein 1969a, 95–96), and in Barakaevskaya Cave in the Northern Caucasus (37 percent) (Lyubin and Autlev 1994). Site disturbance processes also might have increased the percentages of notches and denticulates in two caves in the Prut Valley region (i.e., Buzduzhany I, Starye Duruitory [Ketraru 1973; Praslov 1984b, 103]). Despite the fact that they often appear to be the product of natural processes, microwear analysis of notches and denticulates in Western Europe indicates that many functioned as tools. They are more closely associated with the working of wood than most Mousterian artifact types (Beyries 1988, 214–215; Anderson-Gerfaud 1990, 396–397). Such functional specialization might explain why they are less common in Eastern Europe, where hardwood sources were less abundant.

The quantity and diversity of *bifaces* constitute the most distinctive feature of the Mousterian of Eastern Europe. In Western Europe, bifaces are primarily confined to small handaxes in assemblages dating to OIS 3 (traditionally referred to as the "Mousterian of Acheulean Tradition"). In Central Europe, bifacial tools comprise a wider array of forms and date as early as OIS 5, especially across the North German-Polish Plain ("Micoquian"); bifacial foliate points become common after the beginning of the Last Glacial (OIS 4–3) in assemblages assigned to various industries (e.g., "Altmühlian") (Bosinski 1967; Allsworth-Jones 1986, 1990; Kozlowski 1988; Svoboda et al. 1996).

In Eastern Europe, bifacially worked tools are found throughout the entire Mousterian sequence, beginning with the Last Interglacial warm peak (OIS 5e) at Khotylevo and Sukhaya Mechetka on the central and southern plain. Bifaces are found in a number of Early Glacial localities (OIS 5d–5a), such as Zaskal'naya V and VI in the Crimea, and Stinka, Nosovo I, and probably Korshevo I on the East European Plain. During the Early-Middle Pleniglacial (OIS 4–3), they are most common in the Crimean sites (e.g., Kabazi II, Starosel'e). The East European bifaces exhibit a wide range of variation, including small handaxes, knives, scrapers, triangles, foliate points, and others (Gladilin 1976; Praslov 1984b). Perhaps for aesthetic reasons, investigators lavish much attention on bifacial artifacts—even in sites where only isolated and fragmentary specimens are present—but they rarely represent more than 10 percent of the tools in an assemblage (fig. 4.15).

There are sharp regional differences in the distribution of bifacial tools in Eastern Europe, but it is difficult to explain these differences because of uncertainties about why bifaces were made and how they were used. Although they probably performed a variety of functions, it appears increasingly likely that they were used for butchering large mammals. Experiments indicate that bifaces are especially suited to the process of skinning and dismembering a carcass (Walker 1978; Jones 1980), and their link to butchering is supported by association with butchered mammal remains and by microwear analysis at some Lower Paleolithic sites

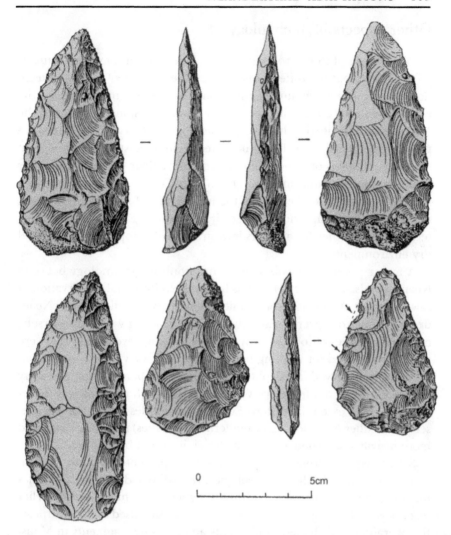

Figure 4.15. Mousterian bifacial tools from Sukhaya Mechetka (after Zamyatnin 1961, fig. 13).

(e.g., Villa 1990; Keeley 1993). Perhaps more important for understanding the contrasts between Western and Eastern Europe is the observation that—with the possible exception of the more finely flaked foliate points—bifaces seem to have been tools that could be held and used in the hand without wooden handles. As in the case of Levallois techniques, the distribution of bifaces might be related to the reduced availability of hardwoods.

Other Aspects of Technology

If comparisons of stone tools between Neanderthals and modern humans tell us little about the differences between the two, other aspects of technology suggest a more fundamental contrast in ability to manipulate the environment. Reconstructing technology other than stone tools among the Neanderthals is difficult, however, because very little of it is preserved in the archaeological record. Nonlithic technology must be inferred from microwear analysis of stone artifacts, occupation floor patterns, and other indirect sources of information. The inherent risks in such an approach need to be confronted because of the importance of assessing this aspect of Neanderthal adaptation—especially in Eastern Europe, where lithic technology provides few insights into strategies for coping with cold and dry environments.

A major potential bias in comparing nonlithic technology between Neanderthals and modern humans lies in the differential preservation of raw materials. As already noted, microwear analysis indicates that Neanderthal stone tools were primarily used for working wood. Conversely, microwear studies of tools used by modern humans reveals that they were more heavily used for working bone, antler, and ivory. While wood is almost never preserved in Paleolithic sites, objects of bone, antler, and ivory have been recovered from many localities. Although this bias may be reduced to some extent in Eastern Europe where hides (which are not preserved in either Neanderthal or modern human sites) may have been used more heavily than wood as a raw material, it cannot be ignored.

Several types of wooden implements are represented directly or indirectly in the Neanderthal archaeological record. Wooden hafts or handles for stone tools may be inferred from the microwear studies noted earlier (e.g., Beyries 1988); although bone might have been used for this purpose, the overall lack of shaped implements among bone fragments in Mousterian sites suggests otherwise. A preserved wooden spear was recovered from Lehringen (Germany) many years ago (Movius 1950), and older specimens are known from the Middle Pleistocene sites of Schöningen (Germany) and Clacton (England) (Oakley 1949; Thieme 1997). Of special interest are several shaped wooden objects and pseudomorphs, which may represent shovels or possibly dishes, as well as a possible tripod, from the Mousterian cave of Abric Romani in northern Spain (Carbonell and Castro-Curel 1992; Castro-Curel and Carbonell 1995). Given their emphasis on wood-working, the Neanderthals may have manufactured a variety of other wooden tools and weapons.

Microwear analysis of stone tools also indicates that hides were worked (not merely removed from mammal carcasses), and—as already noted—

hide-working may have been more common in Eastern than Western Europe. The use of hides is especially important for cold-climate adaptation among modern humans, and it may be assumed that protection from low temperatures was their most important function. There is significant negative evidence that hides were not used by the Neanderthals to produce tailored clothing comparable to that of modern hunter-gatherers of arctic regions. Microwear analysis of East European scrapers reveals that they were employed only for the initial phases of hide preparation—and not the more advanced phases of clothing production (traces of which are observable among modern human Upper Paleolithic scrapers) (Shchelinskii 1974; Anderson-Gerfaud 1990, 405). Even more important is the complete absence of bone needles in Mousterian sites, despite preservation and recovery of small bone fragments from many localities. By contrast, eyed bone needles appear in the earliest modern human sites in Eastern Europe and Siberia (Hoffecker 1999a; Goebel 2000).

Neanderthal use of hides for cold protection thus appears to have been confined to relatively simple items such as blankets and ponchos, although occasionally they may have been perforated for attachment. Isolated examples of stone awls or perforators are present in a few East European sites, including Ketrosy, Rozhok I, and Kiik-Koba (Praslov 1968, 87–89; Anisyutkin 1981a, 48; Kolosov et al. 1993, 123), and bone awls are reported from Chokurcha I and Prolom II in the Crimea (Kolosov et al. 1993, 115–116; Stepanchuk 1993, 33). Similar tools are occasionally found in Mousterian sites of Western Europe as well (Mellars 1996, 122–124). Microwear analysis of several stone awls from Ketrosy supports the notion that they were used to puncture soft material (Shchelinskii 1981, 56–57). However, if hides were sometimes tied together with twine or thongs, they probably did not provide the insulation of tightly sewn fur clothing.

Hides also might have been used to cover artificial shelters (i.e., skin tents), but there is negative evidence against such technology in the Neanderthal archaeological record as well. The data from the open-air sites of the East European Plain, where both the need for artificial shelters and potential for their recognition would be exceptionally high, is especially significant. The most likely example of an artificial shelter is represented in Level 4 at Molodova I on the Middle Dnestr River (Chernysh 1982, fig. 8). Possible traces of shelters are described from several levels at Ripiceni-Izvor in the neighboring Prut Valley (Paunescu 1965, 16–19) (fig. 4.16). But these examples lack clear evidence of former structures in the form of sufficient quantities and suitable types of bone debris, postmold arrangements, and/or central hearths (which are present in many Upper Paleolithic sites). At the most, they would appear to represent windbreaks or simple lean-tos (Paunescu 1978; Soffer 1989, 736). Moreover, the

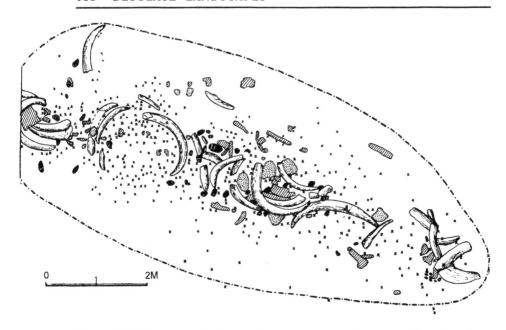

Figure 4.16. Arrangement of mammoth bones and associated occupation debris from Mousterian level at the open-air site of Ripiceni-Izvor in the Prut Valley (after Păunescu 1965, 17, fig. 10).

majority of mapped occupation floors in Mousterian open-air sites reveal no traces of any type of artificial shelter (e.g., Molodova V, Korman' IV, Ketrosy, Sukhaya Mechetka, Rozhok I, and others).

The remains of hearths are found in many Mousterian caves and open-air sites of Eastern Europe, reflecting the extensive use of controlled fire. In fact, this technology is well established in Europe and elsewhere prior to the appearance of the Neanderthals (see chapter 3). Hearths in the East European open-air sites vary considerably in diameter and thickness, ranging from small scatters of charcoal fragments to ash lenses measuring several square meters in area and up to five centimeters in depth (e.g., Zamyatnin 1961, 13; Praslov 1968, 72–86; Chernysh 1977, 13; 1982, 20–34). Most are less than one meter in diameter and two centimeters in thickness (and in many cases, their dimensions were probably reduced by wind and slope erosion). The deep hearth pits that are present in some Upper Paleolithic sites (e.g., Kostenki II [Boriskovskii 1963, 145, fig. 105]) are absent in the Mousterian sites.

Perhaps especially important is the lack of evidence for intensive use of bone fuel in these sites. Along with fragments of wood charcoal, some burned bone has been recovered from many Mousterian open-air sites on the East European Plain (e.g., Molodova I, Korman' IV, Rozkok I [Praslov

1968, 74–86; Chernysh 1977, 13–14; 1982, 23]), and in caves of the Crimea and Northern Caucasus (e.g., Zaskal'naya V, Mezmaiskaya Cave [Kolosov 1983, 47–95; Golovanova et al. 1999, 79]). But the large quantities of bone ash reported from many Upper Paleolithic sites of the central East European Plain are absent. Also absent are large pits containing bones and bone fragments—found in the same Upper Paleolithic sites and probably used to keep bone fresh for burning (see chapter 6).

The nonlithic aspects of Neanderthal technology—to the extent that they are archaeologically visible—provide for some measure of overall complexity. The most complicated Neanderthal implement known to date is the hafted spear or scraping/cutting tool, which comprised a stone blade, adhesive, and shaft/handle (i.e., three components or "technounits" [Oswalt 1976, 38–44]). There is no evidence for the more complex types of weapons or instruments found among recent hunter-gatherers, especially in cold environments (Mithen 1996, 125–127). Assuming a variety of single-component stone and wooden instruments and weapons—along with a number of hafted implements—the total number of technounits for Neanderthal instruments, weapons, and facilities probably fell below one hundred (perhaps as low as fifty). In addition to the apparent lack of tailored clothing and insulated shelters (i.e., technology for cold protection), there is no evidence for untended facilities or storage devices, which represent special technological adaptations to cold environments among recent hunter-gatherers (Torrence 1983; Oswalt 1987; see chapter 1).

Diet and Foraging Strategy

The European Neanderthals appear to have consumed large quantities of meat. Stable isotope analyses of Neanderthal bones recovered from Western and Central Europe indicate a predominance of meat in the diet (Bocherens et al. 1999; Richards et al. 2000), which is consistent with the expectations of hunter-gatherers in cold environments (i.e., effective temperature below 12.0 degrees C [Kelly 1995, 66–73]). Although stable isotope analyses of Neanderthal skeletal remains from Eastern Europe have not been published, the percentage of meat in the diet seems likely to have been equally high or greater in this part of the continent. The daily caloric demands of the large and muscular Neanderthals in low temperature settings must have been high (perhaps comparable to modern arctic people or roughly 3,000 calories per day), and may have been increased further by heavy physical stress and limited technological aids to heat conservation and cold protection.

Despite a heavy reliance on meat, plant foods might have provided a small but significant component of the Neanderthal diet. Even a relatively

Measuring Technological Complexity among the Neanderthals

Although it is possible to identify many individual implements that were made and used by the Neanderthals, it is difficult to arrive at a comprehensive picture of their technology. The principal reason for this is the almost complete lack of preserved wood in the Neanderthal archaeological record. The analysis of stone tool microwear patterns indicates a substantial amount of wood-working and some hafting to wooden handles (Anderson-Gerfaud 1990), and occasional cases of wood preservation confirm the production of various wooden tools and weapons (e.g., Movius 1950). Other implements may have been fashioned from soft plant materials and animal hides. Thus, any application of Wendell Oswalt's (1976) measure of the complexity of food-getting technology among the Neanderthals entails many assumptions and much guesswork. It has been attempted here in order to arrive at a rough approximation for the purposes of comparison with recent hunter-gatherers in cold environments (see chapter 1).

Many of the stone tools of the Neanderthals [each of which counts as one "technounit" following Oswalt (1976)], seem to have been used for making other tools rather than obtaining food (based on microwear data). At least some appear to represent resharpened forms of another tool type [e.g., double straight scraper = recycled single straight scraper (Dibble 1987)]. Nevertheless, some stone tool forms were almost certainly used for food-getting activities (e.g., bifaces, knives, grinding stones), and an estimated range of 20 to 30 implements

modest percentage of gathered foods could have reduced the high mobility demands of a heavy meat diet—especially in a cold terrestrial environment (Kelly 1995, 130–132). However, evidence for plant food consumption is limited and ambiguous. In Eastern Europe, grinding stones that are widely thought to have been used for processing plant foods have been recovered from a number of Mousterian sites in the Dnestr Valley (e.g., Chernysh 1982, 52–53; 1987, 22–23). But microwear analysis of a similar artifact from Barakaevskaya Cave in the Northern Caucasus indicated use on mineral pigment and not vegetal matter (Shchelinskii 1994). Palynology offers another potential source of data. It has been suggested that the density, diversity, and composition of pollen encountered in the occupation layer at Ketrosy (Middle Dnestr Valley) reflect Neanderthal plant collection and use (Levkovskaya 1981, 134–135).

Most of our information regarding diet and foraging practices prior to the end of the Pleistocene is derived from the study of animal remains. In the early 1980s, pointed critiques of past interpretations of faunal remains in Lower and Middle Paleolithic sites helped stimulate new research

> (i.e., 20–30 technounits) seems reasonable. Evidence of hafting indicates more complicated tools and weapons, each of which would count as three technounits (stone blade + adhesive + wooden handle). An estimated minimum of five and maximum of ten different types of hafted stone-tipped spears and cutting implements yields a range of 15 to 30 total technounits in this category.
>
> Wooden tools have the lowest archaeological visibility in Neanderthal sites and are the most difficult to assess. Future discoveries may reveal these to have been significantly more diverse and complex than currently envisioned, and it is safe to assume that many food-getting items have yet to be found or properly identified (e.g., digging sticks, clubs). Also, while most of these tools and weapons would count as one technounit, it is possible that Neanderthals made some multicomponent devices [e.g., the suspected tripod at Abric Romani (Castro-Curel and Carbonell 1995, 378)]. Accordingly, it is best to allocate a wide range of estimated technounits for this category (i.e., total of 15–40 technounits).
>
> In contrast to modern humans (see chapters 5 and 6), there is no direct or indirect evidence for the use of untended facilities (e.g., traps and snares) or storage devices (e.g., pits) among the Neanderthals. Thus, the estimated total technounits for food-getting technology is confined to the categories described above and falls between 50 and 100. If correct, this estimate would place the Neanderthals below the level of complexity of recent hunter-gatherers in cold environments (Oswalt 1976, 192, fig. 9.3; Torrence 1983, 19, fig. 3.2).

in this area (e.g., Binford 1981; Stiner 1994). Much of this research involved the application of taphonomic methods to the analysis of mammal assemblages in an effort to distinguish between vertebrate remains collected by humans and those accumulated and modified by other processes (e.g., carnivores, stream action). Taphonomic methods have been used to help differentiate between remains accumulated by human hunting and scavenging, and to address questions concerning planning and scheduling of resource use (e.g., Pike-Tay et al. 1999).

All of this research has been undertaken within the context of a spirited debate regarding the character of Neanderthal foraging strategy and the extent to which it differed from that of modern humans. Some researchers have suggested that the Neanderthals occupied a different ecological niche than their modern human successors (Stiner 1994, 371–387). They have postulated a heavier reliance on scavenging—as opposed to hunting—large mammals, and generally narrower range of exploited animal taxa. Binford and others have argued that Neanderthal foraging was based on an "opportunistic encounter" strategy that entailed limited planning and scheduling of resource use (e.g., Binford 1984, 1985; Soffer 1989).

Taphonomic studies of site faunas in Western Europe indicate that the Neanderthals were competent hunters of large mammals (e.g., Chase 1986; Gaudzinski 1996). Although they may have exploited available scavenging opportunities (e.g., Stiner 1994), regular hunting of large mammals was probably necessary to provide a high meat diet in their environmental setting. This seems especially likely in Eastern Europe, where taphonomic analyses of large mammal assemblages—although limited in number—also reveal hunting (e.g., Burke 1999b; Hoffecker and Cleghorn 2000). Reflecting the more steppic character of local environments, Neanderthal site faunas in Eastern Europe are dominated by grazers like mammoth, steppe bison, horse, and saiga. With the exception of horse, these taxa are less common in Western Europe—particularly during OIS 5—where woodland forms like red deer are more abundant (Chase 1986).

Neanderthal sites in Europe and the Near East exhibit a certain degree of functional differentiation, and apparently include examples of multiple-activity sites, special-function sites, and stone quarry locations (Henry 1995; Mellars 1996, 245–268). Some of this functional differentiation is expressed in terms of contrasts in associated faunal remains and possible differences in the season of occupation, which may reflect the scheduled exploitation of seasonally available resources. Such a foraging strategy must have been critical to settlement of colder environments with pronounced seasonal fluctuations in climate and resources. In Eastern Europe, functional and seasonal differentiation among sites is most evident in the Crimea and Northern Caucasus during OIS 4–3 (Chabai and Marks 1998; Hoffecker and Cleghorn 2000). And in both Western and Eastern Europe, many Neanderthal cave and open-air sites reveal a pattern of central-place foraging (i.e., collection of resources at a central location), which is likely tied in part to their emphasis on hunting and may not have been practiced by their predecessors (see chapter 3).

Perhaps the most significant contrast between Neanderthal and modern human foraging strategy lies in mobility and group size. Throughout their range, the Neanderthals seem to have foraged across comparatively small territories (as inferred from the distances of transported raw materials from their sources). Evidence for movements in excess of 100 kilometers is very rare and apparently confined to colder periods in Central Europe (Roebroeks et al. 1988; Mellars 1996, 161–168; Féblot-Augustins 1999). There is corresponding evidence (although very limited) for relatively small group size in both Western and Eastern Europe. In the few cases where it is possible to measure the area of an occupation, estimated group size is also small by recent hunter-gatherer standards (roughly ten individuals [Mellars 1996, 269–295; Hoffecker and Baryshnikov 1998, 205–206]).

East European Plain

Faunal remains are generally less well preserved in the open-air sites of the East European Plain than in the caves and rockshelters of the Crimea and Northern Caucasus. Bones and teeth in open-air settings were probably exposed to more subaerial weathering and buried in a less favorable geochemical medium. At some of these sites, no identifiable faunal remains were preserved (e.g., Stinka, Nosovo I, Rikhta [Anisyutkin 1969, 7; Praslov 1972, 75; Smirnov 1979, 11]), while weathered remains are reported from other localities (e.g., Sukhaya Mechetka, Ketrosy [Vereshchagin and Kolbutov 1957, 81; David 1981, 135]). The latter are likely to reflect some bias toward larger taxa and heavier body parts because they are more resistant to the effects of weathering. The best-preserved assemblage may be from Rozhok I on the coast of the Sea of Azov (Praslov 1968).

Mousterian open-air sites on the East European Plain are typically found in deposits of loess and/or colluvium on low river terraces. This type of topographic and depositional setting is not likely to be the locus of natural concentrations of large mammal debris, and it may be assumed that most of the large mammal remains were accumulated at these sites by their human occupants. However, the remains of carnivores, which are typically present in small numbers, may represent animals drawn to the site by the organic refuse (i.e., garbage). Smaller mammals probably burrowed into the disturbed soil following human occupation, and at some sites complete skeletons were recovered from fossil burrows (e.g., David 1981, 140). Some of the more fragmentary small mammal remains may represent the prey of predatory birds that nested at abandoned site locations (see table 4.6).

In the southwest region, remains of mammoth dominate faunal assemblages at most open-air sites. Mammoth is the most common large mammal at Molodova I and V, Ketrosy, and Ripiceni-Izvor (Paunescu 1965; David 1981, 136, table 1; Chernysh 1982, 88, table 4; Alekseeva 1987, 154, table 1), and is present at Korman' IV (Tatarinov 1977, 113, table 1). The relative abundance of this species is probably influenced by the high preservability and easy identification of the massive bones, teeth, and tusks. It is not clear if mammoths were hunted or even scavenged by the occupants of these sites, and it should be noted that evidence for Neanderthal hunting of mammoths is generally scarce. (Perhaps the most likely

Table 4.6. Large Mammal Remains in Mousterian Sites of the East European Plain

Taxon	Sites and level					
	Molodova I Layer 4	Molodova V Layer 11	Korman' IV Layer 12	Ketrosy excavation II	Khotylevo I	Sukhaya Mechetka
Woolly mammoth (*Mammuthus primigenius*)	abundant	xx/5	2/2[1]	192/4	common	51/?
Woolly rhinoceros (*Coelodonta antiquitatis*)	present	x/1	8/2	6/1	present	–/–
Horse (*Equus caballus*)	present	x/1	66/2	51/2	common	42/?
Red deer (*Cervus elaphus*)	present	x/1	–/–	6/1	common	15/?
Elk (*Alces alces*)	present	–/–	–/–	–/–	absent	–/–
Giant deer (*Megaceros giganteus*)	absent	–/–	5/1	2/1	absent	–/–
Reindeer (*Rangifer tarandus*)	absent	x/1	–/–	–/–	present	1/1
Bison (*Bison priscus*)	present	x/1	57/3	10/2	common	366/?
Wolf (*Canis lupus*)	present	–/–	–/–	–/–	present	1/1
Bear (*Ursus arctos*)	present	x/1	–/–	–/–	present	–/–
Lion (*Panthera leo*)	absent	x/1	–/–	–/–	absent	–/–

Sources: Vereshchagin and Kolbutov 1957, 84, table 1; Tatarinov 1977, 113, table 1; Zavernyaev 1978, 29; David 1981, 136, table 1; Chernysh 1982, 88, table 4; 1987,23.
[a]Number of Identified Specimens/Minimum Number of Individuals

instance of mammoth hunting is reported from La Cotte de St. Brelade on the island of Jersey [Scott 1989].)

Many of the mammoth bones and tusks in the East European Plain sites may have been collected for use as raw material, windbreaks, and other nondietary purposes (Vereshchagin and Baryshnikov 1984, 492–493; Alekseeva 1987, 156). Collection seems likely to account for the mammoth remains at the Il'skaya open-air sites in the Northern Caucasus, which exhibit more weathering and carnivore damage than the remains of other taxa (Baryshnikov and Hoffecker 1994, 10–11). Mammoth bones and tusks were almost certainly collected by modern humans in the central East European Plain during the Last Glacial Maximum (Klein 1973, 53), although their use for fuel and material is more clearly evident in that context than in Neanderthal sites (see chapter 6).

The sites in the southwest plain contain numerous remains of other large mammals, including steppe bison, horse, and red deer. Less common species include woolly rhinoceros, giant deer, and reindeer. They are widely believed to represent the remains of hunted prey, and not those of scavenged carcasses or collected bone (e.g., Tatarinov 1977; Chernysh 1982, 56). There is evidence from other parts of Europe that Neanderthals hunted most of these large mammals (Chase 1986; Baryshnikov et al. 1996). But with the exception of possible stone tool cut marks on some of the limb bones from Ketrosy (David 1981, 135), there are few available data to support or refute this interpretation at the East European Plain sites.

Evidence of functional and possible seasonal differentiation among sites in the southwestern region is limited. Many sites appear to represent multiple-activity loci and relatively long-term habitation sites (e.g., Molodova I and V, Korman' IV, Stinka, Ripiceni-Izvor). Most occupation floors at these sites contain large and diverse assemblages of stone artifacts and faunal remains (except where weathering processes seem to have destroyed most or all of the latter) that reflect a variety of activities. Extended occupation is suggested by thick hearth lenses and in a few cases possible traces of crude shelters. These occupation floors represent the closest analogue to a modern hunter-gatherer "base camp" in the Neanderthal archaeological record. Although both the quantity and diversity of occupation debris have almost certainly been enhanced by multiple episodes of occupation on each floor, much of the diversity is reflected in individual concentrations (which are more likely to represent single occupation episodes). The site of Ketrosy, which contains only one major occupation level, has been interpreted as a short-term "hunting camp" (e.g., Shchelinskii 1981, 58), but does not differ significantly from the other sites in terms of the quantity and diversity of artifacts and faunal debris (Anisyutkin 1981a).

Sites in the southwest plain that appear to contain traces of short-term occupations and a limited range of activities are chiefly confined to natural shelters. In Layer 4 at the cave of Trinka I (located on a tributary of the Prut River), the artifact assemblage is smaller than the number of identified large mammal remains (primarily cave-dwelling carnivores) and includes a high proportion of tools (29 percent). This site was probably a carnivore den used only sporadically by Neanderthal groups (Anisyutkin et al. 1986). A similar pattern is evident at the former cave of Vykhvatintsy (Dnestr Valley), where a small assemblage of artifacts was found in association with a large quantity of mammal remains, including a high percentage of carnivores (24 percent) (Sergeev 1950; Ketraru 1973, 21–24). Also, it should be noted that archaeological surveys in the major river valleys have encountered some open-air localities containing very small quantities of Mousterian artifacts that may represent ephemeral occupations (Chernysh 1965, 106–114; 1973, 46–52; Ketraru 1973, 11–62).

Information regarding faunal assemblages from sites in the central plain is especially limited. The only sizable large mammal assemblage is reported from Khotylevo I on the Desna River, where mammoth, red deer, horse, and steppe bison are the most common taxa (Zavernyaev 1978, 29). Few details are available. Because many of the artifacts at Khotylevo lie in stream sediments and are redeposited (Zavernayev 1978, 31–35), the relationship between them and the faunal remains is unclear. Many of the latter may have been concentrated at the site location by the river or a side-valley stream. Only isolated large mammal remains are described from Betovo (also on the Desna River), including mammoth and woolly rhinoceros (Tarasov 1977, 23–24).

Steppe bison predominates in the large mammal assemblages from Sukhaya Mechetka and Rozhok I and II in the southern part of the East European Plain (Vereshchagin and Kolbutov 1957, 84, table 1; Praslov 1968, 71). Steppe bison is the most common large mammal at Kodak, situated along the Lower Dnepr River, although—as at Khotylevo I—the relationship between the faunal remains and artifacts at this locality is problematic (Boriskovskii and Praslov 1964, 22). This species also predominates heavily at Il'skaya I in the Northern Caucasus foothills that adjoin the southern margin of the plain (Hoffecker et al. 1991, 121, table 2), and is abundant in sites occupied by modern humans during the Last Glacial Maximum (OIS 2) in the region (see chapter 6). The dominance of steppe bison in Pleistocene large mammal assemblages of the south East European Plain makes this region unique for Northern Eurasia. These sites lie within the modern steppe belt of southern Russia and the Ukraine, and they reflect the consistent presence of arid climates and open grassland in the area.

Analysis of the highly fragmented remains at Il'skaya I indicates that bison was probably hunted by the occupants of this site (Hoffecker et al. 1991). This is based chiefly on the prevalence of prime-age adults among the assemblage; scavenging (or bone collecting) is more likely to yield an attritional age profile dominated by old adults and juveniles (Klein 1982) (fig. 4.17). Prime-age adults are also predominant among the bison from Sukhaya Mechetka, but the sample at this site is very small (J. F. Hoffecker, unpub. notes, 1988). Few data are available from the Rozhok sites, but a high percentage of juveniles is reported from the lowest level at Rozhok I (Praslov 1968, 84). Hunting of steppe bison is inferred from caves in the Northern Caucasus such as Mezmaiskaya, which yields a prime-dominated age profile and numerous stone tool percussion and cut marks on bone fragments (Hoffecker and Cleghorn 2000). However, none of these sites provides evidence that Neanderthals hunted bison in large groups or herds.

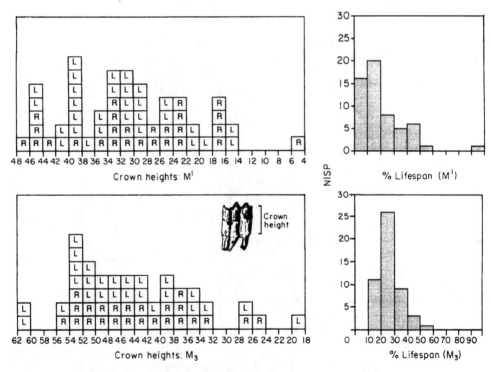

Figure 4.17. Age mortality profile for steppe bison based on molar crown-height measurements from Il'skaya I in the Northern Caucasus (after Hoffecker et al. 1991, 136, fig. 10).

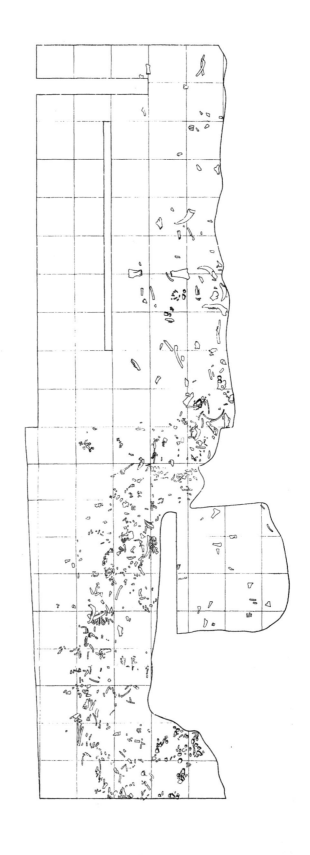

Sites in the central and southern parts of the East European Plain exhibit more functional differentiation than those of the southwest region. A multiple-activity and probable habitation site ("base camp") is present at Sukhaya Mechetka on the Lower Volga River (Zamyatnin 1961). However, a different type of site appears to be represented at Khotylevo on the Desna River, where the ratio of cores to tools is 2.7:1 and the percentage of tools among artifacts is less than one percent (Zavernyaev 1978, 37–51). Khotylevo seems to represent a "workshop" locality with a heavy emphasis on production of tool blanks (Praslov 1984b, 108). Another site probably used for special functions is Rozhok I on the coast of the Sea of Azov, in which large mammal remains outnumber the artifacts on all six occupation levels; Layer 6 is especially striking because the small artifact assemblage includes a high proportion of tools (24 percent) and there are few traces of hearths (Praslov 1968, 86–89). This locality seems likely to have been occupied on a short-term basis for the processing of large mammal carcasses (chiefly bison) (fig. 4.18).

Crimea and Northern Caucasus

The caves and rockshelters of the southern uplands contain types of faunal assemblages different from those of the East European Plain. The preservation of remains is generally better in natural shelters, although there are a few exceptions to this pattern (e.g., Monasheskaya Cave in the Northern Caucasus [Hoffecker and Baryshnikov 1998, 194]). The continual infusion of calcium carbonate into the sediment—derived from exfoliation of the cave ceiling and walls—creates a favorable geochemical medium for bones and teeth. However, natural shelters attract a greater number and variety of nonhuman occupants than open-air localities, and the process of sorting out their contribution to the faunal assemblage is often a difficult one (Brain 1981).

All of the Crimean and Northern Caucasus caves were inhabited by carnivores, including bears, wolves, hyenas, leopards, and/or others. In fact, there is evidence from various parts of Europe that the Neanderthals shared their caves with carnivores more often than modern humans (although the contrast with the latter may not be evident until OIS 2 [Straus 1982]). Many of the carnivores that occupied these sites accumulated substantial quantities of prey remains—including large mammals—in them. Owls and other predatory birds also nested in the caves, and they account for large quantities of small vertebrate remains in the faunal assemblages (e.g., Baryshnikov et al. 1996, 326–327).

Figure 4.18. *(opposite)* Occupation floor mapped from Horizon 6 at Rozhok I on the coast of the Sea of Azov (modified from Praslov 1968, fig. 49).

Table 4.7. Large Mammal Remains in Mousterian Sites of the Crimea

Taxon	Zaskal'naya V Layer 2	Prolom II Layer 2	Kiik-Koba upper level	Chokurcha I	Shaitan-Koba	Starosel'e Level 3
Woolly mammoth (*Mammuthus primigenius*)	122/7[a]	28/11	42/2	3041/?	1/1	–/–
Woolly rhinoceros (*Coelodonta antiquitatis*)	–/–	23/7	5/1	4/?	–/–	–/–
Horse (*Equus caballus*)	8/3	214/13	103/6	147/?	11/1	
Wild ass (*Equus hydruntinus*)	–/–	72/12	3/2	730/?	50/5	380/8
Bison (*Bison priscus*)	–/–	72/7	–/–	813/?	2/1	8/1
Saiga (*Saiga tatarica*)	120/5	1730/42	144/5	69/?	177/5	25/2
Red deer (*Cervus elaphus*)	–/–	17/5	16/1	59/?	7/1	7/1
Giant deer (*Megaceros giganteus*)	–/–	6/2	236/8	3/?	–/–	–/–
Reindeer (*Rangifer tarandus*)	–/–	53/7	–/–	7/?	Present	1/1
Wolf (*Canis lupus*)	3/1	73/10	3/1	9/?	1/1	7/1
Arctic fox (*Alopex lagopus*)	–/–	–/–	12/3(?)	–/–	5/1	–/–
Bear (*Ursus* sp.)	2/1	79/12	2/1	2/?	–/–	2/1
Hyaena (*Crocuta* sp.)	–/–	139/16	10/1	4/?	2/1	6/1

Sources: Vereshchagin and Baryshnikov 1980, 32, table 1; Kolosov 1972, 110; Kolosov et al. 1993, 73–74; Burke 1999a, 2, table 1-1.
[a]Number of Identified Specimens/Minimum Number of Individuals

As in other parts of Eastern Europe, a regional faunal complex may be recognized in the Crimean shelters. The most abundant large mammals in these sites are wild ass and/or horse, saiga, and giant deer; also relatively common are mammoth and red deer (Klein 1969a, 82; Kolosov et al. 1993) (see table 4.7). At some sites, one of these species is predominant. Mammoth constitutes 62 percent of the large mammal assemblage from Chokurcha I (Vereshchagin and Baryshnikov 1980, 32, table 1), while saiga constitutes 67–70 percent of the assemblages from Shaitan-Koba and Adzhi-Koba (Kolosov 1972, 110–111; Baryshnikov et al. 1990, 38). Over 98 percent of the large mammal remains from the 1952–1956 excavations at Starosel'e were wild ass (Formozov 1958, 53). However, the consistent dominance of a particular species is not found here, and both steppe and woodland forms are well represented at most sites (along with some tundra species), reflecting a diverse array of plant communities in the region.

Carnivore remains are also common in the Crimean caves and rockshelters, especially hyena, wolf, and various types of fox. Less abundant species include bear, lion, and lynx (Kolosov et al. 1993). The carnivore component of the large mammal assemblages is often significantly higher than it is in the East European Plain open-air sites, although this pattern is undoubtedly influenced by preservation because most of the carnivore taxa are smaller than the ungulates. In many occupation layers, carnivores account for 10 percent or more of the assemblage, and among the large samples from the upper layers at Prolom II, they constitute 24–26 percent of the total (Kolosov et al. 1993, 96–97). In general, the percentage of carnivores is higher in layers containing a low density of occupation debris.

The strong presence of carnivores underscores the probability that many of the large mammal remains in these sites were not gathered by their Neanderthal occupants. The uppermost level at Chokurcha I is believed to represent a hyena den containing a small quantity of artifacts (Kolosov et al. 1993, 115–116). And a recent taphonomic study of the Prolom II assemblage concluded that most of the ungulate remains in this former rockshelter were accumulated by hyenas (Enloe et al. 2000). The mass of wild ass remains at Starosel'e has been interpreted in the past as evidence of a herd kill in the narrow canyons that surround the site (e.g., Vereshchagin 1967, 373). More recent research on the Starosel'e fauna indicates that the earlier interpretation is probably correct, but that scavenging of animals killed by flash floods in the canyons may have occurred as well (Burke 2000).

On the basis of evidence from other parts of Western Eurasia (Chase 1986; Stiner 1994), it is reasonable to assume that many of the large mammal remains in sites that contain low percentages of carnivores and high densities of occupation debris were hunted (or at least scavenged) by

Table 4.8. Large Mammal Remains in Mousterian Sites of the Northern Caucasus

Taxon	Il'skaya I	Il'skaya II Layer 4b	Monasheskaya Cave	Barakaevskaya Cave	Mezmaiskaya Cave, Layer 2b
Woolly mammoth (*Mammuthus primigenius*)	7/2[a]	135/4	–/–	–/–	–/–
Horse (*Equus* sp.)	27/4	1/1	1/1	8/4	–/–
Red deer (*Cervus elaphus*)	16/2	1/1	3/?	5/2	11/2
Giant deer (*Megaceros giganteus*)	25/3	–/–	4/?	14/4	–/–
Bison (*Bison priscus*)	1334/51	8/2	29/?	290/11	216/9
Sheep (*Ovis orientalis*)	–/–	–/–	19/?	23/?	10/2
Goat (*Capra caucasica*)	–/–	–/–	6/?	3/1	13/2
Saiga (*Saiga tatarica*)	2/1	–/–	–/–	23/?	–/–
Wolf (*Canis lupus*)	15/3	1/1	–/–	2/1	1/1
Bear (*Ursus spelaeus*)	3/1	–/–	–/–	–/–	5/2
Hyaena (*Crocuta spelaea*)	12/3	–/–	–/–	5/4	–/–

Sources: Hoffecker et al. 1991, 121, table 2; Baryshnikov and Hoffecker 1994, 3, table 1; Baryshnikov et al. 1996, 319, table 3; Hoffecker and Baryshnikov 1998, 193–195, tables 1–3.
[a]Number of identified specimens/minimum number of individuals.

Neanderthals. The most likely examples include Shaitan-Koba (Kolosov 1972) and Zaskal'naya V and VI (Kolosov et al. 1993, 69–89). But taphonomic analyses of the assemblages are needed to properly test this assumption.

There is growing evidence of functional and seasonal differentiation among sites in the Crimea (Kolosov et al. 1993; Chabai et al. 1995). Kiik-Koba and Zaskal'naya V and VI appear to represent multiple-activity locations and relatively long-term occupations (base-camp pattern). More briefly occupied limited-function sites seem to be present at the open-air localities Sary-Kaya I and Krasnaya Balka, which contain small artifact assemblages with a high proportion of tools (30–60 percent) and lack traces of hearths. Both Starosel'e (Layers 1, 2, and 4) and Kabazi V are interpreted as short-term camps occupied during the late summer/fall for procurement of wild ass and saiga (Chabai and Marks 1998, 365; Burke 1999b, 2000).

Faunal assemblages from sites in the northwestern Caucasus contain large quantities of bison, goat, and sheep (see table 4.8). Steppe bison is the most common species at Il'skaya I, Barakevskaya Cave, Monasheskaya Cave, and Mezmaiskaya Cave (Hoffecker et al. 1991, 121, table 2; Baryshnikov et al. 1996, 319, table 3; Hoffecker and Baryshnikov 1998, 193–195). Goat and sheep become increasingly common in sites located over 500 meters above sea level (asl). These taxa represent 45 percent of large mammal remains at Mezmaiskaya (1,300 meters asl), and goat predominates in the upper layers at Myshtulagty lagat (Weasel Cave) (1,125 meters asl) in the central north Caucasus (Hidjrati et al. 1997). Mammoth is found only at lower elevations, and is especially common at Il'skaya II near the base of the foothills (Baryshnikov and Hoffecker 1994, 3, table 1). These variations almost certainly reflect the altitudinal zonation of plant communities in this region, which were shifted downward during cold phases.

Carnivores are not abundant in the Northern Caucasus sites, and account for 5 percent or less of the large mammal assemblage in most occupation layers (Baryshnikov and Hoffecker 1994; Hoffecker and Baryshnikov 1998). Exceptions of this pattern are found at Dakhovskaya and Matuzka Caves, where they represent 50–96 percent of the assemblages, and the density of occupation debris is low (Formozov 1965, 33–36). Taphonomic studies of the Matuzka remains indicate that it was primarily a cave bear den (*Ursus deningeri*), while Dakhovskaya Cave was probably used by several carnivores, including foxes, wolves, and lions.

Analysis of the large mammal remains from Mezmaiskaya Cave reveals that steppe bison, goat, and sheep were all probably hunted and butchered by its Neanderthal occupants (Baryshnikov et al. 1996). This conclusion

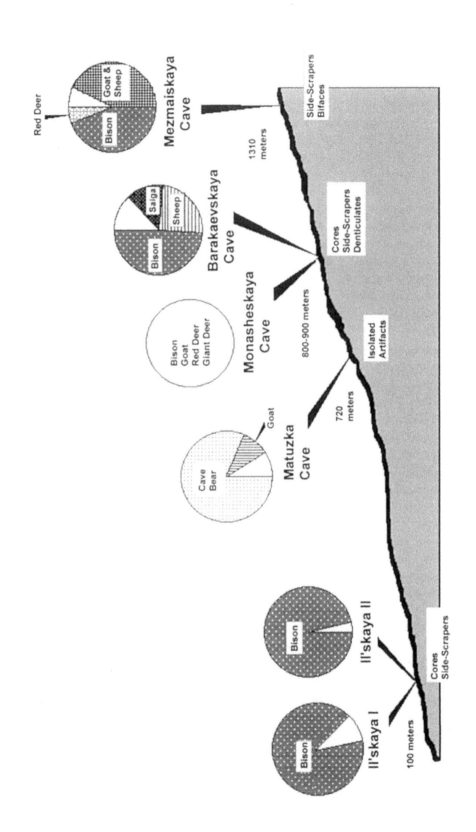

is supported by prime-dominated age profiles and numerous examples of stone tool percussion and cut marks on the bones. Breakage patterns indicate that most bones were fractured in a fresh condition. Hunting of bison is also inferred at Il'skaya I and Barakaevskaya Cave, but there is less supporting evidence at these sites (Hoffecker et al. 1991; Hoffecker and Baryshnikov 1998).

In the Northern Caucasus, both functional and seasonal differentiation of occupations has been documented for Early and Middle Pleniglacial sites (Hoffecker and Baryshnikov 1998; Hoffecker and Cleghorn 2000). Large multiple-activity sites with evidence of tool production and resharpening, and a diverse array of tools and large mammal remains are found at lower and medium elevations (Il'skaya I, Monasheskaya Cave, and Barakaevskaya Cave). There is some evidence of cold-season occupation at Il'skaya I (Hoffecker et al. 1991). A limited-function site is present in Matuzka Cave (720 meters asl), which contains small artifact assemblages and little evidence of tool production or resharpening. This site was used by cave bears in the winter, and presumably occupied by Neanderthals during the warmer months. An intermediate site type is represented at Mezmaiaksya Cave (1,300 meters asl), which was also probably used during the warm season for hunting bison, sheep, and goat (Baryshnikov et al. 1996) (fig. 4.19).

Neanderthal Socioecology

Many anthropologists suspect that Neanderthal social organization was fundamentally different from that of modern hunter-gatherers (e.g., Klein 1973, 122–126; Wobst 1976; Gamble 1999). Much of the speculation on Neanderthal society is based on the almost complete lack of archaeological evidence for the use of symbols (and inferred lack of symbolic language), which play such an important role in modern human social interactions and alliance formation. As noted in chapter 1, the character of Neanderthal organization has major potential ecological significance, because of the central role of organization in modern hunter-gatherer adaptations—especially in very cold and dry environments.

The open-air sites of the East European Plain provide an exceptional opportunity to examine the structure of Neanderthal sites (caves and rockshelters are less likely to yield clear spatial patterns of remains). The mapped occupation floors of these sites reveal patterns unlike those of

Figure 4.19. *(opposite)* Early and Middle Pleniglacial sites in the northwestern Caucasus: variations in elevation, artifacts, and large mammal remains (after Hoffecker and Baryshnikov 1998, 207, fig. 7).

recent hunter-gatherer occupations. While the organizational implications of these alien patterns are unclear, they do support the notion that Neanderthal society was somehow different. The study of comparative socioecology among the higher primates offers some insights into the probable organization of Neanderthal social groups (e.g., Foley 1989).

Another pertinent aspect of the archaeological record is the evidence—noted earlier—that the sizes of Neanderthal territories and groups were relatively small. If correct, this may also reflect a fundamental difference in the organization of their societies. A lack of ability to establish large social networks over wide areas could have been a critical constraint to occupation of the more extreme cold and dry environments of northern Eurasia (Gamble 1986; Whallon 1989). Large social networks are found among recent hunter-gatherers in arctic and desert settings, where resources are often scarce and unpredictable (e.g., Lee 1972; Kelly 1995), and there is evidence for them in the archaeological record of modern humans.

Symbols and Language

Several researchers have explored the possibility that the Neanderthals lacked the morphological basis for fully modern human speech (Laitman et al. 1979; Lieberman et al. 1992). Noting the relatively flat or unflexed cranial base and its implications for the position of the larynx, they concluded that the Neanderthals were unable to produce some vowel and consonant sounds. However, this limitation would not preclude Neanderthal language based on a smaller range of speech sounds.

More significant is the negative evidence in the archaeological record pertaining to the use of symbols. Ornaments and art objects are present in many modern human Upper Paleolithic sites in Western Eurasia, but are extremely rare in Neanderthal sites. The latter have yielded a small number of simple ornaments (e.g., perforated mammal teeth) and crude art objects. Most of the purported art objects are mammal bone fragments bearing incisions that may have been caused by natural processes or during butchering (Mellars 1996, 369–375; D'Errico and Villa 1997). The near absence of ornamentation and art in their archaeological record has often been interpreted as an indication that the Neanderthals made little use of symbols, including modern human language based on symbols (e.g., Chase and Dibble 1987).

The most important evidence for some form of Neanderthal symbolic expression is intentional burial of the dead, which—at least in a modern human context—is invariably tied to ritual and ideology. Although they have been periodically disputed since the early twentieth century, there appear to be a number of intentional burials of Neanderthals of varying ages in caves of Europe and the Near East (Gargett 1989; Trinkaus and Shipman 1992). Funerary objects or "grave goods" have been reported from a number of these burials; most if not all of these appear equally likely to represent artifacts or faunal remains fortuitously associated with Neanderthal skeletons (Chase and Dibble 1987, 271–276). Dismissing this evidence for ritual practices associated with the graves, some researchers have wondered whether intentional burial by itself necessarily indicates an ideology (e.g., Klein 1999, 467–470).

The Neanderthal sites in Eastern Europe reflect a pattern similar to that of other parts of Western Eurasia with respect to material evidence of symbolic expression. Incised bones interpreted as art objects or ornaments have been recovered from Layers 2 and 4 at Molodova I in the Dnestr Valley, and from the caves of Chokurcha II and Prolom II in the Crimea (Chernysh 1982, 53–65; Stepanchuk 1993, 34–35). With the possible exception of the items from Prolom II (which might be simple decorated pieces or ornaments), none of these is a convincing art object, and the incisions may represent damage caused by natural processes or butchering.

Neanderthal burials are rare in Eastern Europe, reflecting the smaller number of caves and rockshelters, but are documented in both the Crimea and Northern Caucasus. The remains of a partially disturbed adult burial were recovered from a subrectangular pit in the lowest layer of Kiik-Koba (Bonch-Osmolovskii 1940). Another burial comprising the very poorly preserved skeleton of an infant was reported from the same level. More recently, the partial skeleton of an infant was found in Mezmaiskaya Cave (Golovanova et al. 1999, 80–84). Although pits were not observed in the latter two cases, the presence of significant portions of the skeleton in anatomical order suggests placement in a grave, because other skeletal remains in these caves have been dispersed. All remaining Neanderthal bones and teeth were found in isolated contexts, and may or may not have been derived originally from burials.

Occupation Floors and Site Structure

Analysis of occupation floors and speculation about Paleolithic society have a long and rich history in East European archaeology. During the early 1930s, when West European archaeologists were absorbed with problems of cultural stratigraphy, Soviet scholars made the perceived social revolution of the Upper Paleolithic a central issue in prehistory. This was chiefly a consequence of the politically driven effort to apply Marxist-Leninist ideas to archaeology and to reject the traditional approach of prehistorians in western capitalist nations. The social evolutionary schemes of Lewis Henry Morgan (1878), which had been incorporated into Engels's

The Origin of the Family, Private Property, and the State (1884), were applied to the interpretation of Paleolithic sites (Howe 1976; Trigger 1989). The Neanderthals were linked to the "promiscuous horde" stage of organization, while exogamous clan society (based on matrilineal descent) was attributed to modern humans of the Upper Paleolithic (Boriskovskii 1932; Efimenko 1938).

Although wider political developments in the Soviet Union during 1928–1934 were the prime catalyst for the new theoretical framework, the character of the East European archaeological record was an important factor (Bahn 1996, 210–215). The lack of caves in much of Eastern Europe compelled a focus on open-air sites buried in loess or colluvium with minimal disturbance to remains. The recognition of former dwelling structures on several Upper Paleolithic occupation floors during the 1920s inspired Soviet archaeologists to excavate broad horizontal areas and record complex patterns of features at many open-air sites across the East European Plain. The latter were used to support the social evolutionary framework. A massive feature complex recorded at Kostenki in 1931–1936 was interpreted as evidence of matrilineal clan organization in the Upper Paleolithic (Efimenko 1938, 448–451). However, little information was available on Neanderthal occupation floors in East European open-air sites until after World War II.

The social evolutionary framework applied to the Paleolithic in the 1930s was eventually rejected by most archaeologists in Eastern Europe (Grigor'ev 1968). Nevertheless, the continued broad-scale excavation of occupation floors at open-air sites on the East European Plain yielded an unparalleled body of data regarding the spatial organization of remains at Middle and Upper Paleolithic sites. Despite our inability to deduce the organization of Neanderthal groups from their occupation floor patterns, the East European record contains some valuable information on this issue.

In the southwest region of the East European Plain, occupation floor maps have been published for Molodova I (Mousterian Layers 1–5), Molodova V (Layers 11, 12a, 12), and Ketrosy in the Middle Dnestr Valley (Chernysh 1959, 1982, 1987; Anisyutkin 1981a). The exposed areas range from 225 square meters at Ketrosy to up to 1,200 square meters for Layer 4 at Molodova I, and the combined total for these sites equals 4,356 square meters (although the published maps cover a subset of this total). Several hundred square meters of occupation floor were mapped and published for Ripiceni-Izvor in the Prut Valley (Păunescu 1965, plate I).

In the southern part of the East European Plain, over 600 square meters of occupation floor area were mapped at Sukhaya Mechetka on the Lower

Volga River (Zamyatnin 1961, fig. 4) (fig. 4.20). Floor areas ranging from roughly 40 to 85 square meters (total of over 400 square meters) are published for Layers 1–6 at Rozhok I on the Sea of Azov coast (Praslov 1968, 72–88). A smaller occupation area of roughly 35 square meters (lacking faunal debris) was mapped at the nearby site of Nosovo I (Praslov 1972, 77, fig. 2). Finally, artifacts, features, and faunal remains in two trenches occupying a combined area of approximately 45 square meters were published for Il'skaya I in the lower foothills of the Northern Caucasus (Gorodtsov 1941, 24, fig. 10).

The mapped occupation floors on the East European Plain reveal a recurrent pattern of randomly distributed artifact and/or faunal debris concentrations of varying size and density. Former hearths are associated with some debris concentrations, but also found in isolated contexts; a consistent pattern of linkage between the two features seems to be lacking. As noted earlier, several sites in the southwest region (Molodova I and V, Ripiceni-Izvor) exhibit circular or linear arrangements of large mammal bones that may represent windbreaks or other simple structures, although this remains problematic. A semicircular arrangement of rocks was reported earlier from Il'skaya I (Gorodtsov 1941, 23), but this is now thought to represent an intrusive modern feature (Formozov 1965, 38).

The occupation floors reveal a low degree of structure or organization in the use of space. The contrast with open-air sites occupied by modern humans during the later Upper Paleolithic (OIS 2) is especially striking. Many of the latter contain highly structured arrangements of former dwellings, hearths, pits, and debris concentrations with parallels to the organization of modern hunter-gatherer camps (see chapter 6). While the contrast is accentuated by technological differences—the Neanderthals apparently did not construct dwellings or storage pits at their sites—the lack of organization of working areas (represented by debris concentrations) around hearths is significant (Mellars 1996, 313–314). Minimal site structure has also been noted in Mousterian sites of Western Europe (e.g., Farizy and David 1992), but the large sample of occupation floors mapped in Eastern Europe provides the best opportunity to examine this aspect of the Neanderthal archaeological record. These data offer little if any basis for inferences concerning the organization of their occupants, but they do suggest that it was somehow different from that of modern human foragers.

In the Crimea and the Northern Caucasus, occupation floors in caves and rockshelters have occasionally been mapped and published (e.g., Zaskal'naya V, Layer 2 [Kolosov 1983, 46–47, fig. 14], Barakaevskaya Cave [Lyubin 1994, 196–199, figs. 15–18]). The mapped floors of intensively used natural shelters typically exhibit a jumbled mass of artifacts and other

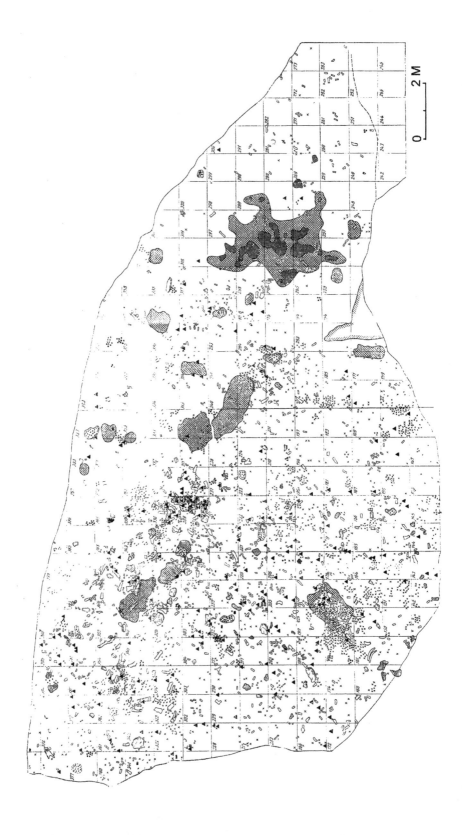

debris, apparently reflecting a spatially confined setting. Although the pattern is less pronounced at relatively open rockshelters such as Kabazi I in the Crimea (Formozov 1959, 151, fig. 7), where spatial constraints are limited, these sites generally yield much less information than open-air sites on the use of space.

However, natural shelters occasionally offer insights concerning the size of the groups that occupied them. The best example of this in Eastern Europe is Barakaevskaya Cave in the Northern Caucasus, which contains a tightly circumscribed area of approximately 35 square meters (Lyubin 1994). Assuming that all members of the groups that occupied this site resided within the narrow confines of the cave, they seem unlikely to have exceeded ten to twelve individuals (Hoffecker and Baryshnikov 1998, 206). Similar conclusions have been reached regarding maximum group size in several small caves in Western Europe (Mellars 1996, 270–292), and suggest that Neanderthal group size may have been consistently smaller than that of typical residential bands among recent hunter-gatherers (about 25 individuals [Kelly 1995, 210–213]).

In the absence of recognizable evidence concerning the organizational structure of Neanderthal societies, the comparative socioecology of higher primates may offer some general insights. It seems likely that early *Homo* organization would have followed a broader hominoid pattern of core groups composed of closely related males (Foley 1989, 487). Cooperative interactions among the core group would have been based on their high degree of genetic relatedness (i.e., kin-selected), and this would have limited its size. As representatives of *Homo* colonized Eurasia and evolved new adaptations to cooler, drier, and more seasonal environments, this generic hominoid pattern was probably modified. The Neanderthal emphasis on a high meat diet and hunting economy may have generated a more pronounced sexual division of labor (assuming that the foraging activities of mature females were constrained by child-rearing demands) (Mellars 1996, 361–362). This would have intensified male-female cooperative bonds with respect to food-sharing and provisioning offspring.

Neanderthal Adaptation in Eastern Europe: Summary and Conclusions

The East European Neanderthals did not develop their unique adaptation in response to local environments, but evolved in Western Europe prior to their colonization of cooler and drier landscapes farther east. Fully

Figure 4.20. *(opposite)* Occupation floor mapped from the western excavations at Sukhaya Mechetka (Volgograd) on the Lower Volga River (modified from Zamyatnin 1961, fig. 4).

developed Neanderthals were present in Western Europe by the final glacial cycle of the Middle Pleistocene (OIS 7–6), and probably reflect the combined influence of genetic drift and adaptation to cold environments. They did not settle widely in Eastern Europe until the beginning of the Late Pleistocene (OIS 5), and their ability to cope with cold and dry environments was almost certainly a prerequisite for occupation of this part of northern Eurasia (Hublin 1998; Hoffecker 1999a).

Eastern Europe may reveal as much or more about Neanderthal adaptations than any other region in northern Eurasia. It was the margin of their range, and the shifting distribution of their sites during OIS 5–3 suggests that their ability to occupy cold environments—while superior to their predecessors—had limitations. The Neanderthals seem to have been forced to abandon much of the East European Plain during the colder phases of the Early and Middle Pleniglacial (OIS 4–3). Comparison of their archaeological record with that of the modern humans who subsequently colonized these environments reveals potentially critical differences.

Studies of recent hunter-gatherers provide insights to the requirements of cold and dry environments during the Pleistocene. Many aspects of the morphology and behavior of recent hunter-gatherer peoples are correlated with latitude and climate. Populations that have inhabited high latitudes for at least a few thousand years possess a variety of morphological and physiological adaptations to cold temperatures (e.g., shortened distal limb segments) (Trinkaus 1981). The percentage of meat in a hunter-gatherer diet is closely tied to latitude, and groups in arctic and subarctic regions subsist primarily on a protein-rich diet of hunted meat and fish (Kelly 1995). Also, foraging peoples in northern environments (especially in continental areas) tend to inhabit larger territories and move greater distances, reflecting the mobile and dispersed character of their resource base (Binford 1980). Finally, hunter-gatherers in colder regions possess more sophisticated technology both in terms of the diversity of tools and complexity of individual implements (Oswalt 1976; Torrence 1983). The Neanderthals clearly possessed both morphological and dietary adaptations to cold environments, but—in comparison to modern humans—they seemed to have been restricted to a relatively simple technology and limited mobility.

The "classic Neanderthals" of Western Europe exhibit many skeletal traits that conform to the predictions of Bergmann's and Allen's ecological rules regarding body size and shape in cold climates. They possessed large heads, thick trunks, and shortened extremities, and at least some of these features represent an extreme cold-climate (or "hyperpolar") adaptation when scaled against modern humans (Holliday 1997). It may be assumed that—like modern humans—the Neanderthals also evolved a number of physiological responses to cold temperatures (Coon 1962). Contrary to earlier speculation, it is now clear that the East European Neanderthals were similar to the "classic Neanderthals" of Western Europe, and their cold-adapted morphology may have been critical for surviving winter temperatures in Eastern Europe, especially given their apparent lack of advanced technology for cold protection (see fig. 4.12).

Like recent hunter-gatherers in cold environments, the European Neanderthals had also developed a heavy meat diet and the foraging adaptations that were necessary to support it. The predominance of meat in the diet is indicated by stable isotope analyses of bone collagen from Neanderthal skeletal remains in Western and Central Europe (Bocherens et al. 1999; Richards et al. 2000). Heavy reliance on meat may also be inferred from evidence for the hunting of large mammals from Neanderthal sites in Europe and the Near East (e.g., Chase 1986). Steppe bison, wild ass, saiga, and other large grazers are more abundant in the East European sites, reflecting local environmental conditions (e.g., Vereshchagin and Kolbutov 1957; Baryshnikov et al. 1996; Burke 2000).

The regular hunting of large mammals may have been broadly coincident with the appearance of fully developed West European Neanderthals in the late Middle Pleistocene, and this too might have been a prerequisite for settlement in Eastern Europe. Particularly in Eastern Europe, the high caloric demands of cold climates, combined with the limited availability of plant foods in many areas, probably required a heavy meat diet and a hunting economy (Hoffecker 1999a; Hoffecker and Cleghorn 2000) (fig. 4.21).

In contrast to what has been determined about their morphology and diet, the complexity of Neanderthal technology—although probably superior to their predecessors—appears to have been very low relative to that of recent hunter-gatherers in northern latitudes. Like the regular hunting of large mammals, the development of Levallois stone tool production techniques seems to be broadly coincident with the appearance of the Neanderthals in Western Europe. These techniques are probably tied to the practice of hafting stone tools (Schick and Toth 1993), for which there is supporting lithic microwear evidence (e.g., Anderson-Gerfaud 1990). The hafting of tools would have increased their efficiency with corresponding reduction of the energy demands on their users. The Neanderthals also may have fashioned a wider variety of wood implements than earlier Middle Pleistocene hominids (lithic microwear analyses indicate substantial working of wood [e.g., Beyries 1988]). The increased variety and efficiency of Neanderthal technology might have contributed to their ability to colonize colder and drier environments, although it should be noted

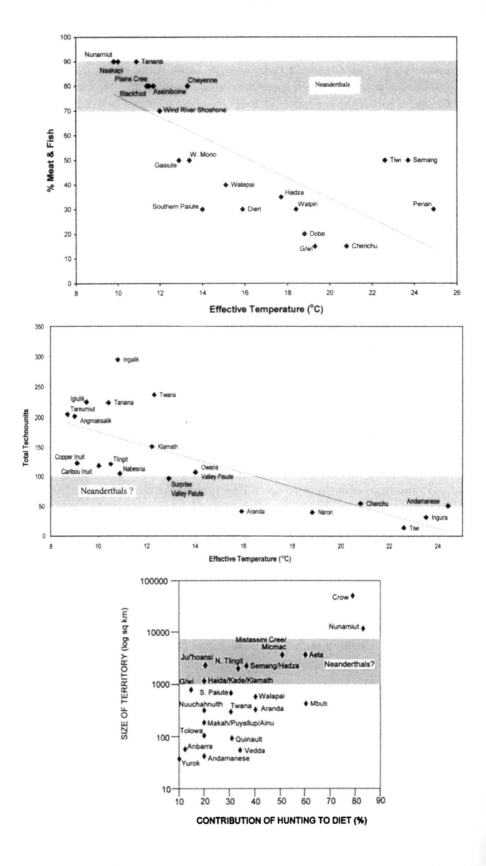

that their stone tool assemblages from sites in Eastern Europe reflect a comparatively low use of Levallois techniques and may indicate a greater emphasis on hide-working (e.g., Gladilin 1976; Praslov 1984b).

Despite the apparent advances over their predecessors, Neanderthal technological complexity falls at the low end of the scale for recent hunter-gatherers. Both in terms of the number of types and component parts of individual implements, the complexity of Neanderthal tools and weapons is significantly lower than that of hunter-gatherers in northern latitudes (and more typical of modern groups in temperate or equatorial regions [Oswalt 1976]). Technological complexity in colder environments seems to reflect the need for greater foraging efficiency in settings where many resources are available only for limited periods of time (Torrence 1983). More specifically, the Neanderthals seem to have lacked untended facilities (e.g., traps and snares) and devices for food storage, which are common technological strategies for coping with resource fluctuations and high mobility requirements among hunter-gatherers in high latitudes (fig. 4.22).

Equally important is the apparent lack of technology—found among modern humans—for cold protection and heat conservation. Although microwear analyses of stone tools indicate that the Neanderthals often scraped hides (especially in East European sites), which were presumably used as blankets and simple clothing, there is no compelling evidence for tailored fur clothing or insulated shelters. Perhaps the critical function of the latter is that they provide protection from extreme low temperatures in a form that permits humans to forage and perform other important economic tasks (e.g., tool manufacture).

The Neanderthals also seem to have been characterized by relatively limited movements and small territories in comparison to recent hunter-gatherers in northern latitudes. If the distances of raw materials in Paleolithic sites from their original sources provide a measure of movements (or at least the dimensions of social networks), Neanderthal groups were rarely moving more than 100 kilometers (Roebroeks et al. 1988; Mellars 1996), and would seem to have been confined to territories of less than 10,000 square kilometers. As in the case of technological complexity, this level of mobility is more typical of recent hunter-gatherers in lower latitudes

Figure 4.21. *(opposite)* Effective temperature and percentage of meat and fish in diet among recent hunter-gatherers (based on data in Kelly 1995, 67–69, table 3.1), showing projected values for Neanderthals; **(4–22)** Effective temperature and technological complexity (measured in total "technounits" for food-getting technology) among selected recent hunter-gatherers (based on data in Oswalt 1976, 173, table 9.1), showing projected values for Neanderthals; **(4–23)** Contributions of hunting to diet and size of territory among recent hunter-gatherers (modified from Kelly 1995, 131, fig. 4.8), showing projected values for Neanderthals.

subsisting on a diet high in plant foods (Kelly 1995, 111–132). There is also evidence from both Western and Eastern Europe that the size of Neanderthal groups was smaller than that of typical modern hunter-gatherer bands (e.g., Hoffecker and Baryshnikov 1998), which seems consistent with small territories (fig. 4.23).

Among modern hunter-gatherers, mobility and networks are closely tied to social organization. Although the organization of Neanderthal societies remains unknown, there is enough negative evidence to suggest that it was different in some way from that of modern humans. The almost complete absence of material expressions of ideology and the disconcerting lack of structure in the spatial organization of their sites imply some fundamental difference with the tribal networks and local bands of modern hunter-gatherers (Klein 1973; Gamble 1986, 1999) and Eastern Europe provides much supporting (negative) data for these observations, especially with respect to occupation floors at open-air sites. Neanderthal societies, which seem likely to have been based on a core group of closely related individuals (e.g., Foley 1989), may have lacked the flexibility of modern hunter-gatherer organizations to establish large networks through exogamous marriage alliances across the landscape.

Overall, the pattern of Neanderthal adaptation seems to have been broadly similar across Europe. Despite local environmental differences with Western Europe, there is relatively little evidence of an East European adaptation. The morphology of the East European Neanderthals does not seem to represent a more extreme form of cold adaptation (although the sample available for comparison remains small), and their diet was probably no more heavily based on meat than that of their West European counterparts. Neanderthals in Eastern Europe exhibit some minor differences in their stone tools (perhaps reflecting less hafting of implements due to shortages of hardwood), and hunted large mammal species that were more common in colder and drier habitats (e.g., steppe bison).

This relative uniformity in their adaptation seems to correspond to a certain lack of flexibility in responding to new conditions. When temperatures declined during the Early Pleniglacial (OIS 4), the Neanderthals were apparently unable to cope with periglacial loess-steppe environments on the East European Plain (Hoffecker 1987; Soffer 1989). Much of the latter seems to have been abandoned by Neanderthals at this time, although some areas (notably the southwest region) were reoccupied during the milder Middle Pleniglacial (OIS 3). By contrast, modern humans successfully colonized the periglacial loess-steppe during the terminal phases of OIS 3 and the subsequent Last Glacial Maximum (OIS 2). Unlike the Neanderthals, early modern humans in Eastern Europe left archaeological traces of an adaptation very similar to that of recent hunter-gatherers in arctic environments (see chapters 5 and 6). They developed a complex technology that included untended facilities and storage devices, as well as tailored clothing and insulated shelters, and the long-distance movements of their raw materials indicate the presence of widespread networks over large areas.

CHAPTER 5

The Transition to Modern Humans

Modern humans appear in Eastern Europe at least 35,000 years BP and probably somewhat earlier. They also appear at this time in other parts of northern Eurasia. The timing of the transition remains obscured by the uncertainties of radiocarbon dating in this time range. Modern humans represent a southern form of *Homo*, which had evolved in lower latitudes during the preceding 200,000 years and subsequently dispersed northward. Their appearance in Europe is linked to the more or less simultaneous disappearance of the Neanderthals.

Although the transition to modern humans is marked by significant changes in morphology, the latter are overshadowed by evidence for more important changes in behavior. Indisputable examples of art, burial ritual, and other material manifestations of symbols and ideology appear at this time in the archaeological record. Many researchers believe that their appearance signals the birth of fully modern human language (e.g., Mellars 1996; Klein 1999).

The archaeological record also reveals a quantum leap in technological innovation and complexity, which undoubtedly played a critical role in the rapid modern human colonization of new habitats. The appearance of modern humans in northern Eurasia is accompanied by evidence for a variety of new technologies, and many of them reflect responses to cold environments. At least some of these new technologies exhibit an unprecedented level of complexity that may be tied to the ability to manipulate and communicate with symbols (Dennell 1983; Mithen 1996).

A fundamental change in social behavior and organization also probably occurred at the time of the transition, although its consequences are not clearly evident in the archaeological record until the beginning of the Last Glacial Maximum (ca. 25,000 years BP). The appearance of modern humans in northern Eurasia is associated with long-distance movements of raw materials on an unprecedented scale, which may indicate the development of widely dispersed networks similar to those of recent hunter-gatherers. The ability to establish large social networks is also related to the occupation of cold environments, and also tied to the use of symbols in a social context (Gamble 1986; Whallon 1989).

If controversies attend the study of the Neanderthals, even more vociferous disputes surround the transition to modern humans. Some of these disputes are tied to the Neanderthal controversies. Paleoanthropologists who perceive relatively limited contrasts between Neanderthals and modern humans often see continuity in the transition, while those who view the contrast as significant often interpret the transition as replacement. The West European record probably offers the most convincing example of modern human replacement of the Neanderthals, but even here the issue is complicated by the presence of a Neanderthal Upper Paleolithic industry containing ornaments and bone tools (i.e., Chatelperronian) (Mellars 1996). In other parts of Eurasia, the record of the transition is more complex.

In Eastern Europe—where Soviet archaeologists of the 1930s described the transition as a "great leap" (*bol'shoi skachok*)—the contrast between Neanderthal and modern human adaptations is a stark one (Klein 1973). Much of Eastern Europe represented a marginal environment for modern humans as well as for Neanderthals, and in some respects the adaptive demands were even greater on the former. Recently derived from southern latitudes, modern humans arrived in Pleniglacial Europe with a skeletal morphology best suited for tropical climates (Trinkaus 1981). But they were ultimately more successful than their predecessors in colonizing periglacial habitat, and their strikingly different response to its demands is especially apparent in Eastern Europe. The starkness of the contrast in this part of the world helps illuminate the nature of the transition (Soffer 1989; Hoffecker 1999a).

However, understanding the process of the transition to modern humans is a formidable challenge in Eastern Europe. The reasons for this include: (a) the scarcity of human fossil remains, (b) continuing uncertainties regarding the dating of key sites, and (c) the complexity and uniqueness of the archaeological record. It is difficult to fit Eastern Europe into the global picture of modern human origins and dispersal. There is no compelling evidence of morphologically transitional forms, and species replacement represents the most plausible explanation. But the archaeological record is almost completely different from that of Western Europe, and less easily explained as the rapid replacement of one group of toolmakers by another. At the least, it suggests a protracted and complex transition process.

Origins and Dispersal of Modern Humans

Like their Neanderthal predecessors, the first East European modern humans did not evolve locally, but probably arrived with a developed suite of morphological and behavioral traits from another part of the Old World. But unlike the Neanderthals, they came from southern latitudes and warmer climates, and did not represent a specialized northern variant of *Homo*. The appearance of modern humans in Eastern Europe was part of a global colonization event by a species with an unprecedented degree of adaptive flexibility, and is best understood in that context (Fagan 1990; Stringer and McKie 1996).

In many respects, the evolution of modern humans (*Homo sapiens*) in Africa offers a parallel to the emergence of the Neanderthals in Western Europe. On the basis of his analysis of Middle and Late Pleistocene hominid fossils, German paleoanthropologist Günter Bräuer proposed an African origin for modern humans in the early 1980s (Bräuer 1984, 1989). Subsequent discoveries and research have provided further support for this thesis (Rightmire 1986; Klein 1999). The earliest representatives of *H. sapiens* are identified in East African sites dating to roughly 600,000–400,000 years ago (e.g., Bodo, Ndutu, Broken Hill). Almost fully developed modern humans are present in Africa (e.g., Omo, Klasies River Mouth, Dar-es-Soltan) during OIS 7-5 (250,000–70,000 years ago), and the same "early modern humans" are also found in the Near East during OIS 5 (e.g., Skhul, Qafzeh). As noted earlier, they may have spread farther north into parts of Eastern Europe during this period of milder climates (see chapters 3 and 4).

The morphological characters that distinguish modern humans from other hominids include a vertical frontal bone, reduced browridge, and an occipital that lacks a both an occipital torus and a suprainiac fossa. The region above the canine root is depressed (not inflated), the mandible typically possesses a chin, and the face is positioned directly below the forepart of the braincase (not in front of it). Lacking the prognathism of the Neanderthals, modern humans also lack a retromolar space between the third molar and ascending ramus of the mandible. Although less diagnostic, the postcranial skeleton also displays some characteristic features. The bones are generally less robust than those of earlier hominids, and the cortical bone of the limb bones is thinner. In contrast to that of Neanderthals, the lateral border of the scapula exhibits a ventral groove or sulcus (bisulcate pattern in early modern humans). The distal limb segments are long relative to the proximal segments. The terminal phalanges are small and elongated, especially in comparison to those of the Neanderthals (Klein 1999, 498–503).

Few of the anatomical features of modern humans that distinguish them from Neanderthals possess obvious adaptive significance. As in the case of the latter, many of them are likely to reflect isolation and genetic drift. The altered shape of the cranial vault may reflect development of specific parts of the brain (e.g., frontal lobe) related to characteristic modern human behavior, but archaeological evidence for such behavior does not materialize until later. The lengthened distal limb segments and elongated phalanges are almost certainly adaptations to warm climates, and follow the predictions of Allen's rule with respect to heat loss and the shape of extremities (Trinkaus 1981, 193–201).

It is easier to trace the origin of modern human anatomy than the origin of modern human behavior. There is little evidence for the behavioral changes mentioned earlier in the archaeological record until well after the appearance of anatomically modern people (but see McBrearty and Brooks 2000 for an alternative view). Formal bone tools, comparable to those of modern human Late Stone Age/Upper Paleolithic industries, are currently dated to roughly 100,000 years ago at two African sites (Katanda and Blombos Cave), but their age is problematic (Klein 2000, 28–29). More securely dated examples of ornaments, art, and bone tools are not present in modern human sites until after 50,000 years ago, and are not associated with any observable morphological change. Many paleoanthropologists believe that they reflect the appearance of fully developed symbolic language. Noting the absence of morphological change, Richard Klein has attributed the inferred behavioral shift to altered neurological structure triggered by random mutation ("neural hypothesis") (Klein 1999, 514–517). The African evidence for behavioral change does not significantly antedate the evidence from Eurasia (which dates to greater than 40,000 years ago), although this may be a function of the limitations of current dating techniques.

After the onset of cooler climates during OIS 4 (roughly 70,000 years ago), European Neanderthals appear to have expanded southward into the Near East (Bar-Yosef 1989), possibly replacing the early modern human population in the region. Neanderthal cold-climate adaptations may have conferred some competitive advantage on them during this interval. Not until after the beginning of OIS 3 (after 55,000 years ago), when climates in northern latitudes ameliorated to some degree, is there evidence of behaviorally modern humans outside Africa. Fully modern *Homo sapiens* may have been present in Australia prior to 50,000 years BP, and might represent an early OIS 3 dispersal event (Lahr and Foley 1994). Their appearance in latitudes above 30 degrees North seems to have occurred during 45,000–40,000 years BP. Modern humans and the Upper Paleolithic Ahmarian industry were present in the Levant by 38,000–37,000 years BP, and archaeological evidence suggests they probably antedate 40,000 years BP (Bar-Yosef 1989, 1998).

Modern humans appear in northern Eurasia (above 40 degrees North) at roughly 40,000 years BP, although the age of the oldest sites probably varies by region. In Western and Central Europe, the earliest modern human occupations are recognized by the presence of distinctive Upper Paleolithic artifacts of the Aurignacian industry, which contains a high proportion of blade tools, bone implements, and some figurative art. Aurignacian assemblages are dated to as early as ca. 40,000 years BP in both southwestern and south-central Europe (Mellars 1996). All skeletal remains found in Aurignacian sites represent modern humans (Gambier 1989). East of the Carpathians, Aurignacian sites are scarce, but other fully developed Upper Paleolithic industries containing blade technology, bone implements, and occasional ornaments and art are also present by ca. 40,000 years BP (Hoffecker 1999a; Goebel 2000).

Much controversy remains concerning the process by which modern human morphology became established throughout the world. Although an African source seems to be supported not only by the fossil record but also by molecular and linguistic data (e.g., Ingman et al. 2000), many paleoanthropologists perceive at least some degree of morphological and behavioral continuity in regions outside Africa. They argue that only gene flow and exchange—not species replacement—can explain these continuities (e.g., Wolpoff 1999). The process may have varied among regions, and if Europe represents the most plausible example of replacement, East Asia may constitute the best case for some continuity and genic exchange (e.g., Hapgood 1989).

More important than the process of morphological change in Eurasia is the apparent linkage between the spread of modern human morphology and behavior. At present, archaeological evidence for the use of symbols and associated changes in technology is most firmly dated to about 50,000 years ago, and coincides with modern human dispersal out of Africa. The archaeological record indicates that people expanded into previously unoccupied areas and new habitats at this time (e.g., Australia, Siberia), and much of this expansion was apparently achieved with both technological innovations and organizational adaptations—especially to marginal environments. By the close of OIS 2 (14,000–10,000 years BP), most of the available terrestrial habitat on the planet had been colonized by a single hominid species employing technologies—and perhaps organizational structures—specifically adapted for local conditions (Soffer and Gamble 1990). The use of symbols roughly 50,000 years ago appears to be linked to an unprecedented ability for rapid colonization of new habitat.

Early Modern Human Sites in Eastern Europe

The East European sites considered here occupy a relatively brief interval of geologic time. They are assigned to the later part of OIS 3, and dated to roughly 45,000–25,000 years BP. This period was characterized by generally cool but oscillating climates—including several mildly warm episodes—that preceded the Last Glacial Maximum (OIS 2) (see chapter 2). The dating of these sites is significantly more precise than those of earlier periods. This is a function of their young age, which lies at least partly within the range of radiocarbon dating, although many occupations fall at or beyond the effective reach of this method. Also important is the presence of a widespread stratigraphic marker in the form of a buried soil that formed at the close of OIS 3 and dates to 30,000–25,000 years BP. A number of open-air sites contain occupation levels within and below this buried soil horizon.

Human skeletal remains recovered from the late OIS 3 sites in Eastern Europe are all assigned to modern humans (*Homo sapiens sapiens*), although Neanderthals are present in terminal Mousterian sites of the Crimea and Northern Caucasus that are as late as ca. 30,000 years BP. However, modern human remains that can be firmly dated to OIS 3 are confined to the central East European Plain (see later discussion). Isolated modern human bones have been recovered from Upper Paleolithic levels at several sites in other parts of Eastern Europe, but they may all postdate OIS 3 (i.e., after 25,000 years BP). Moreover, although assigned to the Upper Paleolithic, the archaeological remains in the late OIS 3 sites exhibit an enormous amount of diversity and include lithic assemblages similar to those of the Szeletian industry of Central Europe (e.g., Brynzeny I [Ketraru 1973, 73]). The latter appears increasingly likely to have been produced by Neanderthals, and may exhibit some parallels with the Chatelperronian industry of Western Europe (Allsworth-Jones 1990). In the absence of human skeletal remains from these sites, the possibility that they were occupied by Neanderthals cannot be discounted, although there is reason to believe that this is unlikely (see later discussion).

The distribution of known sites indicates that much of Eastern Europe was probably settled by modern humans during the latter part of OIS 3, but the timing of their initial appearance may have varied from one region to another. The earliest dated sites seem to lie on the East European Plain, especially on the Middle Don River at Kostenki (at least 40,000 years BP) (Sinitsyn et al. 1997). Relatively early occupations may also be present in the Middle Dnestr Valley at localities such as Molodova V (35,000 years BP or earlier) (Ivanova 1987). Curiously, there are no late OIS 3 sites in the Dnepr-Desna Basin, which might reflect scarcity of fuel during this interval (Soffer 1985; Hoffecker 1988). However, several sites are found in northern Russia and possibly the far northeast (Kanivets 1976; Bader 1978). Sites in the Crimea and Northern Caucasus are few in number, and currently appear to be no older than 32,000–30,000 years BP. There is corresponding evidence for a late Neanderthal presence in these regions (Marks and Chabai 1998; Golovanova et al. 1999) (fig. 5.1).

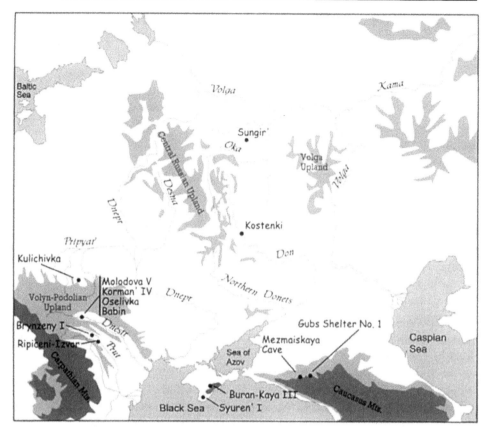

Figure 5.1. Map of early Upper Paleolithic sites (OIS 3) in Eastern Europe.

East European Plain: Southwest Region

The most important localities in this region are concentrated in the Middle Dnestr Valley and include *Molodova I*, *Molodova V*, and *Korman' IV* (Chernysh 1959, 1977, 1982, 1987). These sites are found on the Second Terrace, and all of them contain Mousterian occupation horizons buried in loessic colluvium that overlies stream gravels (see chapter 4). Upper Paleolithic artifact assemblages dating to late OIS 3 are associated with a buried soil horizon formed at the end of this interval (30,000–25,000 years BP) at all of these localities. At Korman' IV (Layer 8) and Molodova V (Layers 10a–10b), small accumulations of artifacts underlie the occupations dating to the end of OIS 3 (Chernysh 1977, 23–24; 1987, 25–26). The older assemblages are buried in colluvial loam and presumably date to roughly 35,000 years BP (Ivanova 1982, 234, fig. 15).

The occupation layers associated with the terminal OIS 3 soil at Molodova V (Layers 10–8) contain lithic assemblages of 500–1,400 artifacts thought to be especially characteristic of the early Upper Paleolithic industry in the region (Chernysh 1987, 27–35) (fig. 5.2). They comprise prismatic blade cores, large blades, and many tools on blades (primarily burins, knives, and end-scrapers). Nonlithic implements are rare (perhaps partly due to limited bone preservation), but several ornaments are present in Layer 8. During 1954–1958, the late A. P. Chernysh mapped occupation floor areas on these levels of 630–750 square meters, providing a rare glimpse of open-air site organization in this time range (Chernysh 1959, 68–76). The floors contain concentrations of lithic and bone debris associated with former hearths. The bone debris is composed primarily of horse and reindeer (Alekseeva 1987, 158).

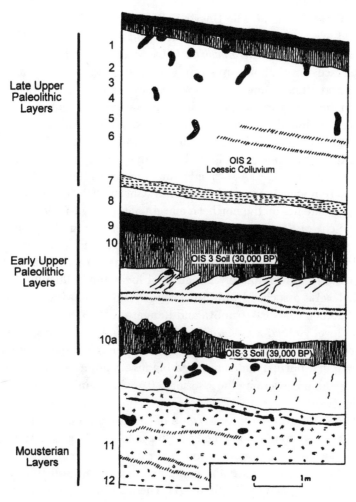

Figure 5.2. Stratigraphy of Molodova V in the Middle Dnestr Valley (redrawn from Ivanova 1987, 104, fig. 8).

The contemporaneous artifact assemblages at Molodova I (Layer 3) and Korman' IV (Layer 7) present a somewhat different picture. The assemblages are small (146–332 items) and contain cores and flakes typical of the Mousterian industry. However, blades and blade tools are common in these assemblages, and the latter are similar to those at Molodova V (burins, knives, end-scrapers) (Chernysh 1977, 24–26; 1982, 70–72). At the nearby site of *Ataki I*, which is also located on the Second Terrace, the terminal OIS 3 soil yielded a small assemblage comprising a similar blend of Mousterian cores and Upper Paleolithic tools, but these artifacts may be redeposited and mixed (Chernysh 1968, 104). The older assemblages that underlie the terminal OIS 3 soil at Molodova V and Korman' IV are also small and of mixed or "transitional" character (Chernysh 1977, 23–24; 1987, 25–26).

Other sites in the Middle Dnestr Valley that may date to the late OIS 3 period are located on higher terraces and include *Babin I* (Layer 3), *Oselivka I* (Layer 3), and *Voronovitsa I* (Layer 2). Surficial deposits on the high terraces are shallow—apparently reflecting the repeated movement of loess and other sediment onto lower valley slopes—and the occupation layers in these sites are not buried in deep stratigraphic context. Although the layers assigned to this time range are contained in dark loam that might exhibit traces of soil formation, their assignment to OIS 3 is based primarily on the typological similarities of the artifact assemblages to Layers 10–8 at Molodova V. Associated faunal remains at these sites are dominated by reindeer, horse, and mammoth (Chernysh 1959, 16–59; 1971).

A major site lies on the northern margin of the Volyn-Podolian Upland near the city of Kremenets at *Kulichivka*. This site is found on a low bedrock promontory along the Ikva River, a small tributary of the Styr' River that flows northward into the Pripyat' Basin. The main occupation layers are buried in approximately 2 meters of loess and colluvium, and the middle level is associated with the terminal OIS 3 soil with a supporting date of 25,000 years BP (Ivanova and Rengarten 1975; Savich 1987). The middle level yielded an assemblage of over 50,000 artifacts, including numerous blades (14 percent) and typical Upper Paleolithic tools (chiefly end-scrapers, burins, and retouched blades). Mammoth, reindeer, and horse predominate among the associated faunal remains (Savich 1975, 44–47) (fig. 5.3).

Other sites in the southwest region of the East European Plain dating to late OIS 3 are found along the Prut River and its tributaries. At the large open-air locality of *Ripiceni-Izvor* on the Prut River (see chapter 4), Upper Paleolithic assemblages are also contained in a buried soil and dated to 28,420 years BP (Paunescu 1965, 1987). These assemblages comprise a low percentage of blades and many typical Mousterian tools (denticulates,

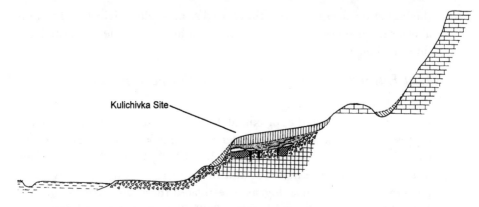

Figure 5.3. Geomorphic setting of the Kulichivka site in the northern Volyn-Podolian Upland (modified from Ivanova and Rengarten 1975, 52–53, fig. 1).

notches, bifacial points) along with some Upper Paleolithic forms (retouched blades, end-scrapers). Associated faunal remains include horse and red deer.

The small rockshelter of *Brynzeny I*, which is located in a nearby tributary valley (Rakovets River), contains an occupation level (Layer 3) assigned to the early Upper Paleolithic (Ketraru 1973, 69–74). A large assemblage (over 8,000 artifacts) is buried at the base of a shallow sequence of loam and rubble deposits that yielded dates on mammal bones and teeth of between 14,700 and 26,600 years BP (Hedges et al. 1996). The wide range of radiocarbon dates may indicate that younger remains are mixed with a late OIS 3 occupation. The stone tools include both typical Mousterian and Upper Paleolithic types such as points, side-scrapers, small bifaces, end-scrapers, and burins. Among the nonstone items is a decorated pendant of mammoth ivory, which represents a rare example of early Upper Paleolithic art from this region (Ketraru 1973, 73, fig. 29). Many thousands of vertebrate remains were found in this occupation level, and they reflect an overwhelming predominance of horse and reindeer (David and Ketraru 1970, 23–25). Farther south in the main valley lies the open-air site of *Korpach*, which contains a similar assemblage buried roughly 4 meters below the surface in a soil horizon, dated to 25,250 years BP (Grigor'eva 1983).

Overall, the distribution of late OIS 3 sites in the southwest region of the East European Plain exhibits a pattern comparable to that of the preceding period. In many cases, early Upper Paleolithic assemblages overlie those of the Mousterian. Modern humans appear to have followed a land-use strategy similar to that of the Neanderthals. Multiple-activity and habitation sites were occupied on the low terraces in the main valleys, while the higher terraces were sometimes used as multiple-activity loci. Presumably a similar set of variables drew modern humans to the same site locations as their predecessors.

East European Plain: Central and Northern Regions

Located on the west bank of the Middle Don River, the sites near the village of *Kostenki* (located approximately 40 kilometers south of Voronezh) represent the most important Upper Paleolithic localities in Eastern Europe (Klein 1969b; Praslov and Rogachev 1982). Archaeological remains were discovered here in 1879, and during the ensuing decades many prominent Soviet archaeologists investigated these sites (e.g., Rogachev 1957; Efimenko 1958; Boriskovskii 1963). The excavation of large occupation floor areas and mapping of complex features at Kostenki had a significant impact on Paleolithic research (Praslov 1982, 7–13).

The Kostenki sites comprise 21 localities, and at least 9 of them contain occupations dating to OIS 3 (antedating 25,000 years BP). They are found in an area where the Don River flows along the eastern margin of the Central Russian Upland. The latter is composed of sedimentary bedrock covered with Middle Pleistocene glacial till that lies about 130–140 meters above the modern floodplain. Below this level rests the Third Terrace (35–40 meters), which is composed of alluvium capped with a forest soil, both of which are correlated with the Last Interglacial climatic optimum (Lazukov 1982, 17–20). No archaeological remains are reported from these deposits (fig. 5.4).

The base of the Second Terrace (20 meters) is composed of sandy alluvium deposited sometime between the Last Interglacial and the late Middle Pleniglacial. This alluvium is interstratified with colluvial deposits reworked from higher slopes (Lazukov 1982, 21). Above the alluvium lies a sequence of colluvial loams and organic horizons derived from soils that date to at least 40,000 years BP (Anikovich 1993; Sinitsyn 1996). The earliest Upper Paleolithic occupation horizons are found in organic colluvium ("Lower Humic Bed") near the base of this sequence. They underlie a volcanic ash lens recently dated to 38,000–33,000 years BP (Sinitsyn et al. 1997, 27–28). Later occupation levels are buried in a series of humic colluvial layers ("Upper Humic Bed") above the ash lens that appear to date to the terminal phase of OIS 3 (32,000–27,000 years BP). An intact buried soil of this age has been discovered at Kostenki I. Loessic colluvium deposited during the Last Glacial Maximum (OIS 2) overlies the Upper Humic Bed and buried soil (Hoffecker 1987, 274–275).

The Kostenki localities occupied during OIS 3 are situated both in the main river valley and along large side-valley ravines (fig. 5.5). The early Upper Paleolithic occupation levels yield a highly diverse array of artifact

THE TRANSITION TO MODERN HUMANS   149

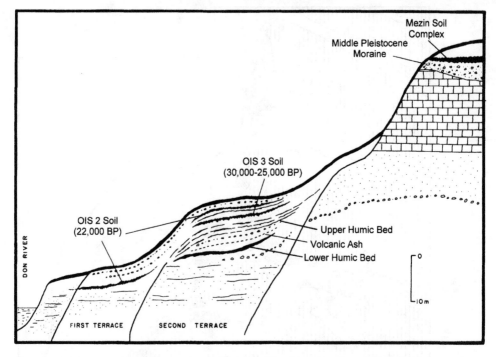

Figure 5.4. Geomorphic setting and stratigraphy of the Kostenki sites on the Middle Don River (redrawn from Praslov 1985, 26, fig. 6).

assemblages (Anikovich 1993). Some contain artifacts made primarily from imported flint with a high proportion of blades and typical Upper Paleolithic tool forms such as burins, end-scrapers, and retouched blades (e.g., Kostenki XII, Layer 2, and Kostenki XVII, Layer 2). Others contain assemblages made with local stone that contain few blades but possess many typical Mousterian tool types, including side-scrapers and bifaces (e.g., Kostenki I, Layer 5, and Kostenki XII, Layer 3). Assemblages in the Upper Humic Bed are especially varied, and include a unique industry comprising a mixture of classic Mousterian and Upper Paleolithic tool types (side-scrapers, points, and end-scrapers) and a rich inventory of nonstone implements (e.g., Kostenki XIV, Layer 2, and Kostenki XV). Associated mammal remains are heavily dominated by horse, and small fur-bearing taxa (Vereshchagin and Kuz'mina 1977, 100–110).

The early Upper Paleolithic sites at Kostenki may include several types of occupations. Most appear to represent multiple-activity loci and long-term habitation sites. A good example is Kostenki VIII, Layer 2 (buried in the Upper Humic Bed), which yielded roughly 23,000 artifacts, large concentrations of debris, and traces of possible dwelling structures with

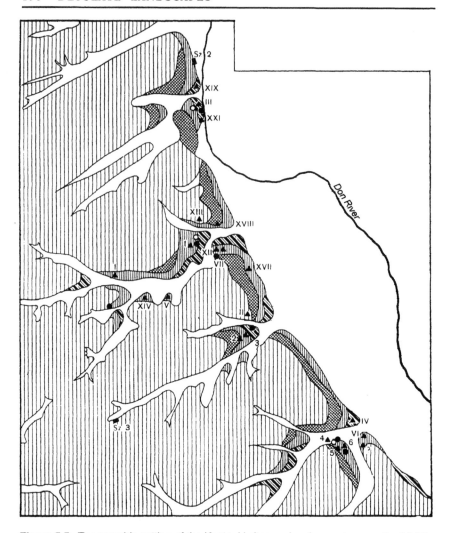

Figure 5.5. Topographic setting of the Kostenki sites and ravine systems on the Middle Don River (modified from Lazukov 1957, fig. 1).

central hearths (Rogachev 1957). Burials were found at three sites (Kostenki XII, XIV, and XV). Evidence of short-term occupations is less common, but may be present at a few localities (e.g., Kostenki I, Layer 4 [Rogachev et al. 1982, "Kostenki 1," 64–65]).

There are no known Neanderthal sites at Kostenki, and—unlike the Middle Dnestr Valley—modern humans appear to have been the first to occupy the area. They camped on the lowest terrace level in the main valley, and along major ravines more than a kilometer from the river. The site locations probably offered a well-drained living surface and some shelter from westerly winds, but their prime attraction seems likely to have been the presence of large mammalian prey (especially horse) and perhaps especially suitable topography for hunting them. Although stone of moderate quality is present, much of their raw material was imported, and this seems unlikely to have been a significant variable.

The site of *Sungir'*, which is located along a tributary of the Oka River at Vladimir in northern Russia at 56 degrees North, is one of the most remarkable Paleolithic sites in the world (Sukachev et al. 1966; Bader 1978). Discovered in 1955, Sungir' was excavated by the late O. N. Bader between 1956 and 1970, who eventually exposed almost 3,400 square meters. The site occupies a promontory composed of late Middle Pleistocene glacial moraine mantled with colluvial loams. The single occupation horizon is associated with a partially redeposited buried soil in the colluvium that probably dates to the terminal phase of OIS 3. Although the occupation debris and soil yielded a wide range of radiocarbon dates, the oldest dates (ca. 25,000 years BP and older) are most likely to be accurate; the burials have recently yielded younger dates (ca. 24,000–23,000 years BP) (Bader 1978, 63–66; Pettitt and Bader 2000) (fig. 5.6).

More than 50,000 stone artifacts and 10,000 nonstone artifacts were recovered from the large occupation area at Sungir'. The lithic inventory consists primarily of typical Upper Paleolithic forms (end-scrapers, burins, and retouched blades), but includes a number of Mousterian types (side-scrapers, bifacial points) (Bader 1978, 114–145). The nonstone artifacts comprise a variety of bone, antler, and ivory implements, and thousands of beads and other ornaments. Despite disturbance from frost action and

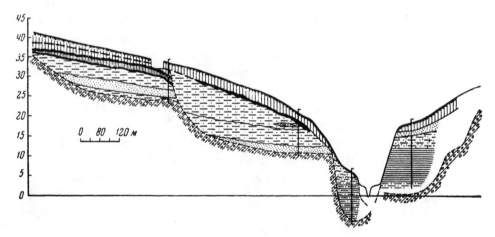

Figure 5.6. Geomorphic setting and stratigraphy of the Sungir' site on a tributary of the Oka River in northern Russia (modified from Tseitlin 1965, fig. 3).

solifluction, many features were mapped on the occupation floor, including hearths, shallow pits, artifact and bone debris concentrations, traces of possible dwellings, and several burials (Bader 1978, 67–113). The associated large mammal remains are chiefly reindeer, horse, and mammoth (Gromov 1966a).

Several sites that appear to date to late OIS 3 are also reported from northeastern Russia as far as latitude 65 degrees North. They include the open-air locality of *Byzovaya*, which is situated on a bedrock terrace covered by alluvial gravels and slope deposits along the Pechora River. Artifacts and faunal remains (chiefly mammoth) were recovered from the alluvial sediment, and dated to ca. 25,000 years BP (Guslitzer and Pavlov 1993, 182). The stone tools were prepared primarily on flakes, and include end-scrapers, side-scrapers, and bifaces, while nonstone implements are absent (Kanivets 1976, 56–71).

Crimea and Northern Caucasus

The distribution of sites occupied by modern humans during OIS 3 in the Crimea and Northern Caucasus exhibits a very different pattern from that of the East European Plain. In both areas, there is evidence for the persistence of a late Neanderthal population and Mousterian industry to ca. 35,000–30,000 years BP (Chabai and Marks 1998; Ovchinnikov et al. 2000). There is a corresponding lack of evidence for modern humans and Upper Paleolithic industries prior to this date. At several sites (e.g., Buran-Kaya III, Mezmaiskaya Cave), fully developed Upper Paleolithic industries directly overlie Mousterian occupation layers (Marks 1998; Golovanova et al. 1999).

Upper Paleolithic sites are generally rare in the Crimea for reasons that remain obscure. The most widely known locality is the rockshelter of *Syuren' I*, which is situated along the Bel'bek River roughly 20 kilometers east of Sevastopol' in the southwestern Crimea. This relatively large natural shelter (43 × 15 meters) was investigated by G. A. Bonch-Osmolovskii during 1926–1929, who exposed 85 square meters (Vekilova 1957). Additional excavations were undertaken in the 1990s (Demidenko 1999). The lowermost occupation layer is deeply buried in deposits of sand and rubble, and has recently yielded radiocarbon dates on bone of 28,450 and 29,950 years BP (Otte et al. 1996). Among the associated large mammal remains, saiga and giant deer are the most common species, but smaller fur-bearing taxa are also abundant (Vekilova 1957, 254). Bonch-Osmolovskii (1934, 131) reported the recovery of an isolated molar from this layer that he assigned to modern humans.

Over 12,000 artifacts were recovered from the lowest layer in Syuren' I, and they have been the object of much debate and controversy (Klein

1965, 59–62). The lithic assemblage contains a high percentage of blades and Upper Paleolithic tool types (end-scrapers, burins, retouched micoblades), which are often compared with the Aurignacian industry of Western and Central Europe. However, a number of typical Mousterian forms (points, side-scrapers, bifaces) are present. The assemblage also contains nonstone implements, ornaments, and art objects (Vekilova 1957, 290–294). On the basis of new investigations at Syuren' I, it has been suggested that the lowest layer comprises a mixture of alternating Neanderthal Mousterian and modern human Upper Paleolithic occupations (Demidenko 1999).

Another early Upper Paleolithic site has been identified recently in the eastern Crimea at *Buran-Kaya III* on the Burul'cha River (Yanevich et al. 1996). Like many of the Mousterian sites in the region, it represents a former rockshelter that has since collapsed. Several Mousterian and Upper Paleolithic occupation layers are buried in roughly 3 meters of silt and rubble that apparently reflect both rockfall and stream deposition. The lowest Upper Paleolithic level has yielded dates of between 32,200 and 36,700 years BP, and contains bifacial points, end-scrapers, and several bone implements (Marks 1998).

In the Northern Caucasus, an early Upper Paleolithic occupation layer has been dated at the Mousterian site of *Mezmaiskaya Cave* (see chapter 4). Layer 1C directly overlies the uppermost Mousterian occupation level, and has been dated (wood charcoal) to 32,010 years BP (Golovanova et al. 1999, 78–79). The artifact assemblage includes end-scrapers, burins, retouched microblades, and a variety of bone implements. An early Upper Paleolithic occupation may also be present at *Gubs Shelter No. 1* in Borisovskoe Gorge on the Gubs River (Amirkhanov 1986, 30–46). The layer containing this occupation appears to represent a buried soil horizon that is tentatively correlated with the terminal OIS 3 soil. The assemblage is dominated by end-scrapers made on blades and flakes. Finally, it may be noted that the assemblage from *Kamennomostskaya Cave* on the Belaya River has long been regarded as early Upper Paleolithic on typological grounds, but remains undated (Formozov 1971) (fig. 5.7).

Early Modern Humans of Eastern Europe: Skeletal Morphology

In contrast to the geographic distribution of Neanderthal skeletal remains in Eastern Europe, modern human remains dating to OIS 3 are almost entirely confined to the central East European Plain. Specifically, they have been found at Sungir' and several of the Kostenki sites, all of which are located within the Oka-Don Lowland (see table 5.1). With the exception

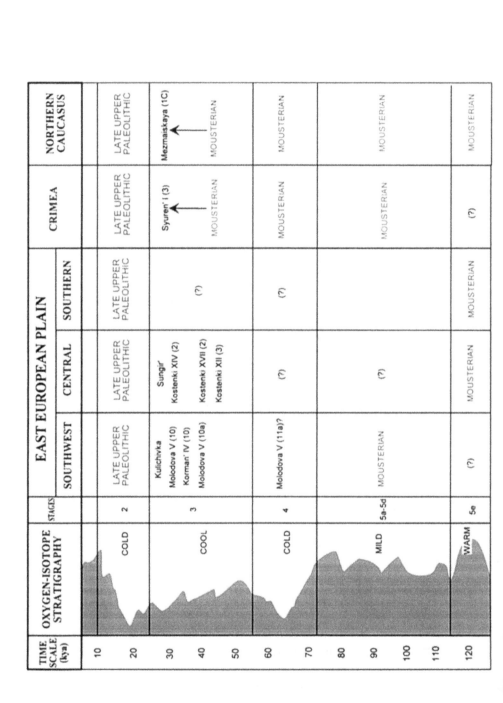

of an isolated tooth from the Crimea (Syuren' I [Bonch-Osmolovskii 1934]), no other skeletal remains can be firmly assigned to Upper Paleolithic sites of OIS 3. This pattern is probably influenced by the general scarcity of early Upper Paleolithic occupations in the southern uplands—especially the Crimea—where Neanderthal remains are most common. Also, with the exception of an isolated tooth from the lowest occupation at Kostenki XVII (Boriskovskii et al. 1982, 186), the modern human sample dates to the terminal phase of OIS 3 (ca. 30,000–25,000 years BP). This may reflect the reduced frost action effects, which destroy many burials in cold regions, during this warmer interval (Hoffecker 1988, 259).

Despite its restricted geographic distribution, the East European Plain sample of skeletal remains provides much information about the modern human inhabitants of the region during the transition period. Cranial parts recovered from Kostenki and Sungir' exhibit the diagnostic features of modern humans, including vertical frontal bone, reduced browridge, and canine fossa. The mandible displays a chin but lacks a retromolar space, and the teeth are relatively small. However, there is considerable variability in the sample. For example, the Kostenki XIV specimen possesses an unusually low cranial capacity (1,165 cubic centimeters), narrow face, and wide nasal cavities (Debets 1955, 43–44; Gerasimova 1982, 253–257). The skull of the child from Kostenki XV is unusually large for its estimated age of six to seven years (Gerasimova 1982, 247). The adult male cranium from Sungir' exhibits a broader face, very wide palate, and greater cranial volume (Debets 1967, 161–162).

The postcranial skeleton also displays typical modern human features, lacking the heavy robusticity, strong muscle attachments, bowed limb bones, and other characteristic Neanderthal features. Especially significant is the length of the distal limb segments relative to the proximal segments on both upper and lower extremities. Both the Kostenki XIV and Sungir' specimens yield high brachial (79–81) and crural (82–84) indices (Alekseev 1978, 253) (fig. 5.8). As in the case of the cranium, there is some variability among the sample. The adult male from Sungir' exhibits unusually wide shoulders, while the adult male from Kostenki XIV exhibits comparatively narrow shoulders (Debets 1967, 160; Gerasimova 1982, 255).

Overall, the modern human remains from the East European Plain fit into the broader sample of European Cro-Magnon fossils (Debets 1955; Alekseev 1978, 182–189). Some anthropologists have suggested that they may be grouped along with remains from Central Europe as a widespread geographic variant of the latter ("Eastern Cro-Magnon" [Yakimov 1957]). There has been much speculation regarding the possible relationships of

Figure 5.7. *(opposite)* Chronology of early Upper Paleolithic sites in Eastern Europe.

Table 5.1. Modern Human Skeletal Remains from Early Upper Paleolithic Sites (OIS 3) in Eastern Europe

Locality	Layer	Date	Age/Sex	Material
East European Plain				
Kostenki I	Layer 3	26,000 yrs BP	unknown	tibiae, pelvis, tooth
Kostenki VIII	Layer 2	27,700 yrs BP	unknown	cranial fragments
Kostenki XII	Layer 1	>25,000 yrs BP	newborn infant	nearly complete skeleton
Kostenki XIV	Layer 3	30,080 yrs BP	male 20–25 years old	nearly complete skeleton
Kostenki XV	——	25,700 yrs BP	6–7 years old	cranial vault, maxilla fragment, mandible
Kostenki XVII	Layer 2	32,200 yrs BP	adult	third molar
Sungir' (1)	——	22,930 yrs BP?	male 55–65 years old	complete skeleton
Sungir'	——	>25,000 yrs BP?	female	skull
Sungir' (3)	——	24,100 yrs BP?	female 9 years old	complete skeleton
Sungir' (2)	——	23,830 yrs BP?	male 12–13 years old	complete skeleton
Sungir'	——	>25,000 yrs BP?	adult	limb bone fragments, phalanges
Crimea and Northern Caucasus				
Syuren' I	Layer 3	29,950 yrs BP	adult(?)	molar

Sources: Bonch-Osmolovskii 1934; Debets 1955, 1967; Gerasimova 1982; Pettitt and Bader 2000.

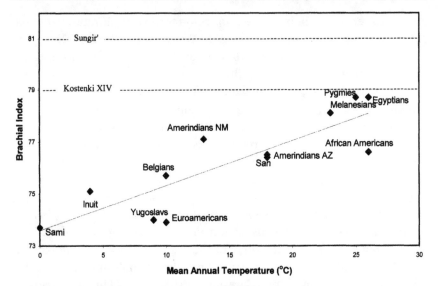

Figure 5.8. Brachial index for modern human populations and mean annual temperature showing brachial indices for Upper Paleolithic specimens (based on data from Alekseev 1978, 253, Trinkaus 1981).

particular specimens to modern human geographic races of the Holocene. For example, the Kostenki XIV skeleton has often been compared with modern African populations, while parallels with modern Asian populations have been noted for the adult male from Sungir' (Debets 1955, 1967). However, modern human geographic races do not appear to have emerged until after the end of the Pleistocene (Klein 1999, 502), and—like the European Cro-Magnon sample as a whole—the diverse morphology of the East European Plain specimens seems to reflect a highly variable population.

In his original analysis of the Kostenki XIV skeleton, G. F. Debets (1955, 43) emphasized the almost complete lack of Neanderthal features. Subsequent discoveries of modern human remains from early Upper Paleolithic localities at Kostenki and Sungir' have revealed the same pattern of discontinuity. The redating of the Starosel'e child and Mezmaiskaya Cave infant (Marks et al. 1997; Ovchinnikov et al. 2000) indicate that the latest Neanderthals in Eastern Europe do not exhibit any significant evolutionary change in the direction of modern humans. In fact, the Kostenki remains appear to be of comparable age to the youngest Neanderthals in the Northern Caucasus. Thus, on the basis of existing evidence, Eastern Europe now reveals the same pattern as Western Europe with respect to the transition from Neanderthals to modern humans. Modern humans appear to represent an intrusive population with clear affinities to earlier *Homo sapiens* in Africa and the Near East that replaced the local Neanderthal population. Both the morphology of the skeletal remains and the mtDNA analyses of Neanderthal specimens from Feldhofer and Mezmaiskaya caves suggest that genic exchange between the two forms was minimal at best (Krings et al. 1997; Ovchinnikov et al. 2000).

Among the morphological contrasts between the East European modern humans and their Neanderthal predecessors, the evidence for climatic adaptation in the postcranial skeleton is particularly striking. The high brachial and crural indices for the modern humans from Kostenki and Sungir' are consistent with a pattern of adaptation to temperate and tropical environments among modern populations (Debets 1967, 161), following the prediction of Allen's rule concerning the length of extremities. The overall shape and size of the body of the adult male from Kostenki XIV (small and thin) seems to conform to both Bergmann's and Allen's rules for warm climate adaptation (Gerasimova 1982, 256). The same pattern is evident among the West European Cro-Magnon sample (Trinkaus 1981), but seems more significant in a north Russian setting like Sungir'—at a latitude comparable to that of Kodiak Island, Alaska.

Modern Human Technology and the Early Upper Paleolithic of Eastern Europe

The technology of behaviorally modern humans represents a quantum leap over the technology of other hominids, and seems to reflect a fundamental change in ability to manipulate the environment. The change is probably linked to a shift in cognition, and plausibly tied to the apparently simultaneous appearance of symbols and other evidence for language in the archaeological record (Mithen 1996). The ability to develop innovative technology in response to local environmental conditions almost certainly played a critical role in the dispersal of modern humans. After 50,000 years BP, behaviorally modern people rapidly colonized boreal forests, deserts, tropical jungles, periglacial steppes, and other environments throughout the Old World (Gamble 1994).

Technological innovation was probably essential to the successful colonization of Eastern Europe (as well as Siberia), where the need to cope with extreme cold is underscored by the warm-climate morphological adaptations that modern humans brought with them into northern latitudes. Increased technological complexity also may have been critical to successful foraging in very cold and dry environments. But here as elsewhere, most of the contrast in technology between Neanderthals and modern humans lies outside the realm of stone tools. As noted earlier (see chapter 4), the Neanderthals were skilled lithic technicians, and probably had little to learn from modern humans in this area. And only limited traces of the novel devices engineered by modern humans in Eastern Europe during OIS 3 are preserved in the archaeological record. The traditional focus of Paleolithic archaeologists on artifacts—especially stone artifacts—has tended to obscure the nature and magnitude of the contrast in technology.

Both the variety and complexity of techniques and devices developed by modern humans in northern Eurasia during the Pleniglacial seem to reflect a fundamental change in ability to manipulate the natural environment (Dennell 1983, 79–96; Mithen 1996, 151–184). Although some of these novel technologies do not appear in the archaeological record until the Last Glacial Maximum (OIS 2), most of them are evident prior to 25,000 years BP. They are particularly well represented on the East European Plain during OIS 3–2, where people seem to have achieved a level of technological complexity comparable to that of recent hunter-gatherers in subarctic and arctic environments (Oswalt 1976; Torrence 1983).

All of the artifacts and features associated with modern humans in Eastern Europe are assigned to the industries of the *Upper Paleolithic*. Although Neanderthals also appear to be present in Eastern Europe during late OIS 3 (in the Crimea and Northern Caucasus), their remains are currently confined to Mousterian sites. However, there is considerable variability in the East European Upper Paleolithic during OIS 3—perhaps more variability than any other region in Western Eurasia at this time. Modern human technology during the transition period must be assessed within the context of these Upper Paleolithic industries, which probably vary for many reasons other than purely technological considerations. As noted earlier, the possibility that some of these artifacts and features might have been produced by local Neanderthals cannot be discounted at present.

Upper Paleolithic industries in Eastern Europe are broadly comparable to those in other parts of northern Eurasia. In general, they exhibit a much heavier emphasis on the production of blades rather than flakes for stone tools, and they contain high percentages of tool types that are typically rare or absent in Lower and Middle Paleolithic industries (e.g., burins, end-scrapers, retouched blades) (Grigor'ev 1970; Klein 1973; Rogachev and Anikovich 1984). The greater reliance on blades as opposed to flakes indicates a more efficient use of stone, reflecting the significantly increased ratio of tool edge to mass provided by the former (Bordaz 1970, 56–57). The high percentages of the chisel-like *burins*, which are especially common in East European sites, is undoubtedly tied to the heavy working of bone, antler, and ivory. *End-scrapers* were often used for working hides (e.g., Semenov 1964, 85–93).

Modern Humans, Symbols, and a Revolution in Technological Ability

The most important revolution in human technology occurred not with the advent of agriculture or power machinery, but with the appearance of symbols in the archaeological record and the spread of modern humans throughout Eurasia. The rather sudden appearance of symbols is widely believed to indicate the birth of symbolic language (e.g., Gamble 1986; Mellars 1996). There seems to have been a quantum leap in technology that reflects a fundamental change in human ability to manipulate the environment, and provided the basis for all subsequent technological development. The association with symbols suggests a relationship between language structure and this new ability (Deetz 1967; Mithen 1996).

Traditional approaches to Paleolithic archaeology have tended to obscure the revolutionary character of the transition to symbol-based technology. The focus on artifacts—especially stone tools—has distracted attention from more significant changes. For the most part, the stone tools made by modern humans as they dispersed across Eurasia simply reflect a shift in emphasis on certain production techniques and tool forms relative to the lithic industry of their predecessors. More important is the abrupt appearance of a variety of specialized implements of bone, antler, and ivory (materials rarely if ever used for tool manufacture by the Neanderthals).

Especially complex technology emerges in the realm of clothing and shelter. The presence of eyed needles and awls in some of the oldest modern human sites of Eastern Europe and Siberia (OIS 3) indicates development of tailored fur clothing (e.g., Praslov and Rogachev 1982; Goebel 2000), which requires a com-

---

Like Upper Paleolithic industries in other parts of northern Eurasia, the East European sites often contain implements manufactured on bone, antler, and ivory. These include needles, awls, rods, points, shovel-shaped tools, and others (Bader 1978; Praslov and Rogachev 1982). At least some of these implements (e.g., needles, awls) were used in the production of tailored clothing, which also required substantial hide preparation with stone end-scrapers, but the functional significance of many nonstone artifacts is unclear.

A number of East European sites dated to OIS 3 contain classic Upper Paleolithic assemblages resembling the Aurignacian of Western-Central Europe and the Near East (e.g., Syuren' I; Kostenki I, Layer 3 [Cohen and Stepanchuk 1999]), although few of them may be considered Aurignacian in a strict sense (Hoffecker 1988). While associated skeletal remains are very rare, these assemblages were almost certainly produced by modern humans. They are found in all major regions of Eastern Europe. Some

plicated sequence of steps for production (e.g., Murdoch 1892, 109–138). Construction of artificial shelters with interior hearths is suggested at several sites dating to OIS 3, and is clearly evident in sites of Last Glacial Maximum age (OIS 2) on the East European Plain (e.g., Pidoplichko 1969; Soffer 1985). Many of the latter also contain traces of deep storage pits that seem to have been dug down to the permafrost layer in order to refrigerate their contents (i.e., similar to "ice cellars" of the modern Inuit).

Technology for the procurement of animals also became more complex and efficient. Spear-throwers were designed to increase the leverage and power of hand-propelled projectiles—at least one example has been recovered from a site in France dating to OIS 2—and bows and arrows may have been in use by this time as well (Klein 1999, 538–542). Large quantities of small and medium mammal remains have been found at many of the East European Plain sites dating as early as OIS 3 (Vereshchagin and Kuz'mina 1977), and probably indicate the use of traps and snares. Trap components may be present at one of the East European Plain sites of the Last Glacial Maximum (Pidoplichko 1976, 165).

The increased diversity and complexity of tools and devices was fueled in part by the development of new production techniques (Dennell 1983, 81–87). The use of a hand-powered rotary drill is evident in one of the earliest modern human sites on the central East European Plain (Semenov 1964, 78–79). A number of sites in Central and Eastern Europe dating to OIS 3 and 2 have yielded traces of fired ceramic technology, including remains of kilns heated to as much as 800 degrees C (Soffer et al. 1993). Recently, evidence of woven textiles has also been reported from Central and East European sites dating to OIS 3 (Soffer et al. 2000).

East European Plain assemblages exhibit a very different pattern, containing a mixture of Mousterian and Upper Paleolithic forms, and many bone tools, ornaments, and art (e.g., Kostenki XIV, Layer 2 [Sinitsyn 1996]). They are clearly associated with modern humans, but bear little resemblance to industries outside Eastern Europe.

The East European sites also include some assemblages with few or no examples of blade technology, comparatively low percentages of typical Upper Paleolithic tool types, and few if any nonstone artifacts, ornaments, or art objects (e.g., Brynzeny I, Layer 3 [Ketraru 1973]; Kostenki XII, Layer 3 [Anikovich 1993]). It is these assemblages that present the greatest challenge to paleoanthropologists attempting to understand the transition to modern humans in this part of Europe. The identity of their makers and their relationship to the other Upper Paleolithic industries of Eastern Europe remain problematic. They exhibit parallels to the Chatelperronian industry of Western Europe (produced by Neanderthals), and especially to the Szeletian industry of Central Europe (possibly produced by Neanderthals) (Amirkhanov et al. 1980; Kozlowski 1988; Allsworth-Jones 1990).

East European Plain: Southwest Region

The oldest known Upper Paleolithic assemblages in the southwest region of the East European Plain are derived from occupation levels that underlie the terminal OIS 3 soil and antedate 30,000 years BP. These occupations may represent the earliest traces of modern human settlement in the region, although human skeletal remains have not been recovered from any of them. They are found at Molodova V (Layers 10a–10b), Korman' IV (Layer 8), and possibly Oselivka I (Layer 4) in the Middle Dnestr Valley, and at Kulichivka (Layer 3) in the northern Volyn-Podolian Upland. The Kulichivka assemblage contains a combination of Levallois blade cores and various tools that include side-scrapers, Levallois points, end-scrapers, and burins (Savich 1985, 1986; Cohen and Stepanchuk 1999, 293). The assemblages from the Middle Dnestr Valley are very small, but reveal a similar mixture of typical Mousterian and Upper Paleolithic stone artifacts (Chernysh 1971, 69–70; 1977, 23–24; 1987, 25–26). Nonstone implements are not reported. The relationship of these early assemblages (which have been characterized as "transitional") to the preceding Mousterian and later Upper Paleolithic remains unclear.

The most widely known early Upper Paleolithic industry in the region is found in the Middle Dnestr Valley, and is often referred to as the "Molodova Culture" (e.g., Rogachev and Anikovich 1984, 173). It is usually described with reference to Layers 10–8 at Molodova V, which are associated with the terminal OIS 3 soil and dated to ca. 30,000–25,000 years BP (Chernysh 1987, 27–35). These layers contain assemblages with large prismatic blade cores, and blades that measure 60–150 millimeters in length. The most common stone tool types are burins (30–40 percent), knives (23–36 percent), and end-scrapers (11–16 percent), and many of them were made on long blades. Less common tool types include points (3–6 percent) and notched bladelets (3–6 percent) (table 5.2; fig. 5.9). Each assemblage also contains a small number of sandstone cobbles interpreted as grinding stones. Formal nonstone tools are absent, although some simple bone ornaments were recovered from the uppermost horizon (Chernysh 1987, 34). Mapping of occupation floor areas at Molodova V revealed concentrations of lithic and bone debris, and many are associated with former hearths (Chernysh 1959, figs. 31–34). Some of the concentrations have been interpreted as traces of artificial shelters, but there is no conclusive evidence of structures (e.g., postmold pattern).

The Molodova industry is recognized at other sites in the Middle Dnestr Valley including Babin I, Layer 1; Oselivka I, Layer 3; Oselivka II; and Voronovitsa I (Chernysh 1973, 19–20), although none of these occupations is firmly dated to OIS 3. On the other hand, the assemblage from Layer 7 at Korman' IV, which is dated by radiocarbon to the end of OIS 3 (Ivanova 1977, 178–179) but might be older on the basis of stratigraphic correlation (Chernysh 1973, fig. 21), exhibits a somewhat different pattern. The cores include double-platform types reminiscent of the Mousterian, and some flakes and blades possess large bulbs of percussion. But blades are common, and some are as much as 100–110 millimeters in length, and the tools comprise typical Upper Paleolithic types similar to those at Molodova (e.g., burins, knives, end-scrapers) (Chernysh 1977, 24–26). Although not usually included as part of the Molodova industry, the Korman' IV assemblage is not fundamentally different.

On the northern margin of the Volyn-Podolian Upland, another local Upper Paleolithic industry has been identified. This industry—labeled the "Lipa Culture"—is represented at sites located near the city of Rovno such as Lipa VI and Kulichivka (Grigor'ev 1970, 45; Rogachev and Anikovich 1984, 175). The large assemblage recovered from the terminal OIS 3 soil at Kulichivka (Layer 2) is the most reliably dated to this time range. A wide variety of cores are present, and blades constitute only 14 percent of the total, although the latter include some very large specimens. The most common tools are retouched blades (26 percent), end-scrapers (25 percent), burins (15 percent), and side-scrapers (5 percent). A small number of cobbles interpreted as grinding stones are also present. Unlike the case of Molodova V, nonstone tools were recovered from Kulichivka and include an antler haft and bone perforators (Savich 1975, 45–47).

Despite the presence of some typical Mousterian cores and tool types, the younger Kulichivka assemblage is generally similar to the Molodova industry and is considered a geographic variant of the latter (Savich 1975, 51; Rogachev and Anikovich 1984, 175). As at Molodova V, traces of artificial structures are reported, but in this case the features are more convincingly interpreted as such. They consist of three shallow oval depressions (measuring 5.5 × 6.0 meters and 7.0 × 7.55 meters), each containing one or two central former hearths lined with small stones. Large stones and bone fragments were found along the margins of the depressions (Savich 1975, 42–43).

The early Upper Paleolithic industries from the Middle Dnestr Valley (Molodova Culture) and northern Volyn-Podolian Upland (Lipa Culture) are broadly comparable to the Aurignacian industry, which is associated with the initial appearance of modern humans in Western and Central

Table 5.2. Artifacts in Early Upper Paleolithic Sites of the East European Plain: Southwest Region

Tool types	Sites and level				
	Molodova V Layer 9	Korman IV Layer 7	Babin I Layer 1	Korpach Layer 4	Brynzeny I Layer 3
End-scrapers	12	4	31	14	x
Composite tools	0	0	6	0	
Burins	23	12	44	21	x
Knives	28	10	69	0	x
Points	5	1	14	0	x
Notched blades	3	1	12	0	
Notches	0	0	0	x	x
Denticulates	0	0	0	x	x
Side-scrapers	0	0	0	16	x
Backed bladelets	2	0	3	5	
Retouched bladelets	0	0	0	45	
Leaf-shaped points	0	0	0	8	x
Total tools:	75	29	187	170	

SOURCES: Ketraru 1973; Chernysh 1959, 20–25; 1977, 67, table 2; 1987, 84–85; Grigor'eva 1983; Allsworth-Jones 1990, 226, table 7.12.

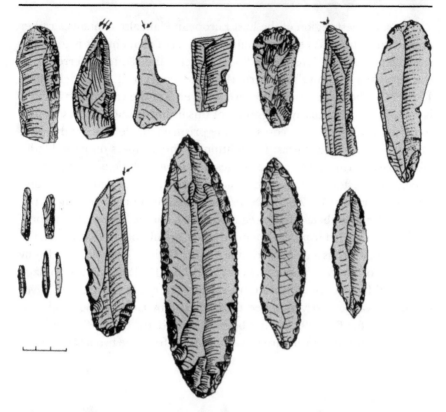

Figure 5.9. Early Upper Paleolithic artifacts from Layer 10 at Molodova V in the Middle Dnestr Valley (redrawn from Chernysh 1959, 71, fig. 33).

Europe. Although none of these sites may be classified as Aurignacian in a strict sense, the prismatic blade technology and array of tool types is similar and can be characterized as Aurignacoid. These industries seem likely to represent part of the wider dispersal of modern humans across Europe during the terminal phases of OIS 3. However—with the exception of some isolated implements from Kulichivka—the nonlithic component is largely absent from the Molodova and Lipa industries. This may be at least partly a function of poor preservation in open-air sites during an interval of soil formation (most nonlithic artifacts from Aurignacian sites in Western and Central Europe were found in caves and rockshelters). The predominance of burins and end-scrapers among the stone tools suggests an emphasis on bone- and hide-working, but more direct evidence of tailored clothing and other complex technology is limited. The most significant indication of the latter are the relatively convincing traces of artificial shelters at Kulichivka.

The southwest region of the East European Plain also contains a rather archaic industry that contains many Mousterian elements. It is found in sites along the Prut River and adjoining areas, and is often referred to as the "Brynzeny Culture" (Ketraru 1973; Rogachev and Anikovich 1984, 172–173). The large assemblage from Layer 3 at the cave of Brynzeny I is considered particularly diagnostic of this industry, and appears to date to the end of OIS 3. The cores include prismatic blade types and typical Mousterian discoidal forms. In addition to end-scrapers on large blades and flakes, burins, and backed blades, the tools include a large number of typical Mousterian types such as side-scrapers, points, denticulates, and bifaces (Ketraru 1973, 70–72). Nonstone tools are not reported (fig. 5.10).

A similar assemblage that is regarded as part of the "Brynzeny Culture" has been recovered from the open-air site of Bobuleshty VI (located on the Reut River, which is a western tributary of the Dnestr), but cannot be firmly assigned to OIS 3 at present (Ketraru 1973, 119–122). An assemblage excavated from Layer 3 in Chuntu rockshelter (located on a Prut River tributary) was formerly assigned to the Brynzeny Culture (Borziyak and Ketraru 1976, 468–469), but has been recently dated to OIS 2 (Borziac et al. 1997). More significant are the assemblages from the two lowest Upper

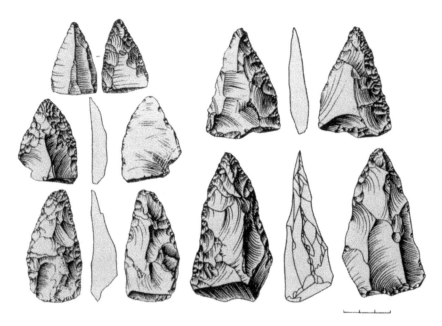

Figure 5.10. Early Upper Paleolithic artifacts from Layer 3 at the rockshelter of Brynzeny I (tributary of the Prut River) assigned to the "Brynzeny Culture" (modified from Ketraru 1973, 72, figs. 27 and 28).

Paleolithic levels (Layers 1a–1b) at the open-air site of Ripiceni-Izvor on the Prut River. These levels are associated with the terminal OIS 3 soil and yielded a date of 28,420 years BP. The proportion of blades is low, and tools include denticulates, notches, bifacial points, and end-scrapers (Păunescu 1987). The nearby open-air site of Korpach also contains an assemblage (Layer 4) dated to the end of OIS 3 with parallels to the Brynzeny industry (Grigor'eva 1983; Allsworth-Jones 1990, 224–225).

Although it is possible that some postdepositional mixing of artifacts from various time periods occurred at Brynzeny I and Bobuleshty VI, this is highly unlikely at Ripiceni-Izvor, and the archaic character of this industry appears to be genuine. Moreover, the geographic concentration of these sites in the Prut Valley and adjoining areas suggests that the assemblages probably do not represent an activity variant of a larger Aurignacoid industry. They exhibit a limited contrast with Mousterian technology that is largely confined to an increased emphasis on the production and use of stone blades and burins. Despite the recovery of artifacts from caves in which conditions are favorable for bone preservation, nonstone implements are lacking. At the same time, the open-air occupation layers have failed to reveal evidence for the construction of artificial shelters.

Many archaeologists have noted the similarities between the Brynzeny Culture (and related assemblages in the Prut Valley area) and the Szeletian industry of Central Europe, and suggested a relationship between the two (e.g., Ketraru 1973, 65; Kozlowski 1988, 206–214; Allsworth-Jones 1990, 226). The Szeletian may represent an industry created by local Neanderthals in response to the appearance of modern humans during OIS 3 (Allsworth-Jones 1986). But it should be noted that Brynzeny I, Layer 3 yielded one of the most sophisticated Upper Paleolithic art objects recovered from the region (Ketraru 1973, 73, fig. 29). Unless this object is intrusive from a younger time period (no late Upper Paleolithic layers are reported from Brynzeny I), the Layer 3 lithic assemblage seems unlikely to have been produced by Neanderthals. However, the question of who made the Brynzeny Culture cannot be fully resolved until diagnostic skeletal remains are found in association with these assemblages.

East European Plain: Central and Northern Regions

The central plain contains the longest and most diverse succession of Upper Paleolithic industries in Eastern Europe. The oldest occupations in this region significantly antedate the terminal OIS 3 soil, and most probably date to 40,000 years BP or earlier. These occupations are found below the volcanic ash lens and within the Lower Humic Bed at Kostenki on the Middle Don River. Assemblages from Kostenki XVII (Layer 2) and Kostenki XII (Layer 2) are grouped together as the "Spitsyn Culture"

Table 5.3. Artifacts in Early Upper Paleolithic Sites of the East European Plain: Central and Northern Regions

Tool types	Kostenki XVII Layer 2	Kostenki I Layer 5	Sungir'	Kostenki XV	Kostenki VIII Layer 2
End-scrapers	22	24	398	85	50
Composite tools	20	0	0	5	0
Borers	0	4	21	~10	29
Burins	160	8	288	12	>500
Knives	0	0	692[a]	0	0
Points	0	1	0	x	3
Retouched blades	>100	0	x	75	x
Notched blades	0	0	0	0	180
Side-scrapers	0	6	75	~10	0
Backed bladelets	3	0	0	0	~900
Leaf-shaped points	0	43	16	0	0
Splintered pieces	10	4	291	53	1
Total stone tools:	330	119	1869	~370	2100
Bone awls	2	1	10	x	x
Bone/ivory points	2	0	3	0	14
Bone needles	0	0	1 (fragment)	10	1 (?)
Ornaments	50	1	>200	30	x
Beads	0	0	~10,000	0	0
Art objects	1(?)	0	4	0	x

Sources: Rogachev 1957; Boriskovskii 1963; Klein 1969b; Bader 1978; Praslov and Rogachev 1982.
[a] Classified as "cutting tools" (Bader 1978, 137–139).

(Rogachev and Anikovich 1984, 181). Most of the stone artifacts in these assemblages were made of flint imported from a source located 130–150 kilometers to the southwest (Boriskovskii 1963, 168–169). Prismatic blade technology is predominant, and long blade cores were recovered from Kostenki XVII. The most common tool types are burins (26–48 percent), retouched blades (up to 30 percent), and end-scrapers (7–13 percent). Kostenki XVII also yielded nonstone implements, including two bone awls, two bone point fragments, and two fragmentary objects of ivory; various ornaments are also present (Boriskovskii 1963, 80–124; Rogachev and Anikovich 1982a, 137–138). Symmetrical holes bored in the latter indicate the use of a hand-operated rotary drill (Semenov 1964, 78). Although two former hearths and associated debris concentrations were mapped at Kostenki XVII, neither appears to represent the location of an artificial structure (table 5.3; fig. 5.11).

The Spitsyn Culture assemblages contain few of the diagnostic elements of the Aurignacian industry, but they are broadly similar in terms of stone tool technology and implements and are considered "Aurignacoid" (Anikovich 1993, 14). Accordingly, these assemblages may be linked to a younger occupation associated with the terminal OIS 3 soil at Kostenki I (Layer 3), which does contain some characteristic Aurignacian tool types. The artifacts from this occupation include prismatic blade cores, carinated end-scrapers, retouched microblades, one bone awl, and an ivory implement; both ornaments and decorated objects are present (Rogachev et al. 1982, "Kostenki 1," 63–64). Arctic fox predominates heavily among faunal remains in Kostenki I, Layer 3 and may indicate fur-trapping (Vereshchagin and Kuz'mina 1977, 100). All of these assemblages are presumably related in some way to the spread of modern humans with Aurignacian industries across Western and Central Europe (although skeletal remains are confined to an isolated tooth from Kostenki XVII, Layer 2). However, the Spitsyn Culture in the Lower Humic Bed may antedate the early Aurignacian occupations in other parts of Europe.

The Lower Humic Bed at Kostenki also contains several sites that represent a very different type of Upper Paleolithic industry. These assemblages have been found at Kostenki VI and Kostenki XII, Layer 3, and are grouped into the "Strelets Culture" (Rogachev and Anikovich 1984, 179–181). Prismatic blade technology is rare, and typical Mousterian methods of blank production and retouching predominate. The tools include side-scrapers (12–17 percent), bifaces (17–18 percent), atypical end-scrapers (7–17 percent), and triangular points (4 percent). Especially diagnostic of these assemblages are small triangular bifacial points with concave bases. Nonstone implements, ornaments, and art are completely absent (Bradley et al. 1995, 990–991) (fig. 5.12).

  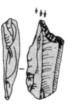 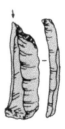

Figure 5.11. Artifiacts of the early Upper Paleolithic "Spitsyn Culture" from Layer 2 at Kostenki XVII on the Middle Don River (redrawn from Rogachev and Anikovich 1984, 247, fig. 83).

Younger occupations found in the Upper Humic Bed at Kostenki also contain remains assigned to the Strelets Culture that are currently dated to 32,000–27,000 years BP (Sinitsyn et al. 1997, 28–29). These include Kostenki I, Layer 5; Kostenki XI, Layer 5; and Kostenki XII, Layer 1a (Anikovich 1993). The artifact assemblages are similar to those in the Lower Humic Bed, but contain more typical Upper Paleolithic end-scrapers and some burins. Small triangular bifacial points were found in all of them. Nonstone tools are lacking, but an object thought to represent a partially finished pendant of marl was recovered from Kostenki I, Layer 5 (Rogachev et al. 1982, "Kostenki 1," 65–66). This occupation also yielded a sharply delimited oval concentration of debris (measuring 6 × 4.5 meters) with a centrally placed hearth that is interpreted as a former dwelling location (Rogachev 1957, 35–36).

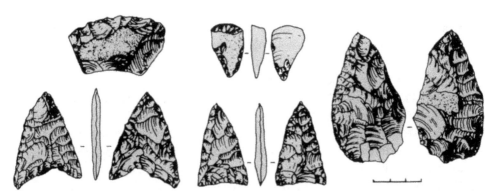

Figure 5.12. Artifacts of the early Upper Paleolithic "Strelets Culture" from Layer 5 at Kostenki I, Kostenki VI, and Layer 3 at Kostenki XII on the Middle Don River (redrawn from Rogachev and Anikovich 1984, 244–245, figs. 80–81).

The site of Sungir', located on a tributary of the Oka River near the city of Vladimir, contains an assemblage that many consider a late manifestation of the Strelets Culture dating to the terminal phase of OIS 3 (ca. 25,000 years BP) (Rogachev and Anikovich 1984, 180–181; Bradley et al. 1995, 993–994). Blade technology is more common at Sungir' than at the earlier sites, and the tools include large percentages of end-scrapers (21 percent) and burins (15 percent). However, many typical elements of the Strelets Culture are present, including side-scrapers (4 percent) and small triangular bifacial points (Bader 1978, 114–145). In contrast to the earlier Strelets Culture assemblages, there is a large and diverse inventory of nonstone tools comprising bone rods, points, awls, and at least one needle fragment. Indirect evidence for traps or snares is provided by intact skeletons of arctic foxes that are missing the third phalanx (presumably removed with the pelts [Gromov 1966b, 102]). In addition to numerous debris concentrations, former hearths, and shallow pits, traces of four artificial structures are thought to be present on the occupation floor. Sungir' also produced the richest array of ornaments, decorated objects, and figurative art in the early Upper Paleolithic of Eastern Europe (Bader 1978, 165–173).

Sites containing assemblages thought to be related to the Strelets Culture are also reported from northeastern Russia along the western slope of the Ural Mountains. They include the open-air sites of Byzovaya on the Pechora River (dated to 25,000 years BP) and Garchi I on the Kama River (undated) (Guslitzer and Pavlov 1993, 182–184). In the southern part of the East European Plain, a large assemblage containing triangular bifacial points and other types found in the later Strelets occupations has been reported recently from Biryuchya Balka on the Northern Donets River (Bradley et al. 1995, 993).

East European archaeologists have long noted the similarities between the Strelets Culture and the Szeletian industry of Central Europe (e.g., Efimenko 1958, 439–441; Anikovich 1993, 14). Other researchers have emphasized their differences, and concluded that they represent independent phenomena (Grigor'ev 1963; Allsworth-Jones 1986, 180). It is important to note that the late stage of the Strelets industry at Sungir' is associated with more typical Upper Paleolithic stone tool technology, figurative art, and modern humans. It thus appears likely that modern humans also made the earlier Strelets assemblages, and the latter probably do not represent a local Neanderthal industry like the Chatelperronian or Szeletian of Western and Central Europe (Cohen and Stepanchuk 1999, 311). In fact, raw material differences can account for some of the differences between the Strelets and Spitsyn sites (Allsworth-Jones 1990, 227–228), and activity differences might account for others. It is possible that the same population of modern humans produced both groups of assemblages at Kostenki.

The central East European Plain contains another early Upper Paleolithic industry that is unique in northern Eurasia. This industry is termed the "Gorodtsov Culture" and is represented in occupation levels within the Upper Humic Bed (32,000–27,000 years BP) at Kostenki XII, Layer 1; Kostenki XIV, Layer 2; Kostenki XV; and Kostenki XVI (Rogachev and Anikovich 1984, 183–185; Sinitsyn 1996). Many of the stone artifacts in the Gorodtsov assemblages were made on flint imported from a source located 130–150 kilometers to the southwest. The tools exhibit a bizarre combination of typical Mousterian types (roughly 30 percent), including side-scrapers, points, denticulates, and others, and a large array of end-scrapers (28–45 percent). The assemblages also contain an exceptionally rich inventory of nonstone implements, including eyed needles, awls, points, shovel-shaped tools, and other objects of unknown function; ornaments and decorated items are also present (fig. 5.13). Over eight hundred bones of hare were found in Kostenki XIV, Layer 2 and may reflect fur-trapping (Vereshchagin and Kuz'mina 1977, 107).

It is especially difficult to integrate the Gorodtsov Culture with other early Upper Paleolithic industries in northern Eurasia. However, it is clearly associated with modern humans, and dated to ca. 30,000 years BP. While the stone technology and tool types are very different from those in the Aurignacoid industries, the Gorodtsov sites reveal the presence of a complex nonlithic technology, including the earliest known examples of eyed needles (for tailored clothing production) in Europe.

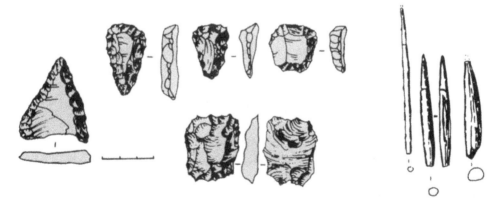

Figure 5.13. Artifacts of the early Upper Paleolithic "Gorodtsov Culture" from Layer 2 at Kostenki XIV (Markina Gora) on the Middle Don River (redrawn from Rogachev and Sinitsyn 1982a, 155, fig. 50; Rogachev and Anikovich 1984, 249, fig. 85).

An assemblage found in the Upper Humic Bed at Kostenki that falls outside the industrial groupings described above was recovered from Kostenki VIII, Layer 2 (Rogachev et al. 1982, 101–108). Almost half of the large collection of stone tools (more than 2,000 specimens) comprise small backed micoblades and points, and it has been compared with the Eastern Gravettian industries that are widespread in Central and Eastern Europe during the early phase of OIS 2 (Anikovich 1993, 14) (see chapter 6). A large array of nonstone tools, ornaments, and decorated pieces were also found in this layer, and the occupation floor contained possible traces of up to five artificial shelters (which may instead simply represent working areas). The latter are represented by oval concentrations of debris (measuring roughly 5–7 meters in diameter) surrounding former hearths (Klein 1969b, 99–101).

Crimea and Northern Caucasus

Few early Upper Paleolithic sites are known in the Crimea and Northern Caucasus, which is at least partly a function of the comparatively late presence of Neanderthals and Mousterian industries in these areas. Modern humans do not seem to have occupied them until after ca. 30,000 years BP. Unlike the East European Plain, Upper Paleolithic occupations directly overlie Mousterian levels at some sites.

In the Crimea, an early Upper Paleolithic occupation has recently been dated to ca. 28,000–30,000 years BP in Layer 3 of Syuren' I rockshelter, which is located near Sevastopol in the southwestern part of the peninsula. The assemblage excavated in 1926–1929 contains many prismatic blade cores and blades, and the tools include retouched microblades (44 percent), retouched blades (24 percent), end-scrapers (10 percent), and burins (9 percent). Among the end-scrapers are carinated forms. Small percentages of Mousterian tool types (chiefly points [3 percent]) are also present. The nonstone inventory includes bone awls and points (Vekilova 1957).

The Syuren' I, Layer 3 assemblage appears to represent a rare example of the Aurignacian industry (strictly speaking) in Eastern Europe (Cohen and Stepanchuk 1999, 284). Many diagnostic elements of the Aurignacian are present, including carinated end-scrapers, backed bladelets, and bone points. It has been suggested that the Mousterian tools in the assemblage are derived from short-term Neanderthal occupations that occurred during the same period (Demidenko 1999). Modern humans are thought to be associated with the assemblage, but skeletal remains are confined to an isolated molar (Bonch-Osmolovskii 1934, 131).

Another early Upper Paleolithic occupation seems to be present at the collapsed rockshelter of Buran-Kaya III on the Burul'cha River in the eastern Crimea. The lowest Upper Paleolithic level at this site (Layer C) has been recently dated to ca. 32,000–37,000 years BP and directly overlies the uppermost Mousterian layer. There is no evidence of prismatic blade technology, and the small sample of tools contains bifaces, end-scrapers on flakes, and trapezoidal microliths; several bone rods were also recovered. Like the Gorodtsov Culture on the Middle Don River, the Buran-Kaya III, Layer C assemblage has no clear relationship to other Upper Paleolithic industries in Eastern Europe (Marks 1998).

In the northwestern Caucasus, an early Upper Paleolithic occupation recently has been identified at Mezmaiskaya Cave, located at 1,300 meters above sea level on a tributary of the Kurdzhips River (Golovanova et al. 1999). Layer 1C, which directly overlies the uppermost Mousterian level and has been dated to 32,000 years BP, yielded prismatic blade and microblade cores and numerous bladelets. Stone tools include end-scrapers on large flakes and blades, burins on flakes, and retouched bladelets and micropoints. Nonstone implements include bone awls, rods, and points. The Mezmaiskaya Cave assemblage may be characterized as Aurignacoid, and was presumably produced by modern humans.

Ornaments and Art in the Early Upper Paleolithic

Throughout northern Eurasia, ornaments and art objects appear in early Upper Paleolithic sites and are viewed widely as an indication of major behavioral change tied to the use of symbols. Ornaments probably should be placed in a separate category from decorated objects and figurative art. Simple ornaments appear in a Neanderthal context—sporadically in the Mousterian and regularly in the Chatelperronian—and may not carry the same implications (Mellars 1996, 372–375). Many nonhuman animal species have evolved elaborate forms of ornamentation without symbols and language. However, the quantity and complexity of ornaments increases dramatically with the appearance of modern humans and Upper Paleolithic industries (White 1989). Decorative and figurative art is wholly confined to the latter, and accompanied by evidence for ritual associated with burials.

Ornaments and art objects have been recovered from a number of early Upper Paleolithic sites in Eastern Europe, and their occurrence is broadly consistent with the archaeological record in other parts of northern Eurasia. However, their temporal distribution varies somewhat from that of Western and Central Europe (but not from Siberia), and may reflect a different pattern of development (Hoffecker 1988, 258–260). Only ornaments are present in the earliest Upper Paleolithic sites. Decorated objects and burials containing evidence of ritual appear later, and figurative art does not materialize until the terminal phase of OIS 3 (ca. 30,000–25,000 years BP). But the number of sites remains small, and the combined effects of preservation bias and sampling error could significantly influence this pattern.

The earliest Upper Paleolithic sites in Eastern Europe are found in the Lower Humic Bed at Kostenki on the Middle Don River (ca. 40,000 years BP or earlier). In Kostenki XVII, Layer 2 (assigned to the Aurignacoid Spitsyn Culture), ornaments are present including perforated arctic fox teeth (n = 37), perforated fossil shell, coral, and belemnite, and perforated stone (n = 7) (Boriskovskii 1963, 102–105). It should be noted that all of these items were made on relatively resistant materials (i.e., teeth, stone). Several of them appear more elaborate than any ornaments recovered from a Neanderthal context (fig. 5.14). No ornaments have been recovered to date from other occupation levels in the Lower Humic Bed, including those assigned to the Strelets Culture.

Younger Upper Paleolithic occupations are found in the Upper Humic Bed at Kostenki, which underlies the terminal OIS 3 soil at Kostenki I and apparently dates ca. 30,000 years BP or slightly earlier (Sinitsyn et al. 1997). Many of these occupations have yielded ornaments in the form of perforated teeth (typically arctic fox), or objects of shell, stone, bone, or ivory interpreted as pendants or beads. Ornaments have been recovered from Kostenki XII, Layer 1; Kostenki XIV, Layer 2; and Kostenki XV (assigned to the Gorodtsov Culture), and Kostenki VIII, Layer 2. A possible pendant (perhaps only partially finished) is present at Kostenki I, Layer

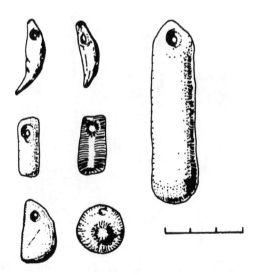

Figure 5.14. Ornaments from Layer 2 at the site of Kostenki XVII on the Middle Don River (redrawn from Boriskovskii et al. 1982, 182, fig. 65).

5 (Strelets Culture). At Kostenki XV, over 150 perforated fox teeth were found in a burial, along with stone and bone artifacts that appear to represent funerary items. Many of the teeth were located near the skull and may represent a headdress or necklace (Rogachev 1957, 111). Some occupations also contain objects of bone decorated with incised geometric patterns (i.e., lines, chevrons, zigzags). These are found only in the Gorodtsov Culture sites (Kostenki XII, Layer 1; Kostenki XIV, Layer 2; and Kostenki XV), and in Kostenki VIII, Layer 2 (Klein 1969b; Praslov and Rogachev 1982) (see fig. 5.15).

Upper Paleolithic occupations associated with the terminal OIS 3 soil are dated to ca. 30,000–25,000 years BP, and are only slighter younger than those in the Upper Humic Bed at Kostenki. Ornaments and decorated objects have been found in sites located in both the southwestern and central regions of the East European Plain (their absence in the Crimea and Northern Caucasus is probably due to the scarcity of early Upper Paleolithic sites in these regions). In the Middle Dnestr Valley, perforated shells, a perforated ivory object, and a decorated ivory fragment were found at Molodova V, Layer 8 (Chernysh 1987, 33). In the northern Volyn-Podolian Upland, decorated antler fragments and a pendant were recovered from Kulichivka, Layer 2 (Savich 1975, 47). All of these items are associated with Aurignacoid industries. However, at Brynzeny I on a Prut River tributary, a spatula-shaped ornament of ivory decorated with dotted

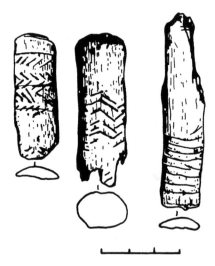

Figure 5.15. Decorated bone objects from Layer 2 at Kostenki XIV (Markina Gora) on the Middle Don River, assigned to the "Gorodtsov Culture" (redrawn from Rogachev and Sinitsyn 1982a, 155, fig. 50).

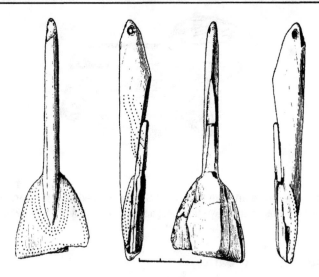

Figure 5.16. Decorated ornament and possible art object from Layer 3 at the rockshelter of Brynzeny I on a tributary of the Prut River (modified from Ketraru 1973, 73, fig. 29).

line patterns, was found associated with the archaic Brynzeny Culture assemblage in Layer 3 (Ketraru 1973, 73, fig. 29) (fig. 5.16). On the Middle Don River at Kostenki I, Layer 3 (Aurignacoid), there are objects of bone and ivory decorated with incised parallel lines and dots (Rogachev et al. 1982, "Kostenki 1," 64). And in the north-central East European Plain at Sungir' (on a tributary of the Oka River), over 10,000 ivory beads were recovered from burials, along with a number of elaborate ornaments representing pendants, bracelets, rings, and others; decorated objects include an ivory disk with radiating lines of incised dots (Bader 1978, 165–173).

Figurative art in the early Upper Paleolithic of Eastern Europe is apparently confined to the terminal phase of OIS 3. Although an ivory object interpreted as an anthropomorphic carving was recovered from Kostenki VIII, Layer 2 (which probably antedates the terminal OIS 3 soil), its significance remains unclear (Rogachev et al. 1982, "Kostenki 8," 106–107). Four small animal figurines of ivory, including two horses and two other animals that may be bison, have been found at Sungir' (Bader 1978, 170–173). The figurines are stylized, two-dimensional representations that differ from the realistic three-dimensional animal figurines recovered from the Aurignacian in Western Europe (e.g., Vogelherd, Germany [Hahn 1972]) (fig. 5.17). Also, the elaborate pendant from Brynzeny I, Layer 3 (tentatively dated to the terminal phase of OIS 3) may represent a stylized fish (Ketraru 1973, 73).

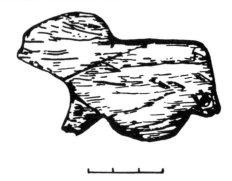

Figure 5.17. Figurative art from the site of Sungir' located on a tributary of the Oka River in northern Russia (redrawn from Rogachev and Anikovich 1984, 270, fig. 106).

Diet and Foraging Patterns

The diet of modern humans in Eastern Europe during OIS 3 probably contained a high percentage of meat. Although this has yet to be confirmed by stable isotope analyses on human bone samples, digestible plant foods are unlikely to have been widely available—except perhaps in the southern upland regions. Moreover, an emphasis on meat consumption is supported by the presence of numerous large mammal bones in East European Plain sites, many of which exhibit stone-tool butchering marks.

Thus, it appears likely that substantial niche overlap existed between modern humans and Neanderthals in Eastern Europe and elsewhere. Where regional comparisons are possible, modern humans of the transitional period seem to have occupied the same locations, performed many of the same activities, and consumed the same large mammals. A high degree of niche overlap could be critical to understanding the transition process, because it would have generated competition and eventual exclusion of one species from areas of shared habitat.

However, while it is difficult to detect differences in the ecological niche occupied by Neanderthals and modern humans, there is evidence for differences in foraging strategy among the latter. The transport of raw materials on the East European Plain during OIS 3 achieved unprecedented distances (up to 500 km), suggesting significantly greater foraging movements and/or widely dispersed cooperative networks. This suggests that modern human groups were able to draw on resources across much larger areas, which may have been critical to occupation of the central plain during the Pleniglacial. As noted earlier, modern humans also developed new technologies that probably increased their foraging efficiency in these environments.

Analysis of Faunal Remains in Archaeological Sites

As with other periods during the Pleistocene, most information concerning diet and foraging patterns in the early Upper Paleolithic is derived from the analysis of faunal remains in archaeological sites. In contrast to earlier periods, a substantial quantity of vertebrate remains have been recovered from open-air sites on the East European Plain, while few data are available from the Crimea and Northern Caucasus. Unfortunately, most faunal assemblages have not been subject to taphonomic or zooarchaeological studies, and the information they provide is still very limited.

Like the earlier Mousterian sites, most early Upper Paleolithic open-air localities on the East European Plain are buried in slope deposits that overlie low river terraces. They occupy a geomorphic setting where large mammal remains are unlikely to be concentrated by either abiotic or non-human biotic processes. Most of the large mammal remains in these sites were probably accumulated by their human occupants, although they need not necessarily represent food debris. On the other hand, the remains of smaller carnivores, rodents, birds, and other small vertebrates often appear to represent animals that inhabited these sites or their food debris (Hoffecker 1988, 261–262).

In the southwest region of the East European Plain, large mammal assemblages in early Upper Paleolithic sites are primarily composed of horse, reindeer, and mammoth. In the Middle Dnestr Valley at Molodova V, remains of horse (28–61 percent) and reindeer (31–61 percent) are heavily predominant in Layers 8–10; less common taxa include mammoth, woolly rhinoceros, and steppe bison (Chernysh 1959, 70–76). Reindeer (60 percent) and horse (30 percent) are also the most abundant species in Layer 3 at Molodova I and reindeer (93 percent) is overwhelmingly predominant at Babin I, Layer 1 (Chernysh 1959, 25; 1982, 72), but at Voronovitsa I, Layer 2, mammoth (48 percent) is more common than horse (35 percent) and reindeer (14 percent) (Chernysh 1959, 44). Mammoth is also the most common taxon at Kulichivka, Layer 2 (northern Volyn-Podolian Upland), followed by reindeer and horse, but this may be influenced by the generally poor state of preservation (Savich 1975, 45–47) (table 5.4).

Unfortunately, few data are available about these large mammal assemblages regarding fragmentation, surficial damage, skeletal parts, and other aspects of their taphonomy. It is widely concluded that they represent the remains of prey hunted by the early Upper Paleolithic occupants of these sites (e.g., Chernysh 1959, 1982), but this conclusion—however reasonable—needs to be buttressed with supporting data.

Table 5.4. Large Mammal Remains in Early Upper Paleolithic Sites of the East European Plain: Southwest Region

Taxon	Molodova I Layer 3	Molodova V Layer 9	Babin I Layer 1	Kulichivka Layer 2	Brynzeny I Layer 3
Woolly mammoth (*Mammuthus primigenius*)	6/–[a]	10/2	1/1	common	32/4
Woolly rhinoceros (*Coelodonta antiquitatis*)	3/–	2/1	–/–	absent	28/6
Horse (*Equus caballus*)	36/–	124/5	38/4	common	8419/194
Roe deer (*Capreolus capreolus*)	–/–	–/–	–/–	–/2	17/4
Red deer (*Cervus elaphus*)	4/–	–/–	–/–	present	34/8
Giant deer (*Megaceros giganteus*)	–/–	–/–	–/–	absent	4/1
Reindeer (*Rangifer tarandus*)	73/–	63/6	535/11	common	3987/117
Bison (*Bison priscus*)	–/–	3/2	–/–	present	326/21
Wolf (*Canis lupus*)	–/–	–/–	1/1	present	18/6
Bear (*Ursus arctos*)	–/–	–/–	–/–	present	2/1
Hyena (*Hyaena spelaea*)	–/–	–/–	–/–	present	4/2

SOURCES: Chernysh 1959, 25, table 1; 1982, 72; 1987, 32; David and Ketraru 1970, 23–25, table 6; Savich 1975, 45–47.
[a]Number of identified specimens/minimum number of individuals

Farther south, along the Prut River and its tributaries, a similar pattern is evident. The earliest Upper Paleolithic levels at Ripiceni-Izvor contain remains of horse and red deer (Păunescu 1965). Over 12,000 large mammal bones and teeth were recovered from Layer 3 at the Brynzeny I rockshelter, among which horse (65 percent) and reindeer (31 percent) are heavily predominant (David and Ketraru 1970, 23–25). Although a variety of carnivores are represented, their total percentage is very small (0.5 percent), and it seems unlikely that the shelter was intensively used as a carnivore den during the period of occupation. The ungulate bones are highly fragmented, and many of them reportedly exhibit stone tool cut marks (David and Ketraru 1970, 29).

In the central East European Plain, the earliest Upper Paleolithic occupations are associated with the Lower Humic Bed at Kostenki (Middle Don River). Mammal remains in these occupations are dominated by horse, which represents 67 percent of the assemblage in Kostenki VI, and 72–89 percent of the assemblages in Kostenki XII, Layers 2–3 (Vereshchagin and Kuz'mina 1977, 103–106). Other large mammals (e.g., mammoth, reindeer, bison) are present in relatively low numbers. An interesting contrast to this pattern is found at Kostenki XVII, Layer 2 in which 70 percent of the mammal assemblage belongs to wolf, and at Kostenki XIV, Layer 4, where hare (86 percent) is overwhelmingly predominant (Vereshchagin and Kuz'mina 1977, 107–108). If the horse remains constitute food debris, the accumulations of wolf and hare bones at Kostenki XVII and XIV seem likely to represent hunting or trapping for furs as well (table 5.5).

The pattern in the younger occupations of the Upper Humic Bed at Kostenki is similar, although the assemblages are significantly larger and more taphonomic data are available. Overall, horse accounts for more than 68 percent of the mammalian remains in the Upper Humic Bed (Vereshchagin and Kuz'mina 1977, 95, table 5), and some sites contain dense concentrations of bone. Over 2,000 bones and teeth representing an estimated minimum (MNI) of 19 individuals were found at Kostenki XIV, Layer 2, and many were discovered in anatomical groups, including sequences of vertebrae (cervical, thoracic, and lumbar) and extremities. Numerous tool cut marks are reported on the bones, especially vertebrae and ribs, and virtually all of the long bones are fractured. All age groups are represented in the sample (Rogachev 1957, 78–80; Rogachev and Sinitsyn 1982a, 149). Another mass of horse remains (1,500 bones and teeth) was encountered at Kostenki XV comprising six anatomical groups of vertebrae and extremities, and many fragments (Rogachev and Sinitsyn 1982b, 163), and anatomical groups are also reported among a smaller assemblage of bones at Kostenki XVI (Rogachev and Sinitsyn 1982c, 174).

Table 5.5. Large Mammal Remains in Early Upper Paleolithic Sites of the East European Plain: Central and Northern Regions

		Sites and level			
Taxon	Kostenki VI	Kostenki XVII Layer 2	Kostenki XIV Layer 2	Kostenki VIII Layer 2	Sungir'
Woolly mammoth (*Mammuthus primigenius*)	19/2[a]	9/1	6/1	67/3	common
Woolly rhinoceros (*Coelodonta antiquitatis*)	1/1	–/–	16/1	18/2	absent
Horse (*Equus caballus*)	144/4	21/2	2083/19	62/3	common
Red deer (*Cervus elaphus*)	2/2	–/–	41/1	1/1	absent
Giant deer (*Megaceros giganteus*)	–/–	–/–	–/–	2/1	absent
Reindeer (*Rangifer tarandus*)	19/2	11/1	11/1	9/2	abundant
Bison (*Bison priscus*)	10/2	12/1	1/1	11/3	1–3/–
Saiga (*Saiga tatarica*)	–/–	–/–	1/1	7/1	1–2/–
Wolf (*Canis lupus*)	15/2	126/2	4/1	204/8	16–17/–
Arctic fox (*Alopex lagopus*)	2/1	–/–	–/–	56/3	common
Lion (*Panthera* sp.)	–/–	–/–	1/1	3/1	1/1

SOURCES: Vereshchagin and Kuz'mina 1977, 103–108; Gromov 1966a, 74.
[a] Number of identified specimens/minimum number of individuals

The concentrations of horse at sites like Kostenki XIV, XV, and XVI most probably reflect the remains of hunted prey, as their topographic setting—upper reaches of broad ravines—is unlikely to be the locus of natural catastrophic deaths or large accumulations of bone. The recurring presence of anatomical groups suggests that the animals were killed at or near these sites, where complete or partial carcasses were butchered. The overall pattern is reminiscent of a Holocene kill-butchery site at which an animal herd was ambushed or driven into a mass kill, and then rapidly—and sometimes incompletely—processed (e.g., Frison 1974). Horses often form bands and larger herds, and the concentrations of their remains in these sites may reflect predation on groups and less than full use of the food products.

The only other vertebrates represented in significant numbers are hare, wolf, and arctic fox, which collectively account for roughly 25 percent of the mammal remains in the Upper Humic Bed (Vereshchagin and Kuz'mina 1977, 94, table 5). Today, all three taxa yield valuable furs. The large Don hare (*Lepus tanaiticus*), which was distributed only in the central and southern East European Plain during the Late Pleistocene (Averianov 1998), is the most abundant species in several occupation levels (i.e., Kostenki VIII, Layer 2; Kostenki XIV, Layers 3–4 [Vereshchagin and Kuz'mina 1977, 104–107]). The massive concentrations of bones in these sites suggest that they were most probably accumulated by their human occupants, and seem likely to have been a source of food and pelts. Wolves are also very common in some assemblages (e.g., Kostenki VIII, Layer 2), and at Kostenki XIV, Layer 3 they include remains of very young individuals (estimated 8–9 months) that indicate winter mortality (Vereshchagin and Kuz'mina 1977, 85).

The site of Sungir', which is located in the north-central East European Plain and was apparently occupied at the end of OIS 3, presents a slightly different picture. Reindeer is the most abundant mammal at this site, although mammoth, horse, and arctic fox are also common (Gromov 1966a; Bader 1978, 183). Reindeer is represented by all skeletal parts, and several groups of vertebrae and limb bones were found in anatomical order. Tool cut marks are reported on many bones (Gromov 1966a, 78; Bader 1978, 49). The presence of frontal bones with attached antlers (and complete absence of frontals with shed antlers) indicates summer or fall mortality. Reindeer probably moved into the northern plain during the warmer months, and—like horse—may have been hunted in herds or groups near Sungir' (Bader 1978, 189–190).

In the Crimea and Northern Caucasus, where early Upper Paleolithic sites are scarce, few faunal data are available. At Syuren' I, Layer 3 in the southwestern Crimea, the most common large mammals are saiga (32 percent), arctic fox (26 percent), and giant deer (14 percent); among small mammals, hare is very common (Vekilova 1957, 254). Saiga is also common in the earlier Mousterian sites, and undoubtedly reflects consistently drier conditions in southernmost Eastern Europe. Syuren' I provides an example of how Neanderthals and modern humans probably competed for the same resources in jointly occupied regions. On the other hand, the abundance of smaller fur-bearing mammals (e.g., arctic fox and hare) is not characteristic of Mousterian sites, and—as in the central East European Plain—reflects a shift in adaptive strategy probably tied to technological protection from cold climates.

Intersite Patterning and Transport of Raw Materials

East European early Upper Paleolithic sites reveal some variability in terms of size and complexity that may reflect functional and seasonal differences. The analysis of the differences may eventually reveal how modern humans scheduled their movements and exploitation of seasonally available resources during OIS 3. Some paleoanthropologists have argued that modern humans planned and scheduled the use of resources with greater precision and efficiency than the Neanderthals (e.g., Pike-Tay et al. 1999). However, the East European data are insufficient to effectively address this issue at present. More significant in its implications for modern human foraging strategy is the evidence for long-distance transport of raw materials.

As in the case of the Mousterian, most occupations appear to represent multiple-activity loci and long-term habitations, although this pattern is likely to be exaggerated by repeated occupation episodes on the same floor. These sites contain large quantities of lithic and faunal debris, hearths, and sometimes possible traces of former structures. Notable examples include Kulichivka, Layer 2; Kostenki VIII, Layer 2; and Sungir' (Rogachev 1957; Savich 1975; Bader 1978). The size of these sites undoubtedly favors their preservation and discovery over smaller localities.

The early Upper Paleolithic also contains a significant number of sites that produced limited quantities of lithic and faunal debris, and few if any features. These localities were probably occupied for relatively brief periods and a limited array of functions. Examples from the Middle Dnestr Valley include Layer 7 at Korman' IV, which yielded only 146 artifacts, isolated faunal debris, and two hearths within an excavated area of 118 square meters (Chernysh 1977, 24), and Layer 3 at Molodova I, which yielded 332 artifacts and 122 identified mammal remains in an excavated area of 168 square meters (Chernysh 1982, 70–72). Even smaller accumulations of debris were found in the levels that underlie the terminal OIS 3 soil at Molodova V (Layers 10a–10b) and Korman' IV (Layer 8)

(Chernysh 1977, 23; 1987, 25–26). Small occupations on the Middle Don River include Kostenki I, Layer 4, which contained only about twenty artifacts and isolated bones of mammoth and horse (Rogachev et al. 1982, "Kostenki 1," 64–65), and Kostenki VI, which yielded roughly 1,000 artifacts and 200 identified mammal remains from an excavated area of 130 square meters (Rogachev and Anikovich 1982b, 90–91). The latter is unusual for a small occupation, because most of them are found in multilevel sites containing other larger occupation levels that confer high archaeological visibility on the site as a whole.

The East European Plain sites also yield evidence for long-distance movements in the form of materials imported up to hundreds of kilometers from their sources. Whether the result of transport or trade (or both), they indicate more widely dispersed groups or networks. Many of the Kostenki sites (e.g., Kostenki XVII, Layer 2) contain large quantities of black flint from sources no closer than 130–150 kilometers on the southern margin of the Central Russian Upland (Boriskovskii 1963, 166–191). More impressive in terms of distance, if not in quantity, are fossil marine shells in Kostenki I, Layer 3, which were imported more than 500 kilometers from the Black Sea coast (Rogachev 1957, 34). There is no evidence of comparable long-distance transport of materials in the Mousterian.

Early Upper Paleolithic Socioecology

Evidence for a fundamental transformation in organization at the time that modern humans dispersed across Eurasia is more elusive than the signs of a quantum leap in technological ability. Nevertheless, some supporting data are present in the early Upper Paleolithic archaeological record, and Eastern Europe is especially suited to the study of this issue. Overall, early Upper Paleolithic occupation floors exhibit a structure similar to that of modern hunter-gatherer encampments. There is some indirect evidence for the formation of large task groups associated with possible herd kills (see earlier discussion). More important is the evidence for significantly increased mobility and/or network size based on the long-distance movement of materials described earlier.

Large-scale excavations at open-air sites of the East European Plain since the late 1920s have provided a wealth of occupation floor maps for this time period that is unequaled elsewhere in northern Eurasia. In general, occupation floors in early Upper Paleolithic sites of Eastern Europe exhibit a reassuring familiarity to archaeologists, and lack the alien quality of the Mousterian floors (see chapter 4). Unlike the latter, debris concentrations are consistently associated with former hearths, and the overall pattern is similar to that of modern hunter-gatherers (e.g., Yellen 1977,

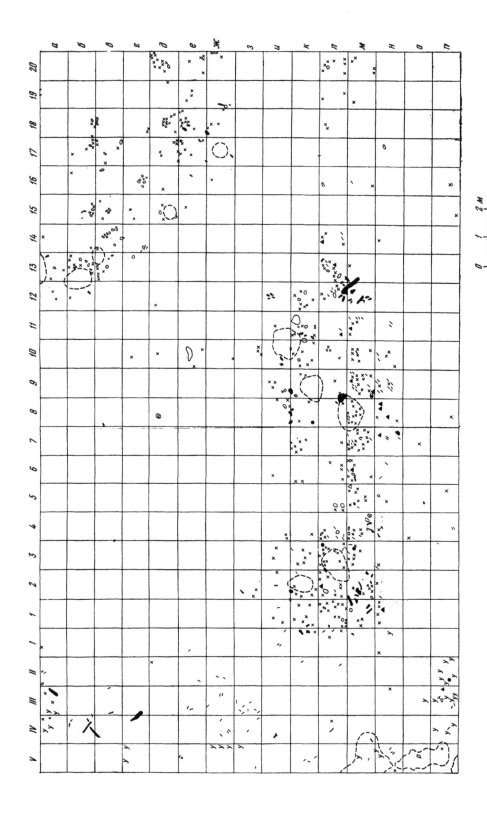

87–91; Binford 1983, 149–156). Only at one site is there credible evidence for multiple dwellings (i.e., Kulichivka, Layer 2 [Savich 1975]), creating a pattern reminiscent of residential camps among many modern hunter-gatherers (e.g., Yellen 1977).

In the southwestern part of the East European Plain, occupation floors at several early Upper Paleolithic sites in the Middle Dnestr Valley were mapped by the late A. P. Chernysh. Floor areas of 630–750 square meters were mapped at Molodova V, Layers 8–10, and areas were also mapped at Babin I, Layer 1 (212 square meters), Voronovitsa I, Layer 2 (130 square meters), and Oselivka I, Layer 3 (55 square meters) (Chernysh 1959, 1961, 1971) (fig. 5.18). Large occupation areas were also mapped at Kulichivka, Layer 2 (northern Volyn-Podolian Upland) and Ripiceni-Izvor (Prut River) (Paunescu 1965; Savich 1975). All of these site floors contain former hearths and debris concentrations. The hearths tend to average 0.5–1.0 square meters in area, and range 2–10 centimeters in thickness. Most hearths are within, adjacent, and/or in close proximity (1–3 meters) to debris concentrations. The latter vary substantially in terms of area and contents (artifacts and bone debris). The most convincing traces of former dwelling structures were mapped at Kulichivka, Layer 2 (Savich 1975, 42–43).

In the central part of the East European Plain, occupation floors have been mapped at several Kostenki localities on the Middle Don River (Klein 1969b, 75–108). These include Kostenki XVII, Layer 2 (66 square meters) in the Lower Humic Bed (Boriskovskii 1963), and Kostenki VIII, Layer 2–3 (530 square meters), Kostenki XIV, Layer 2 (60 square meters), Kostenki XV (70 square meters), Kostenki XVI (50 square meters), and Kostenki XVII, Layer 1 in the Upper Humic Bed (Rogachev 1957; Tarasov 1961; Boriskovskii 1963). These floors also contain former hearths and debris concentrations. The hearths range between .25 and 1.0 square meters in area, and 5–15 centimeters in thickness. They are typically associated with debris concentrations of varying size and contents. Traces of possible former structures were found at Kostenki I, Layer 5, and Kostenki VIII, Layer 2, but they are less convincing than the features interpreted as dwellings at Kulichivka.

The largest known early Upper Paleolithic floor is at Sungir' (roughly 3,400 square meters) on an Oka River tributary, although the cultural layer here is up to 90 centimeters thick and contains features from multiple occupation episodes (Bader 1978, 25–37). More than 40 hearths and "hearth pits" were mapped at Sungir' and, while some were found in one

Figure 5.18. *(opposite)* Early Upper Paleolithic occupation floor of Layer 9 at Molodova V in the Middle Dnestr Valley (after Chernysh 1959, fig. 32).

area of low debris density, most are associated with debris concentrations. The excavator interpreted four of the combined hearth and debris scatters as traces of former structures (Bader 1978, fig. 36), but they may simply represent working areas.

The Transition to Modern Humans in Eastern Europe: Summary and Conclusions

The transition to modern humans in Eastern Europe was part of a global event that resulted in the distribution of *Homo sapiens* populations in terrestrial habitats throughout the world (with the exception of high arctic and alpine regions) by the end of OIS 2. Modern humans had evolved in Africa in the late Middle Pleistocene, but did not begin the process of global dispersal until after ca. 60,000–50,000 years BP (OIS 4–3). Their appearance in regions outside Africa was achieved through colonization of new habitat not previously occupied by hominids, replacement of existing hominid populations in occupied regions, and possibly genic exchange with the latter in some areas. Dispersal coincides with evidence for the use of symbols and possibly fully developed language, and the behavioral changes associated with this development may underlie the technological and organizational adaptations that permitted modern humans to occupy such a wide variety of habitats in a relatively brief period of time (Gamble 1994; Klein 1999).

More specifically, the transition in Eastern Europe was part of the dispersal of modern humans in northern Eurasia (at latitudes above 40 degrees North). These regions were occupied by the cold-adapted Neanderthals at the time that modern humans colonized them during the later phases of OIS 3. The East European record indicates that modern humans probably first occupied periglacial loess-steppe areas that had been abandoned by the Neanderthals and represented an uncontested "empty niche." A similar phenomenon seems to have taken place in parts of north-central Europe (Gamble 1986, 374) and possibly southern Siberia. Modern humans subsequently expanded into areas still occupied by Neanderthal populations in many parts of Europe, and the latter appear to have become extinct—probably with minimal genic exchange—within a few thousand years or less.

The recent dating of skeletal and archaeological remains in the Crimea and the Northern Caucasus indicates that the Neanderthals were present in these areas until at least 30,000 years BP and possibly somewhat later (Chabai and Marks 1998; Ovchinnikov et al. 2000). In terms of soil stratigraphy, these remains may be correlated with the final OIS 3 soil (dated to 30,000–25,000 years BP). Stratified sites on the central East European

Plain at Kostenki (Middle Don River) contain modern human occupations buried below this soil horizon that may be as much as 5,000–10,000 years older (Sinitsyn et al. 1997). There is a corresponding lack of evidence for Neanderthal settlement in this region during OIS 4–3. However, uncertainties remain concerning the dating and human fossil associations of both the latest Mousterian and the earliest Upper Paleolithic occupations on the East European Plain.

The modern humans who colonized northern Eurasia during OIS 3 were at a considerable disadvantage compared to the existing Neanderthal population with respect to the size and shape of their bodies. Derived from southern latitudes, the former were morphologically adapted to warm climates. Their disadvantage must have been especially pronounced in Eastern Europe and other colder and drier parts of Eurasia above 45 degrees North, which had been widely settled by hominids only with the appearance of the cold-adapted Neanderthals (Hoffecker 1999a). On the East European Plain—even during the warm interval at the end of OIS 3—winter temperatures were probably between –20 and –30 degrees C (Velichko et al. 1984, 114).

Sites occupied by modern humans on the East European Plain during OIS 3 reveal evidence of a complex and innovative technology that must have been essential for colonization of this region. This new technology included devices for cold protection and improved foraging efficiency, and probably reached an overall level of complexity comparable to that of recent hunter-gatherers in arctic environments (Dennell 1983; Oswalt 1976, 1987). Occupations on the central plain at Kostenki and Sungir' contain needles and awls almost certainly used for the production of tailored clothing. Large quantities of fur-bearing mammal remains, such as arctic fox and hare—sometimes with evidence for pelt removal—have been found in these sites, and further suggest the invention of traps and snares (Bader 1978; Praslov and Rogachev 1982). Several sites also contain probable traces of artificial shelters with interior hearths, most notably at Kulichivka in the northern Volyn-Podolian Upland (Savich 1975). These types of technological innovations seem to reflect a quantum jump in hominid ability to manipulate the environment that is characteristic of modern humans after the appearance of symbols in the archaeological record (Mithen 1996).

Modern humans also probably depended on a high degree of mobility and organization for successful occupation of areas like the central East European Plain (Gamble 1986; Whallon 1989). Although undoubtedly more productive than modern arctic tundra, the mobility requirements of such areas during OIS 3 must have been high for humans subsisting primarily on a diet of meat. While Neanderthal movements—based on raw-material transport distances—rarely exceeded 100 kilometers, modern humans may have covered distances of up to 500 kilometers. Alternatively, some of these materials may have been moved in part by trade, which suggests the presence of widely dispersed social networks. In either (or both) case(s), modern human groups seem to have been exploiting resources over large areas and on a scale comparable to recent hunter-gatherers in cold and dry environments (Kelly 1995). The development of large social networks is also probably tied to the appearance of symbols in the archaeological record. The creation of symbol-based organization would have permitted modern humans to expand beyond the narrow scope of kin-based social groups (see chapter 1).

Many aspects of the transition to modern humans in Eastern Europe remain obscure. This is due in part to the uncertainties regarding the dating and human fossil associations of some early Upper Paleolithic sites, but it also reflects the difficulty of explaining the extraordinary amount of spatial and temporal variability in those sites. In Western and Central Europe, the early Upper Paleolithic comprises the widespread Aurignacian industry, which is associated with modern humans, and various local "archaic" (or "transitional") industries (e.g., Chatelperronian, Szeletian), which were apparently produced by Neanderthals (Mellars 1996, 392–419). But Eastern Europe follows a different pattern. Sites that may be formally classified as Aurignacian are rare, although broadly similar Aurignacoid industries are widespread (e.g., Molodova, Spitsyn) and most probably produced by modern humans. "Archaic" industries with similarities to the Central European Szeletian (e.g., Brynzeny, Strelets) are present, but there is some reason to believe that these sites may also have been occupied by modern humans (Cohen and Stepanchuk 1999).

If modern humans produced some or all of the local "archaic" industries of the early Upper Paleolithic in Eastern Europe, the process of the transition must have differed significantly from that of Western and Central Europe. The first modern human groups to enter Eastern Europe during OIS 3 might have been considerably more diverse than the Aurignacian people who subsequently spread across other parts of the continent. This could explain the high degree of morphological variability among the occupants of the East European sites. Such diversity, in turn, might reflect Eastern Europe's unique geographic position, and multiple routes of potential access from the Near East (i.e., Balkans, Caucasus, and Central Asia).

The temporal variability in the early Upper Paleolithic of the central East European Plain is also difficult to explain. The sites assigned to the Strelets Culture appear to undergo fundamental changes during OIS 3, as they evolve from Mousteroid assemblages in the Lower Humic Bed at

Kostenki to a typical Upper Paleolithic industry associated with figurative art and elaborate burials (Bradley et al. 1995). The appearance of the Gorodtsov Culture sites in the Upper Humic Bed is equally startling. This industry, which manifests a peculiar mixture of Mousterian tools, abundant end-scrapers, complex nonlithic technology, and simple art, has no parallels with Upper Paleolithic sites outside Eastern Europe (Sinitsyn 1996).

In sum, Eastern Europe may have played a unique role in the transition, but further research is needed to better understand this role. Along with the Balkans, the East European Plain was most probably one of the first areas in Europe to be settled by modern humans. It was probably occupied prior to the spread of the Aurignacian industry across Central and Western Europe, but the relationship between this industry and the generally similar Aurignacoid cultures of Eastern Europe remains unclear. The significance of the archaic industries, which are broadly similar to the Central European Szeletian, also remains unclear (Allsworth-Jones 1990).

CHAPTER 6

People of the Loess Steppe

Climates deteriorated swiftly throughout the northern hemisphere after 25,000 years BP with the beginning of the Last Glacial Maximum. By ca. 18,000 years BP, the Scandinavian glacier had advanced to within a few kilometers of the city of Smolensk, and much of the northwestern plain was covered with ice (Grosswald 1980). Large proglacial lakes and polar deserts formed along glacier margins, and thick deposits of loess blanketed the lowlands. Continuous permafrost extended as far south as 48 degrees North, and estimated winter temperatures on the central East European Plain fell below −30 degrees C (Velichko et al. 1984, 117).

The decline in temperatures brought arctic conditions to middle latitudes, and created an environment without modern parallel. Cold and dry climates, along with frost and wind disturbance, combined with high solar energy and loessic soils to produce a periglacial "loess-steppe" that covered much of the East European Plain. Forest refugia survived in some upland areas. A diverse large mammal fauna inhabited the central and southwest regions, while steppe bison roamed the drier southern plain (Grichuk 1984; Markova 1984).

Modern humans were probably the first hominids to occupy periglacial loess-steppe, and despite the fact that climatic conditions were more severe than at any other time since the Middle Pleistocene, they seem to have thrived in this environment. Their sites are found in most unglaciated areas of Eastern Europe, although they apparently abandoned parts of the central plain during the coldest period. Many occupations are very large and yield substantial quantities of lithic and bone debris. A number of sites contain structural remains and other complex features (Klein 1973; Soffer 1985).

While there is some evidence of morphological adaptation to Last Glacial Maximum environments among European late Upper Paleolithic populations, it is clear that technology played a central role in defining their niche. This is especially apparent in Eastern Europe, where people employed a variety of ingenious devices and techniques, many of which were specifically designed for the periglacial loess-steppe. In treeless areas, they constructed shelters of mammoth bone and heated them with interior bone-fuel hearths. During warmer months, the bone fuel was evidently kept fresh in permafrost-refrigerated pits. Large numbers of fur-bearing mammals were trapped for clothing production. Hunting equipment may have included throwing sticks and bows and arrows (Klein 1999, 535-544).

As during the late OIS 3 period, there is continuing evidence for long-distance movement of materials that may indicate a high degree of mobility. Despite their advanced technology and a seeming abundance of game, late Upper Paleolithic people in Eastern Europe probably foraged across wide areas to obtain an almost exclusively meat diet. At the same time, the size and complexity of many occupations suggests periodic aggregation of relatively large groups. A wealth of spectacular figurative and decorative art, along with numerous ornaments, further testifies to a rich social life. All these observations underscore the important role of organization in coping with periglacial loess-steppe.

The East European archaeological record for the Last Glacial Maximum offers a much more vivid contrast with the adaptations of the Neanderthals than the preceding period, and it is not entirely clear why this is the case. Perhaps—despite the extreme cold—the East European Plain supported a higher biomass of large mammal prey during OIS 2 due to greater aridity and reduced arboreal vegetation (Klein 1969b, 163). Or perhaps—as during the Holocene—novel technological advances were built upon earlier innovations and reached a new cumulative threshold. It is also possible that modern humans had continued to evolve in terms of their capacity for manipulating and communicating symbols, with implications for other aspects of their behavior (e.g., Lindly and Clark 1990).

Modern Humans during the Last Glacial Maximum

The late Upper Paleolithic people of Eastern Europe were a small part of the rich tapestry of human cultures distributed across Africa, Eurasia, and Australia during the Last Glacial Maximum (25,000-13,000 years BP). Perhaps for the first time since the Pliocene, only one hominid species walked the earth. However, the geographic range of hominids had actually expanded with the appearance and dispersal of behaviorally modern humans during OIS 3. By the end of this period, human populations had colonized Australia and the southern portion of Siberia. During the Last Glacial Maximum, the lack of morphological diversity was counterbalanced by a substantial degree of variability in diet and technology.

The Last Glacial Maximum record for temperate and tropical latitudes across Africa, southern Asia, and Australia reveals a particularly wide range of adaptations to various environments (Gamble and Soffer 1990). For example, people in the Nile Valley harvested large quantities of freshwater fish, and processed plant tubers on grinding stones (Close and Wendorf 1990). Along the South African coast, Late Stone Age people hunted a wide range of large mammals with bow and arrow, and may have extracted plant foods with weighted digging sticks (Klein 1994). In southeast Australia, people in the Willandra Lakes region gathered shellfish in bulk quantities, and probably harvested fish with nets (Allen 1990).

The people who occupied Eastern Europe during the Last Glacial Maximum were part of a larger population of modern humans broadly distributed across much of northern Eurasia. In contrast to southern latitudes, environments above 40 degrees North offered a narrower range of potential food sources—especially during glacial periods. As in the case of recent hunter-gatherers in cold terrestrial environments (Kelly 1995), north Eurasian peoples of the Last Glacial Maximum relied heavily on a diet of meat obtained from hunting large mammals. However, they exploited a wider range of small vertebrates in most areas, and often seemed to have employed new technologies to obtain them (Stiner et al. 1999).

Despite the reduced diversity of dietary adaptations across northern Eurasia relative to the temperate and tropical zones, the late Upper Paleolithic record of Europe and Siberia exhibits considerable variability. In Western Europe, where climates were predictably milder, settlement was continuous throughout the coldest phases, and the late Upper Paleolithic industries (i.e., Solutrean, Magdalenian) bear little resemblance to those of Eastern Europe (Laville et al. 1980; Gamble 1986). It is interesting to note that eyed needles—reflecting the production of tailored clothing—appear for the first time in Western Europe at the beginning of the peak cold period (ca. 21,000 years BP), although they were present in Eastern Europe and Siberia during late OIS 3 (see chapter 5).

Late Upper Paleolithic industries in parts of Central Europe display a closer relationship to the East European cultures (Svoboda et al. 1996). This probably reflects both the proximity of Central Europe and the presence of similar loess-steppe environments in the Carpathian Basin. The Eastern Gravettian industry, which is represented in sites dating to ca. 29,000–22,000 years BP, is found in both Central and Eastern Europe. However, climates were milder and arboreal vegetation was more common in the Carpathian Basin, and—in contrast to the central East European Plain—settlement appears to be continuous throughout the Last Glacial Maximum (Svoboda et al. 1996, 133–138).

In Siberia, climatic conditions were comparable to the East European Plain, and a similar reduction in settlement probably occurred during the cold peak (ca. 20,000–18,000 years BP) (Grichuk 1984; Goebel 2000, 218). However, trees were present in most areas inhabited by human populations during the Last Glacial Maximum, and the large mammal community seems to have comprised higher percentages of reindeer, red deer, and other taxa that were less common on the East European Plain. Late

Upper Paleolithic industries of Siberia are broadly similar to those of Eastern Europe, but exhibit differences in technology and art that presumably reflect the combined effects of isolation and local environments.

The skeletal morphology of modern humans in northern Eurasia during the Last Glacial Maximum exhibits some differences with that of the late OIS 3 population. The former are less robust and sexually dimorphic than their predecessors, and their teeth are significantly smaller (Frayer 1978; Wolpoff 1999, 766–777). Curiously, the high brachial and crural indices exhibited by modern human skeletons during OIS 3 (and interpreted as indicative of their southern origins) are observed in the late Upper Paleolithic sample. However, a recent study determined that overall limb length is reduced in the latter, and apparently indicates a cold-adaptation trend (Holliday 1999).

Last Glacial Maximum Sites in Eastern Europe

During the Last Glacial Maximum (25,000–13,000 years BP), sites are found in all major regions of Eastern Europe except for the northern plain. The northwest plain was covered by glacial ice as far south as the city of Minsk (54 degrees North), while northeastern areas—although largely unglaciated—were inundated by a system of massive proglacial lakes (Grosswald 1980, 15–17). Sites are scarce in the eastern plain and Volga Basin (e.g., Beregovaya 1960, 66–68), perhaps due to limited survey and creation of large reservoirs along much of the Volga River. During the peak cold phase (ca. 20,000–18,000 years BP), portions of the central and possibly the southwest East European Plain seem to have been abandoned once again (Dennell 1983, 101–102; Soffer 1985, 173–176). After 16,000 years BP, site density increased on the East European Plain, but occupations did not appear in northern areas until the Late Glacial (13,000–10,000 years BP) (fig. 6.1).

Sites of the Last Glacial Maximum are associated with the thick loess bed deposited during 25,000–13,000 years BP that overlies the terminal OIS 3 soil (Klein 1973, 18–32; Hoffecker 1987, 269). The loess averages 7–8 meters in depth in the central East European Plain, and reflects the periglacial conditions that prevailed at this time (Velichko 1973, 42–43). Two stratigraphic markers that are sometimes observed in the loess are a weakly developed soil dating to ca. 22,000 years BP, and a thin gley horizon dating to ca. 17,000 years BP. Most late Upper Paleolithic occupations are not found in the primary loess, but are buried in colluvial slopewash from the loess on low terraces in stream valleys. High rates of colluviation seem to have been characteristic of periglacial loess-steppe environments.

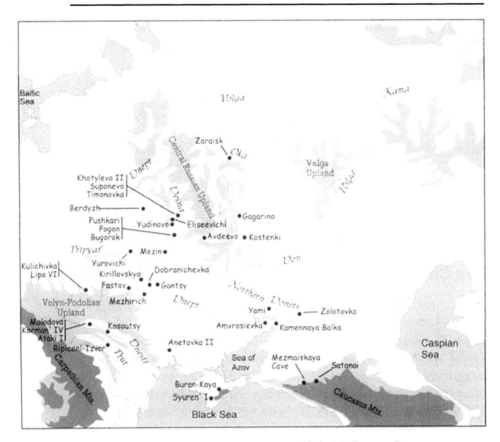

Figure 6.1. Map of late Upper Paleolithic sites (OIS 2) in Eastern Europe.

Late Upper Paleolithic sites on the East European Plain are often dated with reference to their stratigraphic context and position relative to marker horizons such as the terminal OIS 3 soil or gley horizon (e.g., Khotylevo II on the Middle Desna River [Velichko et al. 1981, "Khotylevskie Stoyanki," 60–69]). Radiocarbon dates have been obtained for an increasing number of Last Glacial Maximum sites, all of which lie within the effective range of this method (Svezhentsev 1993; Sinitsyn et al. 1997). However, the scarcity of wood charcoal in areas that were largely treeless during this period (e.g., Dnepr-Desna Basin) has forced investigators to rely heavily on bone, which is more susceptible to contamination by younger carbon. As a consequence, many dates are probably too young (Hoffecker 1988, 245–248). A number of sites have been assigned to the later Upper Paleolithic on the basis of the artifact typology, although this approach is often based on faulty assumptions.

No Neanderthals are known to have survived the final phase of OIS 3, and modern humans are assumed to have produced all the Upper Paleolithic industries of the Last Glacial Maximum. However, as during earlier periods, human skeletal remains are rare in the East European sites. Partial skeletons have been recovered from two localities on the Middle Don River (Kostenki II and XVIII) and one site on the Middle Dnestr (Kosoutsy), but most remains are confined to isolated bones and teeth from widely scattered sites (Klein et al. 1971; Klein 1973, 110, table 5).

East European Plain: Southwest Region

In the Middle Dnestr Valley, many decades of survey and excavation have yielded a particularly rich record of Last Glacial Maximum settlement (Chernysh 1973). As for the preceding periods, the sites of *Molodova V* and *Korman' IV* contain the most complete sequence of stratified occupation levels. The latter rest in a thick bed of loess-derived colluvium and buried soils that overlie the alluvium of the second terrace (see chapters 4 and 5). Both localities contain occupations that date to the early phase of OIS 2 (ca. 24,000–22,000 years BP) preceding the cold maximum. These levels are associated with traces of soil formation that probably correspond to a brief warm oscillation (Chernysh 1977, 26–30; 1987, 35–47).

Molodova V, Layer 7, is dated to 23,700–23,000 years BP, and represents one of the largest Upper Paleolithic occupations in Eastern Europe. This layer (10–15 cm thick) produced over 51,000 artifacts from an excavated area of 680 square meters. The stone tools are primarily on blades and include burins, knives, and end-scrapers, while the nonstone inventory is composed of awls, points, handles, possible digging tools, and other implements. The occupation floor contained more than 50 former hearths, most of which were associated with concentrations of lithic and bone debris (Chernysh 1959, fig. 36). Traces of two possible artificial structures, one of which constitutes a shallow oval depression measuring 4 × 1.6 meters with a centrally placed hearth, were recorded (Chernysh 1959, 77–78). The associated faunal remains are chiefly horse and reindeer.

Korman' IV, Layer 6, is undated, but occurs in a stratigraphic position similar to that of Molodova V, Layer 7, and overlies a level dating to the end of OIS 3. This occupation yielded an assemblage of only 625 artifacts and isolated faunal debris from an excavated area of 210 square meters. The overlying level (Layer 5) is radiocarbon-dated to 18,560–18,000 years BP, but contains a larger assemblage of artifacts (roughly 1,000 items) and faunal remains that are similar to those in Layer 7 at Molodova V (Chernysh 1977, 26–36). This layer is associated with the surface of a buried soil, and the two radiocarbon dates—although obtained on charcoal—might be too young.

At Molodova V, a thin lens containing a small quantity of lithic and faunal debris (Layer 6a) immediately overlies Layer 7 (Chernysh 1987, 47). Above this lies a bed of archaeologically sterile colluvium more than half a meter in thickness that may be correlated with the cold peak of the Last Glacial Maximum (ca. 22,000–18,000 years BP). This gap in the sequence of occupation levels seems to indicate a settlement hiatus during this interval, but a comparable gap in the Korman' IV sequence is absent (Chernysh 1977, 11, fig. 4). However, most if not all of the remaining late Upper Paleolithic occupations in the Middle Dnestr Valley date to the final phase of the Last Glacial Maximum (ca. 18,000–13,000 years BP) that follows the cold peak, and the Late Glacial (13,000–10,000 years BP).

At Molodova V, Layers 6–4 are buried in loessic colluvium and radiocarbon-dated to 16,750–17,000 years BP. These occupation layers were excavated over an area of 875–917 square meters and yielded assemblages of between roughly 3,500 and 7,000 artifacts (Chernysh 1987, 48–60). The stone tool inventory in each assemblage is dominated by burins, while nonstone implements include awls and points. The occupation floors contained numerous former hearths and associated debris concentrations, and traces of artificial shelters may be represented on all levels. Reindeer is the most abundant taxon among the faunal remains. The uppermost Layers, 3–1, are dated to 13,370–10,590 years BP (Late Glacial) and contain artifacts, features, and faunal remains similar to those in Layers 6–4 (Chernysh 1987, 61–82). Layer 3 yielded clear traces of a former structure in the form of a postmold arrangement measuring approximately 3 × 5 meters and enclosing a shallow depression with central hearth (Chernysh 1959, 99, fig. 48).

The site of *Ataki I*, which is located on the second terrace of the Dnestr Valley roughly 8 kilometers upstream from Molodova V, contains four late Upper Paleolithic layers. They are buried in loessic colluvium that overlies the terminal OIS 3 soil and dates to the Last Glacial Maximum (Ivanova 1968). The occupation levels apparently postdate the cold peak, and Layers 3 and 2 yielded radiocarbon dates of 16,600 and 15,375 years BP, respectively (Sinitsyn et al. 1997, 57). The relatively small excavated area (75 square meters) revealed hearths, debris concentrations, and a lithic inventory similar to contemporaneous levels at Molodova V (Chernysh 1968). The faunal remains include reindeer, horse, and bison (fig. 6.2).

The site of *Kosoutsy* is located many kilometers downstream from Molodova V (near the city of Soroki), and contains a lengthy sequence of occupation layers that also date to final phase of the Last Glacial Maximum. These layers are buried in loessic loam on a terrace 10–15 meters above the modern Dnestr River, and have produced more than 20 radio-

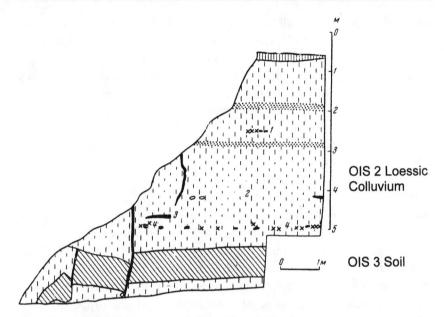

Figure 6.2. Stratigraphic profile of Ataki I in the Middle Dnestr Valley, where occupation layers are buried in loess-derived colluvium deposited during the Last Glacial Maximum (after Chernysh 1968, 104, fig. 2).

carbon dates on wood charcoal that range from 19,620 to 15,520 years BP (Sinitsyn et al. 1997, 59). The lithic assemblages contain burins and end-scrapers, and nonstone artifacts comprise an especially rich inventory of eyed needles, awls, points (including possible arrowheads), possibly digging tools, and cordage (Borziyak and Kovalenko 1989; Soffer et al. 2000). Reindeer dominates the faunal remains, but numerous hare remains were also found in the uppermost five layers (Borziyak 1993, 70–71).

Along the Prut River in northeast Romania, the site of *Ripiceni-Izvor* (see chapters 4 and 5) contains four occupation layers reportedly buried in primary loess overlying a low bedrock terrace (Păunescu 1965). As at Korman' IV, no significant hiatus among the Last Glacial Maximum levels is observed. The stone tools include burins, end-scrapers, and points, and the associated faunal remains comprise horse, red deer, and bison. Also along the Prut River (in Moldova), Chuntu rockshelter contains a small artifact assemblage recently dated to 22,100–18,510 years BP. The associated mammal remains are chiefly reindeer and horse (Borziac et al. 1997).

On the northern margin of the Volyn-Podolian Upland, a late Upper Paleolithic layer is present at *Kulichivka*, which is located on a low bedrock promontory on a tributary of the Styr' River (see chapter 5). Although no radiocarbon dates are available, the artifacts from Layer 1 are similar to Molodova V, Layer 7, and other occupations dating to the early Last Glacial Maximum, and are believed to antedate 21,000–18,000 years BP (Savich 1975, 50). A large lithic assemblage (29,000 artifacts) containing end-scrapers and burins was recovered from an excavated area of 1,626 square meters. Features included a probable former structure, represented by a shallow depression measuring 7.0 × 7.35 meters with a central hearth (Savich 1977, 122, fig. 1). Among the associated faunal remains, reindeer was most abundant, but musk ox was also present, indicating especially cold climates (Savich 1975, 50).

In general, the late Upper Paleolithic sites of the southwest plain are found in the same or analogous settings as those of the preceding periods. They are located on low terraces adjacent to side-valley ravines in major stream valleys. Presumably, a similar set of variables rendered these sites attractive to people during the Last Glacial Maximum (e.g., well-drained surface, proximity to running water, access to raw materials). Most excavated localities represent large multiple-activity occupations with traces of former structures. Limited-activity sites remain scarce, as in the case of the earlier periods.

East European Plain: Central Region

Upper Paleolithic sites in the central plain differ from those of the Middle Dnestr Valley and other parts of the southwest plain. These sites rarely contain more than one or two occupation levels, and are often found on or immediately above the Last Glacial Maximum floodplain in major valleys. Many of them contain enormous quantities of debris, especially in the form of mammoth bones, and the remains of former structures and deep pits are common. Unlike in the sites in the southwest plain, wood charcoal is rare and bone was used heavily for fuel—a phenomenon that has complicated the process of developing a radiocarbon chronology for the region (Velichko 1961a; Pidoplichko 1969; Klein 1973; Soffer 1985).

Most sites can be subdivided into those that were occupied prior to the cold peak (ca. 24,000–20,000 years BP) and those that were occupied after this event (ca. 17,000–10,000 years BP) (Soffer 1985, 173). Although a number of radiocarbon dates fall into the 19,000–16,000 years BP range, the sample of dates from many sites is now sufficiently large that a bimodal distribution around this interval is apparent (Sinitsyn et al. 1997, 29–38). This suggests that settlement of the region was either terminated or significantly reduced during the period of maximum cold (fig. 6.3).

Many of the sites that antedate ca. 19,000 years BP contain artifact assemblages of the *Eastern Gravettian* industry, with characteristic shouldered stone points, along with realistic art—such as Venus figurines—and com-

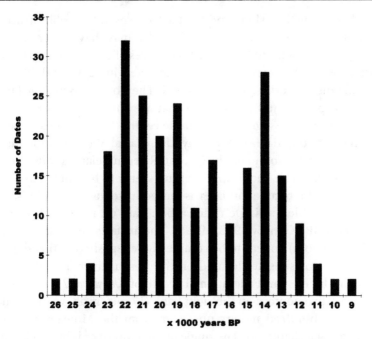

Figure 6.3. Radiocarbon dates from late Upper Paleolithic sites in the central East European Plain (based on data from Sinitsyn et al. 1997, 47–66, table 1).

plex arrangements of large pits. Although isolated examples of mammoth-bone structures may be present (e.g., Berdyzh [Bud'ko 1964]), they are rare at best and may be absent altogether. Sites that postdate the maximum cold (i.e., after ca. 17,000–16,000 years BP) are sometimes subsumed under the term *Epigravettian* (or "Evolved Gravettian"). The Epigravettian sites often contain large quantities of microlithic implements, stylized art, and the remains of mammoth-bone structures with associated storage pits (Rogachev and Anikovich 1984, 197–226; Anikovich 1993, 16).

The site of *Berdyzh* is located on the second terrace of the Sozh River, which is a tributary of the Dnepr River, in Belarus. The main occupation horizon apparently rested on a promontory on the lower slope of an ancient ravine, and was buried in colluvial sands and silts (Bud'ko et al. 1970, 1971). Some artifacts from an earlier occupation seem to have been redeposited over the main level. Mammoth teeth from the main occupation yielded dates of 23,430 and 22,500 years BP (as well as one younger date), and the overlying sediment exhibits signs of frost disturbance that must have occurred during the maximum cold peak.

Over 350 square meters of the main occupation horizon were excavated, revealing roughly 1,800 mammoth bones and tusks and a few bones of other large mammals (Kalechits 1984, 91–92, table 7). Although evidently disturbed by slope and frost action, several possible structures of mammoth bone with interior hearths may have been present (Polikarpovich 1968; Bud'ko et al. 1971). Traces of several pits, up to more than 2 meters in diameter and 40–60 centimeters in depth, and filled with bone and other materials, were also found. The stone tools include burins, shouldered points, and truncations that are similar to those in other Eastern Gravettian assemblages from the central East European Plain.

Another site that dates to the early Last Glacial Maximum (or earlier) is *Yurovichi*, which is found on the Pripyat' River in Belarus (Polikarpovich 1968, 186–190). The geomorphic setting seems to be similar to that of Berdyzh, and the occupation layer is buried in colluvium on the second terrace and yielded a single radiocarbon date on mammoth tooth of 26,470 years BP (Ksenzov 1988, 19–28). The artifacts and faunal remains (chiefly mammoth) appear to be completely redeposited at this site.

Several major sites are located on the Desna River near the city of Bryansk in the Russian Federation. The site of *Khotylevo II* is found on a low bedrock promontory on the right bank (17 meters above the modern river), a few hundred meters upstream from the Mousterian site of Khotylevo I (see chapter 4). The single occupation layer is buried in sandy slope deposits that immediately overlie the terminal OIS 3 soil, and has produced nine radiocarbon dates on bone and teeth ranging between 23,660 and 21,170 years BP (Velichko et al. 1981, "Khotylevskie Stoyanki," 60–69; Sinitsyn et al. 1997, 55). Roughly 500 square meters have been excavated, and the total estimated occupation area is 60,000 square meters. The faunal remains are overwhelmingly dominated by mammoth, although modest numbers of arctic fox remains are present. The occupation floor contains artifact and bone concentrations, and shallow depressions and pits filled with bone, but no reported traces of former mammoth-bone structures (Soffer 1985, 63). The artifact assemblage of over 20,000 items includes burins, end-scrapers, and points, along with a variety of nonstone implements and art—similar to many other early Last Glacial Maximum assemblages in the region (Velichko et al. 1981, "Khotylevskie Stoyanki," 65).

A few kilometers downstream from Bryansk lie the sites of *Timonovka I* and *II* and *Suponevo*, both of which are located on the high terrace of the right bank of the Desna River. The Timonovka localities are situated 38 meters above the modern river, while Suponevo is found on a lower slope only 20–25 meters above the river. These sites are buried in at least partially redeposited loess that postdates the Last Glacial Maximum cold peak, and the occupation layers overlie the gley horizon formed at ca. 17,000 years BP (Velichko 1961a, 154–172) (fig. 6.4). A total of eight radiocarbon dates have been obtained on teeth and bone from the three sites,

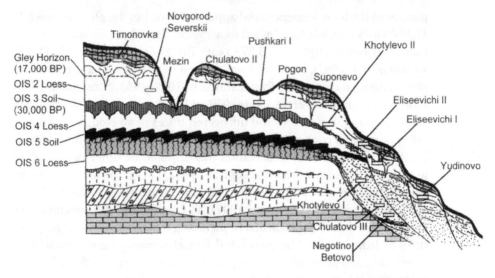

Figure 6.4. Generalized stratigraphic context for the Middle Desna River (modified from Velichko et al. 1977, fig. 2).

ranging from 15,300 to 12,200 years BP (Sinitsyn et al. 1997, 54). Some of the occupation debris at Timonovka I and II had been redeposited in deep ice-wedge casts (Velichko et al. 1977, 56).

Timonovka I and Suponevo were discovered in 1925–1927 and investigated by the noted Russian archaeologist V. A. Gorodtsov during 1928–1933. At the time, Soviet Paleolithic specialists were only beginning to recognize former dwelling structures in open-air sites on the East European Plain. Gorodtsov (1935) reported the discovery of up to six semi-subterranean dwellings at Timonovka I. The reported "earth-houses" measured 11.5 × 3 meters in area and extended to depths of 2.8–3.0 meters below the occupation layer, and contained artifacts and faunal remains. Velichko (1961a, 161–166) subsequently determined that these features were ice-wedge casts, and it is presently unclear whether any former structures were present at this site, although a large quantity of mammoth remains was recovered (Soffer 1985, 102). However, at least one former structure of mammoth bone was encountered at Suponevo, and both Timonovka II and Suponevo contain bone-filled pits, hearths, and debris concentrations (Velichko et al. 1977, 87–93; Sergin 1979).

The *Pushkari* sites are concentrated on a high promontory on the right bank of the Desna River more than 100 kilometers downstream from Bryansk. The promontory is composed of bedrock capped with Middle Pleistocene till and mantled with Late Pleistocene loess (Velichko 1961a, 134–144). The occupation layer at Pushkari I is buried in the upper portion of the loess sequence, and apparently overlies the gley horizon (ca. 17,000 years BP), but has yielded four radiocarbon dates on bone ash and tooth between 21,100 and 16,775 years BP (Sinitsyn et al. 1997, 54–55; Velichko et al. 1997). Pushkari I was the first of ten sites discovered on the promontory, and was initially excavated in 1932–1933. Subsequent excavations were undertaken in 1937–1939 by P. I. Boriskovskii, who exposed roughly 400 square meters and uncovered traces of an elongate structure (12 × 4 meters) containing three centrally placed hearths (Boriskovskii 1953, 181–202). This interpretation was disputed by other investigators (e.g., Voevodskii 1952, 73; Shovkoplyas 1965a, 271–272), but more recent excavations indicate that a smaller structure (roughly 3.5 × 3.5 meters) was probably present (Belyaeva 1997). The immense lithic inventory includes more than 160,000 artifacts, including 2,500 cores and large quantities of lithic waste. The associated faunal remains were primarily of mammoth, and generally not well preserved (Boriskovskii 1953, 226–227; Sablin 1997).

Other sites on the Pushkari promontory include *Pogon* and *Bugorok*. The former contains an occupation horizon that lies near the base of the Last Glacial Maximum loess above the terminal OIS 3 soil, and a single radiocarbon date on bone of 18,690 years BP may be too young (Velichko 1961a, 150–151; Velichko et al. 1997, fig. 2). The Bugorok occupation, which overlaps with Pogon, lies in the upper loess above the gley horizon, and has been disturbed by ice-wedge formation (Velichko 1961a, 151–153). Only small areas of these occupations have been excavated (20–60 square meters), yielding small artifact assemblages and poorly preserved faunal remains (Gvozdover 1947; Voevodskii 1950).

Several sites on the Sudost' River, which is a major tributary of the Desna, present a somewhat different picture. The two occupation layers at *Eliseevichi*, which is located on the second terrace 11–12 meters above the modern river, are buried near the top of a sequence of floodplain deposits and colluvium. At the time of occupation, the site lay close to the active floodplain and was periodically inundated by overbank floods (Velichko 1961a, 177–180) (fig. 6.5). Some material has been redeposited in an ice-wedge cast. Eleven radiocarbon dates on teeth and burned bone range between 17,340 and 12,970 years BP, although there is also an older date on wood charcoal (possibly intrusive) of 20,570 years BP (Sinitsyn et al. 1997, 54). Excavations conducted over a period of many years uncovered at least one probable mammoth-bone structure and several deep pits filled with bone (Sergin 1975). Very large quantities of artifacts and faunal remains were recovered, including many thousands of mammoth and arctic fox bones (Vereshchagin and Kuz'mina 1977, 79–81).

PEOPLE OF THE LOESS STEPPE 205

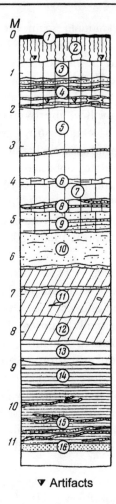

▼ Artifacts

Figure 6.5. Stratigraphic profile for Eliseevichi I on the Sudost' River (modified from Velichko 1961a, 180, fig. 57).

Yudinovo is found downstream from Eliseevichi on the first terrace 7–10 meters above the modern river level. This site was occupied at the time that the first terrace represented the high floodplain, and the artifacts and features are buried in slope deposits that overlie river alluvium (Velichko 1961a, 175–177). A total of 16 radiocarbon dates on teeth, bone, and burned bone range between 18,630 and 12,300 years BP, although most fall between 14,870 and 13,300 years BP (Sinitsyn et al. 1997, 53–54). Traces of four mammoth-bone structures, along with hearths and bone-filled pits, are present (Polikarpovich 1968, 53–71; Sergin 1974, 1977; Abramova 1993, 86–94). The artifacts include many nonstone items and art objects, and the associated faunal remains are primarily mammoth and arctic fox (Vershchagin and Kuz'mina 1977, 81–83).

The site of *Mezin* is located on the Desna River approximately 25 kilometers downstream from the city of Novgorod-Severskii in the Ukraine. The occupation area lies on the lower slope of a large tributary ravine near its confluence with the main valley, approximately 10 meters above the modern river level. The artifacts and features are buried in loess-derived colluvium, and their stratigraphic and chronologic position is not well defined. Three radiocarbon dates obtained on mammoth teeth yielded estimates of 27,500, 21,600, and 15,100 years BP, while two dates on shell yielded ages of 29,700 and 29,100 years BP (Sinitsyn et al. 1997, 56). The artifacts are generally similar to those from sites that postdate the early Last Glacial Maximum (Shovkoplyas 1965a, 293; Rogachev and Anikovich 1984, 199–200).

Mezin was discovered in 1908 by F. K. Volkov, and was excavated over the course of many decades by Ukrainian and Russian archaeologists such as Efimenko, Voevodskii, and Shovkoplyas. Over 1,200 square meters of the occupation area was eventually exposed. The features appear to include at least two former structures of mammoth bone, several bone-filled pits, hearths, and debris concentrations (Shovkoplyas 1965a, 35–95; Pidoplichko 1969, 86–110). More than 113,000 stone artifacts were recovered, with burins predominating heavily among the tools, and the nonstone inventory included a large and diverse inventory of tools and art. The associated faunal remains are primarily mammoth and arctic fox, but large numbers of wolf, horse, reindeer, and musk-ox are also present (Shovkoplyas 1965a, 97).

Near the confluence of the Desna and Dnepr Rivers, within the city of Kiev, is the site of *Kirillovskaya*, which was the first open-air Paleolithic site in Eastern Europe subject to broad-scale excavation. Over 7,800 square meters were exposed during 1893–1902 by V. V. Khvoiko. Although excavation methods were primitive and the spatial distribution of finds was not recorded (Miller 1956, 35), subsequent analysis of notes and collections suggested that three former structures of mammoth bone might have been present (Pidoplichko 1969, 37–38). Most of the identifiable large mammal remains apparently belonged to mammoth.

Along various tributaries of the Dnepr River below Kiev are three major late Upper Paleolithic sites that exhibit similarities to Mezin and Kirillovskaya and probably postdate 18,000 years BP. *Mezhirich* is located at the confluence of the Rosava and Ros' Rivers, approximately 12 kilometers upstream from where the latter empties into the Dnepr. The occupation layer is buried in loessic colluvium on the second terrace level, and yielded a mean radiocarbon age of 15,660 years BP from 14 dates on teeth and burned bone ranging between 19,280 and 11,700 years BP (Korniets et al. 1981, 107–109; Sinitsyn et al. 1997, 55–56). Over 500 square meters have been exposed of an occupation area that may cover roughly 10,000 square meters. Four former structures of mammoth bone, associated with hearths, pits, and debris concentrations, have been uncovered (Pidoplichko 1976; Soffer et al. 1997). The large assemblage of faunal remains is dominated by mammoth and hare, and the artifacts include a rich inventory of nonlithic items (including cordage) and art, in addition to a large collection of burins, and other stone tools (Pidoplichko 1976, 139–214; Korniets et al. 1981; Soffer et al. 2000, 514).

Dobranichevka is found along the Supoi River, adjacent to a large tributary stream, on the second terrace, approximately 16 meters above the modern floodplain. The occupation layer is apparently buried in loess-derived colluvium, and a single radiocarbon date on mammoth tooth of 19,200 years BP has been reported (Soffer 1986). Over 2,000 square meters were excavated, revealing traces of four structures of mammoth bone with associated hearths, pits, and debris concentrations, similar to the Mezhirich occupation (Shovkoplyas 1972; Shovkoplyas et al. 1981). The stone tool assemblage is also similar to Mezhirich, but the inventory of nonlithic artifacts is relatively poor.

Gontsy is located on the Udai River near the city of Lubny. It was one of the first Paleolithic sites discovered in Eastern Europe (1871–1873), and has been investigated intermittently up to the present day. The occupation horizon rests primarily in colluvium on a low terrace 10 meters above the modern river level, and has produced seven radiocarbon dates that cluster between 14,600 and 13,200 years BP (Pidoplichko 1969, 48–51; Sinitsyn et al. 1997, 56). Gontsy seems to have contained several former structures of mammoth bone with associated pits and hearths, along with an artifact inventory similar to that of Mezhirich (Boriskovskii and Praslov 1964, 34–35; Pidoplichko 1969, 53–61; Soffer 1985, 57–61).

On the southwest margin of the Central Russian Upland is the major site of *Avdeevo*, which is located on a small tributary of the Seim River west of the city of Kursk. The occupation area is situated on the first terrace only a few meters above the modern river, and was buried at least in part by floodplain alluvium (Velichko 1961a, 190; Velichko et al. 1981, "Avdeevo," 49) (fig. 6.6). A total of 21 radiocarbon dates on bone, teeth, and burned bone yield a mean age estimate of 19,575 years BP, but most dates fall between 23,400 and 20,100 years BP, and the site is thought to date to the early Last Glacial Maximum. Broad-scale excavations by A. N. Rogachev and others during the 1940s and 1970s exposed more than 1,300 square meters, revealing a complex pattern of features similar to Layer 1 at Kostenki I (see later discussion). The largest group of features

Figure 6.6. Stratigraphic profile of Avdeevo on the Seim River in the Central Russian Upland, illustrating the effects of frost disturbance and rodent burrowing (redrawn from Velichko et al. 1981, "Avdeevo," 50, fig. 12a).

consists of a U-shaped arrangement of 15 debris-filled pits of varying size that Rogachev interpreted as the remains of an elongate structure (or "longhouse") like that of Kostenki I (Rogachev 1953, 146–187). A smaller feature complex, comprising a oval arrangement of pits (approximately 12–14 meters in diameter) surrounding several former hearths, was subsequently uncovered nearby, and has also been interpreted as a former structure (Gvozdover and Grigor'ev 1978, 54; Velichko et al. 1981, "Avdeevo," 53–54). The artifact inventory is also similar to that of Kostenki I and includes characteristic truncations and shouldered points of stone, and numerous nonlithic implements and figurative art.

On the eastern edge of the Central Russian Upland, along the Don River, the Kostenki sites contain a number of occupation layers that date to the Last Glacial Maximum and Late Glacial (see chapter 5). The most famous of these is the uppermost level (Layer 1) at Kostenki I, which was discovered in 1879 and was excavated by many noted East European archaeologists during the twentieth century (Klein 1969b, 25–31; Praslov 1982). Kostenki I is situated near the mouth of a large side-valley ravine on the second terrace level (20 meters). Layer 1 lies above the terminal OIS 3 soil in loessic colluvium, apparently deposited by slopewash under periglacial conditions. More than 40 radiocarbon dates have now been obtained on teeth, burned bone, and wood charcoal (rare), and most of the dates fall between 24,570 and 21,150 years BP, indicating an early Last Glacial Maximum age (Lazukov 1982; Sinitsyn et al. 1997, 47–48).

During 1931–1936, P. P. Efimenko conducted broad-scale horizontal excavations at Kostenki I, exposing more than 700 square meters and a massive complex of features. Unlike Khvoiko during the early excavations at Gontsy, Efimenko carefully recorded the spatial distribution of artifacts and features, and interpretation of the results had far-reaching effects on Paleolithic archaeology in the Soviet Union. The feature complex comprised an extended oval arrangement of 15 large debris-filled pits (36 × 15 meters) surrounding a line of central hearths, and was reconstructed as a single structure or "longhouse" (Efimenko 1938, 1958). This interpretation was closely tied to the Marxist social evolutionary scheme developed for the Paleolithic in the early 1930s by Efimenko, Boriskovskii, and others; the Kostenki longhouse appeared to be representative of the matrilineal clan societies ascribed to the late Upper Paleolithic (Soffer 1985, 10–11). Many archaeologists have questioned the original reconstruction of Efimenko's feature complex, suggesting that one or more smaller dwellings are represented (e.g., Grigor'ev 1967; Klein 1969b, 120–121).

During 1951–1957, A. N. Rogachev uncovered a second feature complex, which was more fully excavated during the 1970s (Rogachev et al. 1982, "Kostenki 1," 45–49). This complex is similar to the first, and consists of an oval arrangement of large debris-filled pits that measures roughly 25 × 18 meters, and contains three central hearths. The discovery of the new feature complexes at Kostenki I and Avdeevo (as well as an apparently similar complex at the recently investigated site of Zaraisk [described later]) indicates that the controversial "longhouse" constitutes a characteristic occupation floor pattern of the Eastern Gravettian. In late prehistoric sites of the North American arctic, analogous floor patterns are found that appear to represent open-air arrangements of hearths and other features created by seasonal aggregations of otherwise dispersed groups (e.g., Schledermann 1978; McGhee 1997).

Kostenki I, Layer 1, yielded an enormous quantity of artifacts that are considered diagnostic of the Eastern Gravettian industry, and more specifically of the "Kostenki-Avdeevo Culture" (Grigor'ev 1993). The stone tools are largely composed of burins, shouldered points, truncations, and backed bladelets. Nonlithic implements include points, awls, needles, mattocks, and others. Among the numerous ornaments, decorated pieces, and figurative art are several Venus figurines (Efimenko 1958, 209–409; Rogachev et al. 1982, "Kostenki 1," 50–61). The associated faunal remains are primarily mammoth, wolf, and hare (Vereshchagin and Kuz'mina 1977, 100).

Several Kostenki sites on the first terrace level (10–15 meters) also appear to date to the early Last Glacial Maximum. At the time of occupation, these sites were close to the river floodplain. Kostenki XXI, Layer 3, is associated with a buried soil layer that dates to ca. 22,000–21,000 years BP and represents a brief warm oscillation prior to the cold maximum

(Praslov and Ivanova 1982, 201–209). The stone tools and nonlithic artifacts are not similar to Kostenki I, Layer 1, but the faunal remains reflect the dominance of mammoth and hare (Vereshchagin and Kuz'mina 1977, 110). Kostenki IV contains two occupation levels that are not separated by intervening sterile sediment and have recently yielded dates on bone of 23,000 and 22,800 years BP (Sinitsyn et al. 1997, 48). Both levels contain artifact assemblages that exhibit few parallels with other sites in the region. Rogachev (1955) recorded two feature complexes on the lower occupation level comprising linear arrangements of hearths similar to the first complex at Kostenki I, Layer 1, but lacking the associated deep pits (Rogachev and Anikovich 1982c). The mammal remains are chiefly hare.

Younger occupations are found on both the second and first terraces. Kostenki XI, Layer 1a is located on the second terrace near the modern surface. Six radiocarbon dates on bone yield a mean age estimate of 16,410 years BP (Sinitsyn et al. 1997, 48–49). The occupation layer contains traces of two former structures of mammoth bone, associated with deep pits filled with bone debris, very similar to the former structures in the Dnepr Basin sites (Rogachev and Popov 1982, 120–124). The faunal remains are primarily mammoth, and the artifact assemblage includes many microlithic tools.

Kostenki II is situated on the slope above the second terrace, and the single occupation layer has produced a wide range of radiocarbon dates; it may be tentatively assigned to the later phases of the Last Glacial Maximum. This site also produced traces of a former structure of mammoth bone, but no reported deep pits. A human burial, encased in a chamber of mammoth bones, was found adjacent to the structural remains (Boriskovskii 1963, 11–64).

Occupations dating to the end of the Last Glacial Maximum and Late Glacial are also found on the first terrace, and include five sites located a few kilometers downstream of Kostenki near the village of *Borshchevo*. The site of Borshchevo I, which has yielded dates of 17,200 and 15,600 years BP, apparently contains another former structure of mammoth bone (Rogachev and Kudryashov 1982; Sinitsyn et al. 1997, 49).

Roughly 100 kilometers north of Kostenki on the east bank of the Don River is another major Upper Paleolithic site at *Gagarino*. This locality was initially investigated by S. N. Zamyatnin during 1926–1929, and subsequently excavated by L. M. Tarasov in 1961–1969. The single occupation layer is buried in colluvium on the second terrace (17–20 meters), and has a mean radiocarbon age of 21,170 years BP on eight dates obtained from mammoth teeth and tusks (Tarasov 1979, 20–28; Sinitsyn et al. 1997, 52). Zamyatnin reported traces of a possible dwelling structure at Gagarino as early as 1927, which encouraged Efimenko's ground-breaking studies of the Kostenki I occupation floor in 1931–1936 (Praslov 1982, 11). The feature at Gagarino is a circular depression approximately 4 meters in diameter that contains a central hearth and several rock slabs; two debris-filled pits were found adjacent to the depression. The artifacts are generally similar to the assemblage from Kostenki I, Layer 1, and include several Venus figurines. Most of the faunal remains are arctic fox (Tarasov 1979).

Farther north (just below latitude 55 degrees North) along a small tributary of the Oka River, lies the recently investigated site of *Zaraisk*. The principal occupation layer at this locality overlies a late OIS 3 soil, and yielded 11 radiocarbon dates between 23,000 and 18,300 years BP (Amirkhanov 1997; Sinitsyn et al. 1997, 61). Only a limited area has been excavated to date, but it is apparent that both the artifacts and features exhibit parallels to Kostenki I, Layer 1, and other Eastern Gravettian sites in the region. Mammoth predominates among the faunal remains.

In general, later Upper Paleolithic sites of the central plain are found in topographic settings similar to those of earlier sites in the region. They are typically located on low terraces, adjacent to side-valley ravines, in the major stream valleys. Most of the known sites contain large and diverse concentrations of occupation debris, as well as pits and other features that suggest multiple-activity loci and long-term habitations. Some sites exhibit an organizational structure similar to that of many recent hunter-gatherer residential base camps (e.g., Mezhirich [Soffer 1985, 392–402]). Presumably, variables similar to those identified for the southwest plain sites continued to attract people to these locations (e.g., well-drained surface, proximity to water, access to raw materials).

However, the central plain sites—especially those that resemble recent hunter-gatherer base camps—are peculiar in their lack of stratification. This observation even applies to the multilayered Kostenki sites, many of which contain only one major occupation level of Last Glacial Maximum age, regardless of the number of underlying levels dating to OIS 3 (Praslov 1982). Site locations seem to have been used intensively for an extended period of time—possibly several months (e.g., Soffer 1985, 408–414)—and then permanently abandoned.

One explanation for this unusual pattern might be exhaustion of local bone fuel sources, reflecting a phenomenon unique to periglacial loess-steppe environments. The central plain sites are also unusual in their almost exclusive reliance on bone fuel, and many investigators have concluded that a significant portion of the bones of mammoths and other large mammals at these sites were collected from natural accumulations (e.g., Gromov 1948; Klein 1973, 53; Vereshchagin and Baryshnikov 1984, 492–493). In periglacial environments, bones in such accumulations may remain fresh and flammable for many years. Natural concentrations of bone in these environments are often found at stream confluences and side-valley ravine mouths (Péwé 1975, 98)—topographic features associated with these late Upper Paleolithic sites. Assuming relatively slow rates of accumulation, several thousand years were probably required to form a large concentration at one locality. Once a concentration was exhausted by the occupants of a nearby site, the location might have lost its value for another few millennia.

East European Plain: Southern Region

Late Upper Paleolithic sites in the southernmost areas of the East European Plain exhibit a distinctive regional character. The faunal remains in these sites are dominated by steppe bison (*Bison priscus*), which is not common in the southwestern and especially the central plain. The abundance of bison in the southern sites (also typical for Mousterian sites in the area [see chapter 4]) reflects the drier climates and less diverse plant communities of the region. Unlike on the central plain, there is no evidence for a settlement hiatus during the maximum cold peak (20,000–18,000 years BP), although the sample of available dates is comparatively small. The artifact assemblages typically contain a large proportion of microlithic tools, but lack the rich and diverse nonlithic inventories of the central plain sites.

The earliest Upper Paleolithic site in this region may be *Yami*, which is located on the Northern Donets River in the Ukraine. The occupation layer directly overlies the terminal OIS 3 soil and is believed to date to the early Last Glacial Maximum (Leonova 1994, 183–188). Another relatively early site on the southern plain is *Anetovka II*, which is found on a tributary of the Lower Bug River, and has a mean radiocarbon age of 19,833 years BP on five dates (Sinitsyn et al. 1997, 58). The occupation area contained over 21,000 bison bones, although more than 400 reindeer remains—chiefly antler—were also present (Bibikova and Starkin 1989).

Amvrosievka is located along a large ravine on a small tributary of the Mius River on the southern margin of the Donets Ridge. The site comprises an occupation area and a massive bed of bison bones on the adjoining ravine slope (fig. 6.7). An estimated 500 individual bison are represented in roughly 300 square meters of the bone bed excavated to date, and a series of nine radiocarbon dates on bone yield a mean age of 18,800 years BP (Krotova and Belan 1993, 125–133; Sinitsyn et al. 1997, 57–58). Features in the occupation area are confined to debris concentrations at varying depths from the surface. The stone tools include end-scrapers, burins, and retouched microblades, and nonlithic items contain a number of laterally grooved bone points. Several simple ornaments and shells were also recovered (Boriskovskii 1953, 336–362).

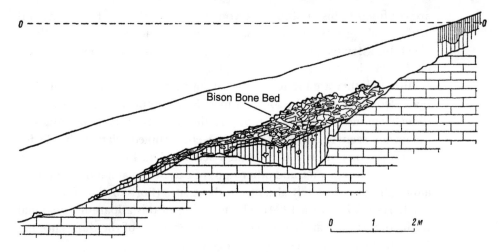

Figure 6.7. Depositional context of the bison bone bed at Amvrosievka, which is located on the southern margin of the Donets Ridge (modified from Boriskovskii and Praslov 1964, pl. 10).

The bone bed at Amvrosievka appears to represent the accumulated remains of several mass kill episodes in which herds of bison were driven into the ravine. The nearby occupation area seems to have been used to butcher and process some of the carcasses (Krotova and Belan 1993, 133–141). The presence of a large bone bed and evidence for a kill-butchery site make Amvrosievka unique in the Paleolithic of Eastern Europe. Parallels have been drawn with Holocene bison kill sites of the North American Plains, although the kill events at Amvrosievka are thought to have occurred during the spring, which is uncommon in the North American sites (Pidoplichko 1953; Frison 1978; Krotova and Belan 1993).

A somewhat different pair of sites is found at *Kamennaya Balka*, on the margin of a deep ravine that empties into the Mertvyi Donets River near the Sea of Azov. Twelve radiocarbon dates on bone and burned bone from the thick occupation layer at Kamennaya Balka II range between 15,400 and 10,000 years BP, and several microhorizons have been distinguished (Leonova 1993, 152–156; Sinitsyn et al. 1997, 57). Mapping of the latter revealed debris concentrations and former hearths. The stone tool inventory is chiefly composed of burins, end-scrapers, and retouched microblades (Gvozdover 1964, 40).

On the Lower Don River near the town of Konstantinovsk, another group of relatively small localities is found at *Zolotovka*. The sites are located on a high terrace roughly 37 meters above the modern river and along the margins of a ravine system. Zolotovka I has yielded dates of 17,400 and 13,600 years BP (Sinitsyn et al. 1997, 58). The faunal remains are confined to bison, and large concentrations of bone fragments and artifacts, along with two hearths, were mapped on the occupation floor. The lithic assemblage includes end-scrapers, burins, and retouched microblades (Praslov and Shchelinskii 1996).

Although they are located in similar topographic settings, the late Upper Paleolithic sites of the southernmost East European Plain consistently differ from most localities in more northerly parts of the plain. Many of the former seem to represent sites that were occupied relatively briefly for a limited range of tasks (i.e., limited-activity loci). Evidence of dwelling structures, storage pits, and other signs of extended habitation are rare—although not completely absent—in these sites (Boriskovskii and Praslov 1964, 10–13; Leonova 1994). The meaning of this apparent functional distinction is unclear, but it may account for much of the difference between them and the sites in other regions.

Crimea and Northern Caucasus

Late Upper Paleolithic sites in the Crimea are more common than those of the preceding period, although most of them probably date to the Late Glacial (13,000–10,000 years BP); few radiocarbon dates are currently available. One occupation layer that may date to the Last Glacial Maximum is the upper level in the rockshelter of *Syuren' I*, located near Sevastopol' in the southwest area of the peninsula (see chapter 5). Artifacts are buried in the upper part of a deep sequence of loam and limestone rubble—well above the levels dated to the end of OIS 3. The lithic tool assemblage is dominated by backed bladelets, and is reportedly similar to the Eastern Gravettian assemblages of the central plain region (Tatartsev 1996, 196–198). The associated large mammal remains are chiefly saiga (Vekilova 1957, 254).

Most other late Upper Paleolithic occupations are thought to date to the Late Glacial or terminal Pleistocene. The stone tools in these occupation layers contain many microlithic forms, including geometric segments. Among them are *Shan-Koba* (Layer 6) and the upper levels at *Buran-Kaya III*; the latter recently yielded bone dates of 11,950 and 11,900 years BP (Rogachev and Anikovich 1984, 222; Sinitsyn et al. 1997, 60).

The temporal distribution of late Upper Paleolithic sites in the northwestern Caucasus appears to be similar to that in the Crimea, although no radiocarbon dates are available and the dating is based on typology and stratigraphic position (Amirkhanov 1986). At *Mezmaiskaya Cave*, which is located in the Kurdzhips Valley southwest of Maikop (see chapters 4 and 5), Layers 1B and 1A directly overlie units dating to the end of OIS 3 and may be of Last Glacial Maximum age. The artifacts contain end-scrapers and backed bladelets, and among the nonlithic items are awls and points. The associated large mammal remains include horse and bison (Golovanova et al. 1999, 80).

The rockshelter at *Satanai*, which is found in Borisovskoe Gorge on the Gubs River, contains an artifact assemblage similar to that of Layers 1B–1A at Mezmaiskaya Cave buried in a thick deposit of rubble and loam. However, unlike the latter, the Satanai lithic inventory includes geometric segments thought to be characteristic of the final Upper Paleolithic (Amirkhanov 1986, 57–80). Almost all of the associated large mammal remains are horse. Another rockshelter in the area that probably contains a late Upper Paleolithic occupation level is *Gubs Shelter No. 1*, which is also located in Borisovskoe Gorge near Satanai (Amirkhanov 1986, 30–52) (fig. 6.8).

Skeletal Remains in Late Upper Paleolithic Sites

As during the preceding period, most modern human remains of the late Upper Paleolithic in Eastern Europe have been recovered from open-air sites on the plain. However, the distribution of remains differs from that of the early Upper Paleolithic. The majority of localities containing skeletal material have yielded isolated fragments and lack complete or nearly complete skeletons from graves. Relatively intact graves containing partial skeletons are known only from Kostenki II and XVIII (Middle Don River) and Kosoutsy (Middle Dnestr Valley), while a former grave is inferred at the Satanai rockshelter in the Northern Caucasus (see table 6.1).

This pattern of distribution probably reflects the effects of permafrost and frost action processes in Last Glacial Maximum and Late Glacial environments. The late Upper Paleolithic inhabitants of the East European Plain probably could not dig graves during winter months. During warmer months, graves could be excavated only to the base of the thaw layer (probably less than a meter below ground surface). Remains buried in shallow graves during the summer would have been exposed to frost heaving—and possible carnivore ravaging—in the autumn and winter. A similar pattern is evident in modern arctic environments (Zimmerman 1985, 24).

Significantly, the least disturbed late Upper Paleolithic grave on the central plain was found within a protective casing of mammoth bones at Kostenki II (Boriskovskii 1963, 48–64). The grave at Kostenki XVIII was only partially protected with mammoth bones, and more heavily disturbed (Rogachev and Belyaeva 1982, 188–189). The child burial at Kosoutsy appears to have been relatively intact (Borziyak and Kovalenko 1989, 210–211). The permafrost zone did not extend into the Northern Caucasus

TIME SCALE (yrs BP)	EAST EUROPEAN PLAIN			CRIMEA	NORTHERN CAUCASUS
	SOUTHWEST	CENTRAL	SOUTHERN		
10,000	Molodova V (1)			Buran-Kaya III	
12,000	Molodova V (3)	Yudinovo Gontsy	Kamennaya Balka		Satanai (?)
14,000	Ataki I (3-2) Molodova V (3-2)	Timonovka I-II Kostenki XI (1a) Mezhirich Mezin (?)	Zolotovka (?)	Shan-Koba (6)?	Gubs Shelter 1(?)
16,000		HIATUS	Amvrosievka		Mezmaiskaya (1A) ?
18,000	Korman' IV (5)				
20,000	Kosoutsy (9)	Pushkari I (?) Gagarino Avdeevo Khotylevo II	Anetovka II		
22,000					
24,000	Molodova V (7)	Kostenki I (1) Berdyzh	Yami (?)	Syuren' I (1) ?	
26,000	Korpach (4)	Yurovichi (?)			
28,000	Molodova V (8)				

during the Last Glacial Maximum (Velichko and Nechayev 1984), and the presence of a former grave is inferred at Satanai, where many ochre-stained bones were found together in association with several ornaments (Amirkhanov 1986, 80).

Because of the fragmentary character of most of the late Upper Paleolithic skeletal sample, conclusions about the morphology of the Last Glacial Maximum population are limited. In general, the sample appears similar to the West European Upper Paleolithic remains (Debets 1955, 48). The adult male skull from Kostenki II is large and dolichocephalic (i.e., long headed), and the juvenile male skull from Kostenki XVIII exhibits similar features (Gerasimova 1982, 248–253). The late Upper Paleolithic remains—although few in number—have yet to display the striking variability of the central East European Plain sample from the preceding transitional period (see chapter 5). The archaeological record reveals a corresponding lack of variability in comparison to the early Upper Paleolithic of this region.

The complete lower left limb from Kostenki II provides some basis for assessing the relative proportions of the extremities (Alekseev 1978, table 49). The pattern is consistent with that observed in the late Upper Paleolithic postcranial sample from Western Europe (Holliday 1999, 559–564). While the crural index for the Kostenki II specimen is very high (86) and comparable to modern tropical populations, the total length of the limb (femoral length + tibial length) is relatively short (813 mm) and slightly below the mean for adult males from late Upper Paleolithic sites in Western Europe (Holliday 1999, 560). This suggests that—as in other parts of Europe—late Upper Paleolithic people on the East European Plain retained the relatively long distal lower limb segments of their early Upper Paleolithic predecessors, but developed some degree of adaptation to cold climates in terms of generally shorter limbs.

Technology in the Late Upper Paleolithic

Despite evidence for some morphological adaptation to cold temperatures, it is apparent that modern humans relied to a much greater degree on technology in coping with the periglacial loess-steppe environments of Eastern Europe. The enormous wealth of the late Upper Paleolithic archaeological record, with its large array of nonlithic artifacts and complex features, provides an unparalleled glimpse of hominid technical ingenuity before the development of agriculture. For reasons that are not entirely clear (and probably include such mundane considerations as sample size

Figure 6.8. *(oppposite)* Chronology of late Upper Paleolithic sites in Eastern Europe.

Table 6.1. Skeletal Remains in Late Upper Paleolithic Sites in Eastern Europe

Locality	Date	Age/Sex	Material
East European Plain			
Korman' IV	OIS 2	(?)	humerus fragment
Kosoutsy	19,020–15,520 yrs. BP	infant	complete skeleton
Starye Duruitory	OIS 2	adult	mandible, teeth
Khotylevo II	23,660–21,170 yrs. BP	child	deciduous teeth, mandible
Chulatovo I	14,700 yrs BP	adult female	frontal, parietal
Eliseevichi	17,340–12,970 yrs. BP	infant	clavicle, ribs, pelvis, femur
Yudinovo	18,630–12,300 yrs. BP	(?)	humerus fragment
Mezin	OIS 2	adult (?)	lower right molar
Novgorod-Severskii	19,800 yrs BP	10–12 years	humerus, cranial fragments
Pushkari I	21,100–16,775 yrs. BP	juvenile	deciduous molar crown
Kostenki II	OIS 2	>50 years male	partial skeleton
Kostenki XVIII	21,020–17,900 yrs. BP	9–10 years	partial skeleton
Northern Caucasus			
Satanai	OIS 2	adult male	disarticulated partial skeleton

Sources: Debets 1955; Klein et al. 1971; Gerasimova 1982; Amirkhanov 1986; Borziyak and Kovalenko 1989.

and differential preservation) the late Upper Paleolithic offers a far more vivid contrast with the Neanderthal record than the preceding period. This contrast is enhanced in no small measure by the demands of full glacial conditions in Eastern Europe.

The archaeological record for OIS 2 illustrates—more fully than that of the transitional period—the unique character of modern human technology after 50,000 years ago. Both direct and indirect evidence from the late Upper Paleolithic reflect an ability to manipulate the natural environment in complex ways that involve interactions among highly diverse elements of that environment. The pattern is apparent in technology related to virtually all aspects of late Upper Paleolithic adaptations including hunting prey, collecting materials, storage of goods, and protection from climate.

Late Upper Paleolithic industries in Eastern Europe exhibit a fully developed Upper Paleolithic appearance. With isolated and problematic exceptions (e.g., Radomyshl' [Shovkoplyas 1965b]), sites containing "archaic" artifact assemblages are absent during the Last Glacial Maximum and Late Glacial. Lithic technology is dominated by blade production, and typical Upper Paleolithic tool forms such as burins and backed blades are supreme. Large and diverse inventories of bone, antler, and ivory implements are common, and ornamentation and art are associated with all major industries. Important insights into the technology of the period may also be found in the features—structures, pits, hearths, and others—mapped on open-air occupation areas scattered across the East European Plain.

The principal subdivision of late Upper Paleolithic industries in Eastern Europe lies between the industries of the early Last Glacial Maximum (ca. 25,000–21,000 years BP) and those of the late Last Glacial Maximum and Late Glacial combined (ca. 17,000–10,000 years BP). The two groups are separated by the period of maximum cold, when settlement of the East European Plain was significantly reduced—especially in the central region. Although there are clear stylistic differences between the industries of the early Last Glacial Maximum and those of the late period, there are no significant contrasts in terms of overall technological skills and achievement.

Early Last Glacial Maximum: Eastern Gravettian

Most East European sites that may be dated to the early Last Glacial Maximum are assigned to the Eastern Gravettian or very similar industries. Although these industries are often assigned to the "early" or a "middle" Upper Paleolithic (e.g., Gamble 1986, 183–185; Rogachev and Anikovich 1984, 197–216), they are included here with other industries of OIS 2 as part of the late Upper Paleolithic. The Eastern Gravettian is viewed as a "technocomplex" that occupies much of Central and Eastern Europe during this period, and contains many local variants. Its geographic distribution is closely tied to loess-steppe environments in these regions. Stone tools typically include large numbers of burins, retouched blades, backed bladelets, and characteristic shouldered points. Nonlithic implements comprise bone points, awls, needles, mattocks, stone tool handles, and others. Ornaments and art are common, and the latter include the famous Venus figurines. Occupation areas are often large, and many contain traces of dwelling structures and numerous pits (Kozlowski 1986; Svoboda et al. 1996). Although the origins of Eastern Gravettian technology and art remain somewhat obscure (Kozlowski 1986, 151–156), they probably lie with the Aurignacian and the earliest modern human population of Central Europe.

The earliest Eastern Gravettian sites are found in the Carpathian Basin of Central Europe, and date to the end of OIS 3. Many elements of the tool assemblages later found in Eastern Europe are present in occupation levels at Willendorf II (Austria) and Dolni Vestonice I and II (Czech Republic) dating to 30,000–25,000 years BP (Svoboda et al. 1996, 131–141). The earliest Venus figurines also appear in this region prior to the beginning of OIS 2. Two sites on the central East European Plain that probably antedate 25,000 years BP and might be tied to this early phase are Kostenki VIII, Layer 2, and Yurovichi (Kozlowski 1986, 166; Anikovich 1993). However, these sites lack Venus figurines and their relationship to the later Eastern Gravettian is more problematic.

After the beginning of the Last Glacial Maximum (ca. 25,000 years BP), shouldered points appear in the Carpathian Basin, and sites containing these especially diagnostic tools and other typical Eastern Gravettian lithic and nonlithic forms—including Venus figurines—are found on the East European Plain. There is much similarity between the central East European Plain and Carpathian Basin artifacts during 24,000–21,000 years BP, and the former are assumed to represent a population and culture derived from the loess-steppe areas of Central Europe (Kozlowski 1986; Grigor'ev 1968, 67; 1993, 63). On the other hand, the East European sites appear to develop some local technological adaptations to the loess-steppe of the central plain where wood fuel was scarce.

Early Last Glacial Maximum sites on the central East European Plain that contain Eastern Gravettian occupations include Berdyzh, Khotylevo II, and Pushkari I in the Dnper-Desna Basin. They also include Avdeevo on the western margin of the Central Russian Upland, and Kostenki I (Layer 1), Kostenki XIII, Kostenki XIV (Layer 1), Kostenki XVIII, and Gagarino on the Middle Don River. The recently investigated site of

Zaraisk may be added to this list, and documents the northernmost outpost of the technocomplex known to date (Rogachev and Anikovich 1984, 206–213; Kozlowski 1986, 165–167; Grigor'ev 1993, 52–57).

Early Last Glacial Maximum assemblages from the southwest region of the East European Plain are less typical and sometimes excluded from the Eastern Gravettian (e.g., Chernysh 1987, 47). Shouldered points are rare and Venus figurines completely absent. However, many of these assemblages are broadly similar to those from the central plain, and may be considered as a local variant (Kozlowski 1986, 165). The distinct regional character of the southwest plain sites may be tied to local environmental conditions (i.e., presence of some forest flora and fauna). Eastern Gravettian sites are largely, if not completely, absent from the southernmost plain and Northern Caucasus, where local conditions were also different from those of the central plain (Amirkhanov 1986; Leonova 1994).

Stone tool production techniques in the Eastern Gravettian reflect a heavy emphasis on the manufacture of blades. Prismatic cores were used to generate large blades, and small (sometimes "secondary") cores were used to make bladelets and microblades (Rogachev and Anikovich 1984, 206). The blades were retouched into shouldered points and other large blade tools, while the smaller blanks were transformed into backed bladelets and micropoints. The overall use of stone was highly efficient—especially the production of the microlithic implements, which maximized the ratio of edge to volume. Such efficiency in the use of stone is not found in the Mousterian, and it is found only in some early Upper Paleolithic industries (e.g., Spitsyn Culture of the Middle Don River) (table 6.2).

The most diagnostic stone tool form of the Eastern Gravettian (particularly in Eastern Europe) is the shouldered point, which is often referred to as the *Kostenki point* in both Central and Eastern Europe (Grigor'ev 1993, 52–54). They come in various shapes and sizes. A large proportion of them exhibit heavy wear around the point and along the sides in the form of polish (e.g., Efimenko 1958, 233–247). Microscopic use-wear analysis indicates that these artifacts were not used as projectile points, but as knifelike tools for cutting hide, meat, and other soft materials (Semenov 1964, 93–94) (fig. 6.9).

Another diagnostic Eastern Gravettian tool is a truncation—a blade or flake ventrally retouched or truncated at both ends. They are usually referred to as *Kostenki truncations* or knives, and their presence is strongly correlated with Kostenki points (Grigor'ev 1993, 52–54). In Kostenki I, Layer 1, they comprise 18 percent of the tool assemblage (Rogachev et al. 1982, "Kostenki 1," 52). The ends of these pieces may have been truncated in order to straighten the natural curvature of the blank and facilitate hafting (Semenov 1964, 64–66). Kostenki truncations also exhibit heavy

Table 6.2. Artifacts in Late Upper Paleolithic Sites of the East European Plain: Eastern Gravettian

Tool types	Molodova V Layer 7	Pushkari I	Avdeevo	Kostenki I Layer 1, Feature 1	Kostenki IV Layer 2	Gargarino
End-scrapers	252	82	94	250	212	104
Composite tools	32	2	27	37	0	30
Borers	8	2	44	41	30	8
Burins	549	91	753	985	158	469
Knives	217	0	0	0	0	0
Backed points	144	188	36	0	185	9
Shouldered points	24	0	73	465	0	39
Truncations	0	163	448	653	0	27
Retouched blades	0	115	524	195	0	131
Side-scrapers	10	3	15	6	0	2
Backed bladelets	156	72	239	368	333	280
Leaf-shaped points	0	0	3	16	0	0
Splintered pieces	0	0	22	1	>120	8
Total stone tools:	2183	>720	~2500	>3300	>7000	1011
Bone/ivory mattocks	4	2	10	4	0	0
Bone/ivory awls	18	0	5	x	32	22
Bone/ivory points	8	0	3	x	1	9
Bone needles	0	0	0	x	0	14
Ornaments	0	1 (fragment)	27	x	126	78
Art objects	0	0	10	x	0	14

Sources: Rogachev 1955; Klein 1969b; Tarasov 1979, 107, table 7; Kozlowski 1986, 158–159, table 3.4; Chernysh 1987, 84–85.

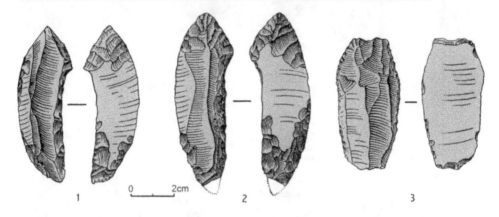

Figure 6.9. Kostenki shouldered points (1–2) and Kostenki truncation (3) from Layer 1 at Kostenki I (Eastern Gravettian) on the Middle Don River (after Efimenko 1958, 232, fig. 53; 241, fig. 56).

polish, and were apparently used for functions similar to those of the Kostenki points (see fig. 6.9).

The most abundant stone tools in the Eastern Gravettian sites are burins, which range between 25 and 30 percent of retouched items in many East European assemblages. These chisel-like tools assume various forms, the most common of which are (a) dihedral, (b) on a retouched truncation, and (c) on the corner of a snapped blade (Klein 1973, 72–74; Kozlowski 1986, 158–159). Microscopic analysis of use-wear striations confirms that burins were consistently used to work bone, antler, and ivory (Semenov 1964, 94–100). Their abundance in East European Plain sites of the Last Glacial Maximum—which exceeds that in Western and Central Europe (Grigor'ev 1993, 55)—undoubtedly reflects the lack of wood on the loess-steppe and corresponding emphasis on these materials. It is interesting to note that percentages of burins are also extremely high in assemblages of the southwest plain, where softwood was available during the early Last Glacial Maximum, but hardwood taxa were rare or absent.

The majority of remaining stone tools in most of these sites are backed bladelets and points (Kozlowski 1986). These represent very small blades that have been blunted on one edge by steep retouching. Like other microliths produced by prehistoric and recent hunter-gatherers, they may have been fitted into slotted bone points as inset blades and points. These, in turn, might reflect the use of spear-throwers or bows and arrows (Klein 1973, 80–82; 1999, 538–542). However, supporting evidence for this in the form of slotted bone points is still largely confined to sites that postdate the early Last Glacial Maximum, with the possible exception of a

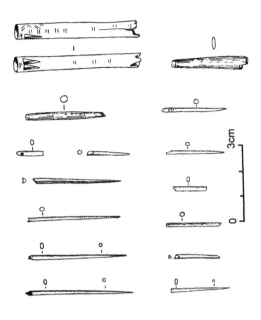
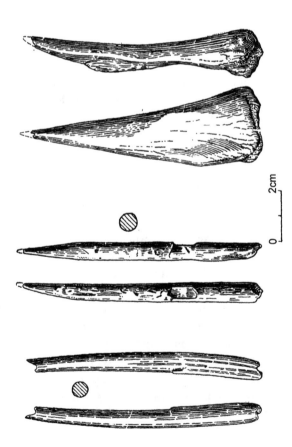

specimen recovered from Layer 5a at Korman' IV that may be tentatively dated to the latter (Chernysh 1977, 42–43).

As in the early Upper Paleolithic sites, the nonlithic artifacts and site features reveal more about the technology of the Eastern Gravettian than the stone tools. Although evidence for composite points with inset blades is lacking, bone, antler, and ivory points are common and may represent the most widely used weapon for large mammal hunting. Specimens measuring 50–75 centimeters in length have been recovered from Avdeevo and Kostenki I (Grigor'ev 1993, 55). Other examples are reported from Gagarino, Khotylevo II, and Molodova V, Layer 7 (Tarasov 1979, 109–110; Velichko et al. 1981, "Khotylevskie Stoyanki," 65; Chernysh 1987, 43–45).

Technology related to the production of insulated tailored clothing is also represented primarily by nonlithic remains. Although small and delicate, eyed needles of bone and ivory have been recovered from several sites including Gagarino (which contained a sample of 25 specimens) and Kostenki IV (Tarasov 1979, 112–114; Rogachev and Anikovich 1982c, 79–83). Gagarino also yielded what appears to have been a decorated needle case of bird bone (Tarasov 1979, 114), similar to ethnographic examples from the Arctic (e.g., Murdoch 1892, 320–321). Bone awls or perforators are also found in many assemblages. In addition to stone end-scrapers, which were used for initial hide preparation, hide burnishers of bone and ivory were found at Avdeevo, Kostenki I, and Gagarino. Microscopic wear patterns on these implements are similar to those of hide-burnishing tools among recent hunter-gatherers (Semenov 1964, 175–179) (fig. 6.10).

Also pertinent to clothing technology are the massive numbers of fur-bearing mammal remains often found at Eastern Gravettian sites. Thousands of bones of wolf, arctic fox, and hare have been recovered from occupations on the central plain (Vereshchagin and Kuz'mina 1977). At Kostenki I, Kostenki VIII, Avdeevo, and Pushkari I, nearly complete skeletons and/or paw bones of wolf were found in anatomical order, reflecting pelt removal. Similar arrangements of arctic fox remains are reported from Kostenki XIII, Gagarino, and Avdeevo (Boriskovskii 1953, 187; Rogachev 1953; Efimenko 1958, 180; Tarasov 1979, 31). The sheer numbers of these taxa suggest large-scale harvesting with traps and snares as among recent hunter-gatherers in northern latitudes (e.g., Coon 1971, 128–133). It is unclear whether any of the more ambiguous items of bone,

Figure 6.10. *(opposite)* Nonstone implements from Kostenki I and Gagarino (Eastern Gravettian) on the Middle Don River (modified from Efimenko 1958, 316, fig. 123; Tarasov 1979, 112, fig. 57).

antler, and ivory in these sites represent components of traps, but the likely use of the latter implies especially complex technology.

The heavy reliance on bone fuel in the treeless regions of the East European Plain reflects a unique loess-steppe adaptation not found in other parts of northern Eurasia during the late Upper Paleolithic. Bone fuel was used almost exclusively in Eastern Gravettian sites of the central plain, where charred bone and bone ash are often found in large quantities. As noted earlier, long-term occupations may have been located near natural accumulations at the mouths of side-valley ravines that would have provided a fuel supply. In addition to such accumulations, site occupants presumably used the fresh bone obtained from their own large mammal kills (most sites lack large numbers of likely large mammal prey remains).

During warmer months, bone fuel supplies were probably kept fresh and flammable in permafrost-cooled storage pits. Large numbers of deep pits—filled with bone and other debris—are found only in the treeless areas of the central plain; such pits are absent in the southwest plain where hearths contain wood charcoal. Although these pits may have also been used for storage of meat (e.g., Binford 1983; Sergin 1983), the strong correlation with bone ash (and large number of nonmeaty body parts found in many pits) suggests that they were most probably used for maintaining fuel reserves (Klein 1969b, 118). They were probably excavated to the base of the thaw layer during the summer (roughly one meter in depth), and designed to provide cold storage like the "ice cellars" of recent hunting peoples in arctic regions. Many pits contain a variety of occupation debris, apparently reflecting complex histories that often seem to have included use as trash dumps (Grigor'ev 1993, 59–60).

The occupants of these sites apparently excavated their deep pits with mattocks that are also unique to the central East European Plain (Grigor'ev 1993, 55). Mattocks fashioned from ivory and bone have been recovered from Kostenki I, Avdeevo, Khotylevo II, and Pushkari I (Efimenko 1958, 291–294; Velichko et al. 1981, "Khotylevskie Stoyanki," 65). Striations on the tips of specimens from Avdeevo and Pushkari I confirm their use as digging implements (Gvozdover 1953; Semenov 1964, 179–181; Klein 1973, 82–83) (fig. 6.11).

Although the Eastern Gravettian sites of the central plain were the first to yield traces of Paleolithic dwelling structures in Europe, their interpretation continues to provoke debate and controversy. As noted earlier, the original reconstruction of a single roofed dwelling or communal "longhouse" at sites such as Kostenki I and Avdeevo has been disputed by many archaeologists (Klein 1969b, 120–121; 1973, 100–104). It appears more likely that these feature complexes represent a group of smaller dwellings or an open-air enclosure surrounding a central line of hearths. The larger

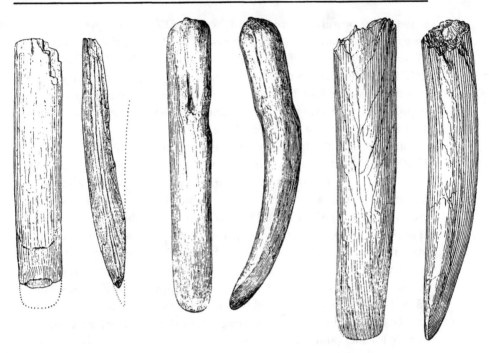

Figure 6.11. Mattocks of mammoth ivory from Layer 1 at Kostenki I (Eastern Gravettian) on the Middle Don River (after Efimenko 1958, 292, fig. 107).

pits (some of which occupy 6–8 square meters) may have been the floors of small semisubterranean houses (e.g., Kozlowski 1986, 176–177).

Evidence for the construction of smaller shelters is more substantial. However, as radiocarbon dates continue to accumulate, it is becoming increasingly apparent that most, if not all, of the mammoth-bone structures on central East European Plain postdate the Last Glacial Maximum cold peak. The superstructure of former shelters in the Eastern Gravettian sites remains unknown (although in the southwest plain wooden poles would have been readily available). The most convincing examples are found at Kulichivka, Layer 1, in the northern Volyn-Podolian Upland (Savich 1975, 122), and at Kostenki IV, Layer 1, and Gagarino on the Middle Don River (Rogachev 1955; Tarasov 1979). All of these shelters are represented by a shallow circular or oval depression, measuring several meters in width and containing a centrally placed hearth and large quantities of occupation debris.

Late Last Glacial Maximum and Late Glacial: Epigravettian

After a major reduction in settlement on the central plain during the maximum cold peak (20,000–18,000 years BP), most unglaciated areas of

Eastern Europe were occupied until the end of the Pleistocene. Sites in this time range are dated to the final phase of the Last Glacial Maximum (ca. 17,000–13,000 years BP) and the Late Glacial (ca. 13,000–10,000 years BP). Throughout this interval, periglacial loess-steppe environments prevailed across the East European Plain, although climates ameliorated slightly during several brief interstadials and began to warm significantly after the beginning of the Late Glacial.

Late Upper Paleolithic sites in Eastern Europe that date to 17,000–10,000 years BP are usually lumped into the classification of Epigravettian (e.g., Kozlowski 1986). As in Central Europe—where it is also used—Epigravettian is a generic term that is "more chronological than cultural" (Svoboda et al. 1996, 143), and subsumes much variability both within and between regions. With the exception of the spectacular structures of mammoth bone on the central plain, the East European Epigravettian sites are similar to those of the early Last Glacial Maximum. Sites on the plain reflect a distinctive adaptation to loess-steppe conditions—as during 25,000–21,000 years BP—and the differences between the two periods may be more stylistic than technological. In treeless areas, bone was used heavily for fuel (and again stored in deep pits), raw material, and shelter construction.

Epigravettian sites are common in the southwest, central, and southern regions of the East European Plain, and are also present in the Crimea and Northern Caucasus. In the Middle Dnestr Valley, some contain multi-level occupation sequences such as Molodova V (Layers 1–6), Korman' IV (Layers 1–4), Ataki I (Layers 1–4), and Kosoutsy (Layers 1–10). In the central plain, they are typically confined to one or two occupation horizons, such as Eliseevichi, Yudinovo, Mezin, Dobranichevka, Gontsy, Mezhirich, and Kostenki II. Sites in the southern plain, which sometimes contain multiple levels, include Kamennaya Balka (Layers 1–3), Amvrosievka, and Zolotovka. Crimean and Northern Caucasus localities include Buran-Kaya III and Satanai (Rogachev and Anikovich 1984, 197–226; Soffer 1985; Amirkhanov 1986; Leonova 1994).

The lithic technology and tools of the Epigravettian are especially similar to those of the preceding Eastern Gravettian (Kozlowski 1986). Prismatic cores were used to produce blades, while smaller cores were used for manufacturing bladelets and microblades. Burins continue to dominate among finished stone tools, and often represent over 50 percent of the tools in individual assemblages. This undoubtedly reflects a sustained emphasis on the working of bone, antler, and ivory in all regions. End-scrapers are usually present in modest numbers (15 percent or less). Backed bladelets, retouched microblades, and backed points are found in most assemblages, and sometimes dominate among tool inventories (e.g.,

Kostenki XI, Layer 1a [Rogachev and Popov 1982, 125]). Backed bladelets and microliths are particularly common in the southern plain sites, where they typically compose about 30 percent of the tools (Leonova 1994, 173) (see table 6.3; fig. 6.12).

Many backed bladelets and microblades were used as inset blades in slotted points of bone, antler, and ivory. Examples of such points have been recovered from Epigravettian sites in most regions, including Molodova V (Layer 3) and Kosoutsy (Layer 3) in the Middle Dnestr Valley (Chernysh 1987, 63–64; Borziyak 1993, 69–73), Mezin on the Desna River (Shovkoplyas 1965a, 207–211), and Anetovka II and Amvrosievka in the south (Leonova 1994). It is unclear whether the abundance of inset-blade points in the Epigravettian relative to the preceding period is a function of sampling and preservation (i.e., more sites and younger sites), or a significant change in hunting technology. Direct evidence of spear-throwers is still lacking for the Epigravettian in Eastern Europe, but examples of the former are known from contemporaneous sites in Western Europe (Klein 1999, 540–542). However, possible arrow or dart points are reported from Eastern Europe (e.g., Mezhirich [Pidoplichko 1976, 180–181]).

Technology related to tailored clothing production is well represented in the Epigravettian, as during the preceding interval. In addition to bone awls, eyed needles have been found at many East European Plain sites (e.g., Kosoutsy, Eliseevichi, Mezhirich, Mezin), along with implements of bone and stone for hide preparation. Sites dating to the late Last Glacial Maximum and Late Glacial often contain particularly large numbers of fur-bearing small mammal remains (e.g., over 15,000 arctic fox remains at Eliseevichi [Vereshchagin and Kuz'mina 1977, 80]). Articulated paw bones of hares, reflecting pelt removal, have been recovered at Gontsy and Mezhirich (Pidoplichko 1969, 52; Gladkikh 1981, 239). At Mezhirich, several carefully shaped items of ivory have been interpreted as components of small mammal traps (Pidoplichko 1976, 164–167).

Sites in treeless regions of the central plain again reveal almost exclusive use of bone fuel in the hearths, and many contain deep pits filled with bone. Over 6 kilograms of bone ash were recovered from Layer 1a at Kostenki XI on the Middle Don River (Rogachev and Popov 1982, 121). Large bone-filled pits, typically measuring 1–2 meters in width and 0.75–1.0 meter in depth, are found at many of the larger Epigravettian sites in the Dnepr-Desna Basin and at Kostenki XI; they are often grouped tightly around the peripheries of former structures at these sites (Rogachev and Popov 1982, 120–124; Soffer 1985, 118–123). Their dimensions and contexts suggest that, as in the Eastern Gravettian sites, they were used to refrigerate fresh bone during warmer months. As in the earlier sites, bone

Table 6.3. Artifacts in Late Upper Paleolithic Sites of the East European Plain: Epigravettian

Tool types	Molodova V Layer 3	Kosoutsy Layer 2	Timonovka I	Mezin	Dobranichevka (1953–1955)	Mezhirich Feature 2	Kostenki XI Layer 1a
End-scrapers	37	18	594	664	248	115	26
Composite tools	10	0	35	143	20	13	0
Borers	0	0	18	159	3	5	10
Burins	183	71	1369	2609	205	268	57
Backed points	8	2	10	0	0	2	0
Shouldered points	1	0	0	0	0	0	18
Truncations	9	0	366	0	32	26	14
Retouched blades	8	0	19	0	13	42	490
Side-scrapers	0	3	5	0	0	5	>10
Backed bladelets	9	95	30	317	82	60	378
Splintered pieces	0	0	0	0	0	0	239
Total stone tools:	267	~200	~2500	4429	637	~560	~1400
Bone/ivory mattocks	1	x	0	10	1	3	0
Bone/ivory awls	2	0	12	70	x	56	0
Bone/ivory points	2	1	0	x	0	5	1
Bone needles	0	x	0	x	0	7	0
Ornaments	0	1	0	>800	0	6	1
Art objects	1	1	23	>20	1	3	0

SOURCES: Shovkoplyas 1965a; Klein 1969b; Pidoplichko 1976; Rogachev and Popov 1982, 125; Kozlowski 1986, 158–159, table 3.4; Chernysh 1987, 63–65; Borziyak and Kovalenko 1989, 210.

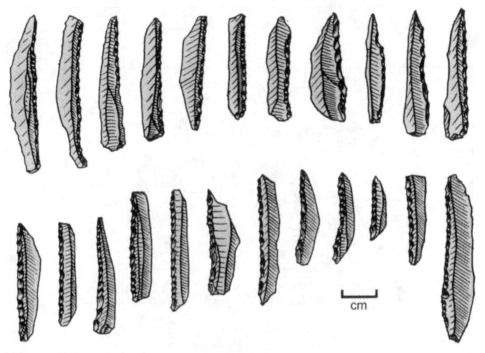

Figure 6.12. Backed bladelets from Mezin (Epigravettian) on the Desna River (modified from Shovkoplyas 1965a, 171, fig. 29).

and ivory mattocks have been recovered from many localities, and the many specimens from Eliseevichi exhibit wear striations characteristic of digging implements (Semenov 1964, 179–181).

The Epigravettian sites of the central East European Plain are justly famous for their large former shelters of mammoth bone, which constitute the most ancient ruins on earth. At least 16 mammoth-bone structures are documented in the Dnepr-Desna Basin sites, and four structures are reported from sites on the Middle Don River (Pidoplichko 1969; Klein 1973, 91–99; Praslov and Rogachev 1982; Gladkih et al. 1984) (fig. 6.13). Many of these sites contain more than one former structure, and at least three of the Dnepr-Desna localities (Dobranichevka, Mezhirich, and Yudinovo) have each yielded the remains of no less than four structures. The structures are circular or oval in plan, and vary in diameter from 3.5 to as much as 8.0 meters. In terms of area, they display a bimodal distribution that contains a small group (roughly 12 square meters) and a large group (roughly 25 square meters). The composition of the bones used for each structure varies, although mammoth crania, mandibles, long-bones, scapulae, and pelves (i.e., largest bones) are most common. Most shelters apparently contained at least one internal hearth.

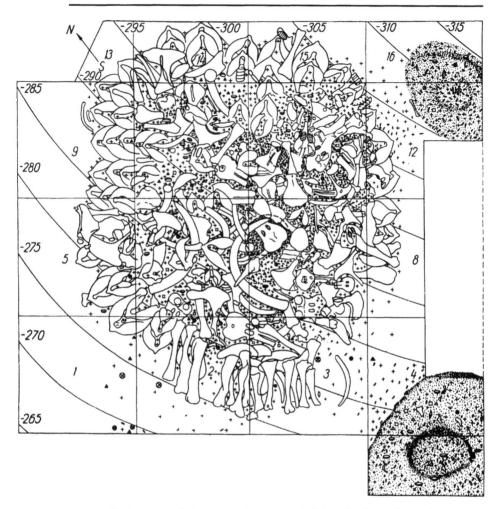

Figure 6.13. Mammoth-bone structure at Mezhirich in the Dnepr Basin (after Pidoplichko 1969, fig. 43).

As noted earlier, the mammoth-bone structures seem to be largely or wholly confined to the period following the cold maximum (i.e., after 18,000 years BP). The reason for their scarcity or absence in the earlier Eastern Gravettian sites is not clear. One explanation for this might be that very large natural concentrations of mammoth bone—which must have been essential for gathering the enormous volume of bone required to build several structures—may have accumulated only during the cold maximum (20,000–18,000 years BP). During this interval, conditions may have been especially favorable for such concentrations, which would have accumulated without disturbance until people returned to the region after 17,000 years BP.

Artificial shelters were constructed at sites in the southwest and southern plain as well. Perhaps the most convincing traces of a former structure in these areas were discovered in Layer 3 at Molodova V (Middle Dnestr Valley), which contains a postmold arrangement surrounding a shallow depression with a central hearth and enclosing an area of roughly 12 square meters (Chernysh 1959 97–99). In the southern plain, the remains of several small structures delineated by postmold patterns were mapped at Osokorovka, Horizon 3 on the Lower Dnepr River (Boriskovskii and Praslov 1964, plate XXI).

Ornaments and Art in the Late Upper Paleolithic

Art is perhaps the single most important indicator of modern human behavior during the Upper Paleolithic. More than the technology and ornamentation of the latter, its presence suggests a rich cultural life similar to that of recent and living peoples. For reasons that are unclear, most Upper Paleolithic art is confined to northern Eurasia (i.e., Europe and Siberia), and a significant part of it is derived from the late Upper Paleolithic of Eastern Europe. With the exception of the paintings on the walls of Kapovaya Cave in the Ural Mountains (Bader 1965), the East European material is limited to mobilary items. Most of them have been found in sites of the central plain, but some objects have been recovered from the southwest and southern plain as well.

The most widely known expressions of late Upper Paleolithic art in Eastern Europe are the naturalistic figurines of the Eastern Gravettian sites (ca. 24,000–21,000 years BP). These include the evocative "Venus" figurines, which have been recovered from Khotylevo II on the Desna River, Avdeevo in the Central Russian Upland, and Gagarino and Kostenki I and XIII on the Middle Don River. The classic form of these figurines is remarkably consistent over broad geographic areas, and depicts an unclothed female with protruding abdomen, pendulous breasts, and exaggerated buttocks. The East European specimens were made with mammoth ivory and marl, and measure roughly 10–15 cm in height; at least some them appear to have been painted (Abramova 1967; Gvozdover 1989) (fig. 6.14).

As noted earlier, the Venus figurines seem to have originated—along with most other elements of the Eastern Gravettian technocomplex—in Central Europe prior to the end of OIS 3. Examples have also been found in Western Europe (Gamble 1982). Some Soviet archaeologists of the 1930s viewed these figurines as further evidence of matrilineal clan society in the Upper Paleolithic (e.g., Boriskovskii 1932, 23), but they probably reveal more about social relationships between groups than about

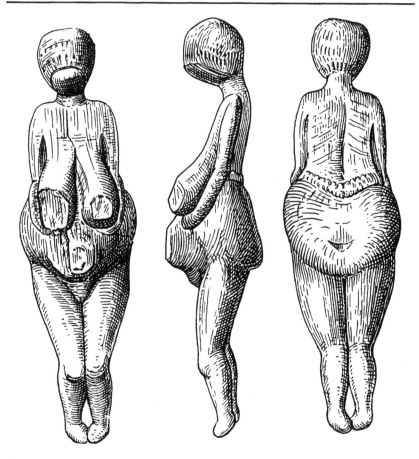

Figure 6.14. Figurative art of the Eastern Gravettian: "Venus" figurine of mammoth ivory from Layer 1 at Kostenki I on the Middle Don River (after Efimenko 1958, 349, fig. 142).

social relationships within them. The widespread distribution of the classic form indicates—more convincingly than any other aspect of material culture—the existence of shared symbols and ethnic identity across a large geographic area for the first time in human prehistory.

The Eastern Gravettian sites of the central plain also contain other forms of representational art, including more stylized human figurines recovered from Khoylevo II, Avdeevo, Kostenki I, and Gagarino (Tarasov 1979, 140–141; Grigor'ev 1993, 56–57). Both naturalistic and stylized animal figurines (mammoths, bison, carnivores, and others) have been recovered from Kostenki I (Layer 1), Kostenki IV (Layers 1–2), and Avdeevo. A series of animal figurines of marl (chiefly mammoths) were found in Layer 2 at Kostenki XI (Praslov and Ivanova 1982, 208), which has been tentatively dated to the early Last Glacial Maximum but might be younger (Sinitsyn et al. 1997, 25). Objects of bone or ivory decorated with simple geometric designs are not especially common in the Eastern Gravettian, but some examples are known from Berdyzh (Dnepr Basin), Avdeevo, Kostenki I, and Kostenki XXI (Layer 3) (Bud'ko et al. 1971, 303; Klein 1973, 83–87; Praslov and Rogachev 1982).

Objects interpreted as simple ornaments have been recovered from many of the larger Eastern Gravettian sites on the southwest and central plain. These include perforated carnivore teeth, shells, or fragments of worked bone, ivory, or marl that represent pendants or beads from Molodova V (Layer 7), Kulichivka (Layer 1), Pushkari I, Avdeevo, Kostenki I (Layer 1), Kostenki IV (Layers 1–2), Kostenki XI (Layer 2), and Gagarino. Avdeevo and Kostenki I also yielded more elaborate ornamentation in the form of what may have been ivory headbands decorated with geometric patterns (Abramova 1967; Savich 1977; Praslov and Rogachev 1982).

Sites of the Epigravettian period (ca. 17,000–10,000 years BP) exhibit a different style of art that is largely based on abstract design (Kozlowski 1986, 183–184). Various objects of bone and ivory engraved or painted with geometric patterns (e.g., diagonal cross-hatching) have been recovered from many sites in the Dnepr-Desna Basin, including Suponevo, Timonvka I, Eliseevichi, Yudinovo, Mezhirich, and Mezin (Abramova 1967, 43–47; Marshack 1979, 271–276). Both Mezin and Mezhirich yielded large mammoth bones decorated with geometric designs painted in red ochre (Shovkoplyas 1965a, plates 54–56; Pidoplichko 1976, 213) (fig. 6.15). The most naturalistic representational art is a female figurine (ivory) found at Eliseevichi, which is slender in form and dissimilar to the earlier Venus sculptures (Polikarpovich 1968, 115–117). A highly stylized anthropomorphic figurine (bone) was found at Kostenki II (Boriskovskii 1963, 48), and stylized carvings of birds, incised with geometric patterns, were recovered from Mezin (Shovkoplyas 1965a, 227–234). The latter site also contained a group of abstract sculptures variously interpreted as anthropomorphic or phallic (Abramova 1967, 39–40).

Isolated examples of art are known from the southwest and southern plain sites dating to the Epigravettian period. In the Middle Dnestr Valley, a highly stylized human figurine of marl was found in Layer 3 at Molodova V, and in Layer 2 at Kosoutsy (Chernysh 1987, 65–66; Borziyak and Kovalenko 1989, 209, fig. 4.1). The fragment of a stylized female figurine of bone is known from Muralovka, near the coast of the Sea of Azov (Praslov and Filippov 1967, 26, fig. 9), which has yielded radiocarbon dates of 19,630 and 18,780 years BP (Leonova 1994, 200).

Many of the Epigravettian sites on the central plain contain ornaments, and some of these are decorated with geometric designs. As during the preceding period, perforated carnivore teeth, shells, and objects of worked

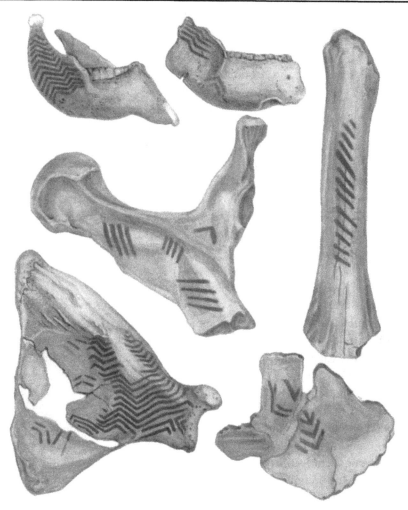

Figure 6.15. Mammoth bones decorated with geometric designs from Mezin on the Desna River (after Pidoplichko 1969, 97, fig. 31).

bone and ivory interpreted as pendants and beads are present in a number of sites (e.g., Suponevo, Eliseevichi, Yudinovo, Mezin, Gontsy, Mezhirich, Kostenki XI [Layer 1a], Borshchevo I). Elaborate bracelets of ivory, engraved with geometric patterns, were found at Mezin (Shovkoplyas 1965a, 237–241), while a carved pin and brooches of ivory were recovered from Mezhirich (Pidoplicho 1976, 210–203). Ornaments are occasionally found in localities in the southwest and southern plain (e.g., Molodova V [Layer 4], Amvrosievka [Boriskovskii and Praslov 1964, plate 10; Chernysh 1987, 60]).

Diet and Foraging during the Last Glacial Maximum

The extreme cold and aridity of climates during OIS 2 probably reduced available plant food for the human inhabitants of the East European Plain to a minimum. At the same time, daily caloric requirements must have been comparable to those of recent hunter-gatherers in the Arctic. The importance of meat in the diet probably reached a maximum level during the late Upper Paleolithic, particularly on the central plain. Vegetal foods probably played a more important role in the diet of people living in areas like the Northern Caucasus, where climates were milder and plant communities more diverse.

As in the case of earlier periods, much taphonomic research needs to be conducted on the faunal remains recovered from late Upper Paleolithic sites in Eastern Europe. Few detailed studies have been completed to date, although in recent years some work has been published on large mammal assemblages from the southern plain (e.g., Bibikova and Starkin 1989; Krotova and Belan 1993). An enormous body of data is available for analysis, because preservation conditions at open-air sites were especially favorable, and bones and teeth were buried rapidly in loess-derived sediment with minimal soil formation and weathering.

Processes of accumulation at late Upper Paleolithic open-air sites on the East European Plain were similar to those of the earlier sites (see chapters 4 and 5). Most large mammal remains were probably accumulated at these sites by their human occupants. However, many of the remains do not necessarily represent food items. As noted earlier, large quantities of mammoth bones seem to have been collected for use as material and fuel. At many sites, equally impressive numbers of smaller fur-bearing mammals were apparently trapped for pelts (and at least sometimes ignored as food sources). Moreover, in the central plain sites—where a significant volume of bone was consumed as fuel—most animal food remains were probably destroyed (e.g., Grigor'ev 1967, 348; Tarasov 1979, 33). This presents researchers with a special problem, and further limits our knowledge of diet and foraging patterns in this region.

Although the periglacial loess-steppe of OIS 2 was probably much more productive than arctic tundra (see chapter 2), several lines of evidence suggest that the inhabitants of this environment were moving and interacting over areas comparable to those of recent hunter-gatherers in high latitudes (Kelly 1995, 116–132). The transport distances of raw materials sometimes exceed 600 kilometers (e.g., Klein 1973, 89), and may indicate foraging movements on a similar scale. The geographic distribution of diagnostic art forms (e.g., Venus figurines) reveals a pattern of movements

Table 6.4. Large Mammal Remains in Late Upper Paleolithic Sites of the East European Plain: Southwest Region

	Sites and level				
Taxon	Molodova V Layer 7	Korman' IV Layer 4	Kosoutsy Layer 3	Kulichivka Layer 1	Chuntu Layer 3
Woolly mammoth (*Mammuthus primigenius*)	105/7[a]	224/1	12/1	–/5	–/–
Woolly rhinoceros (*Coelodonta antiquitatis*)	4/1	–/–	–/–	–/–	8/1
Horse (*Equus caballus*)	709/11	227/7	720/10	–/4	156/6
Roe deer (*Capreolus capreolus*)	–/–	–/–	62/4	–/–	–/–
Red deer (*Cervus elaphus*)	10/1	38/2	3/2	–/–	5/1
Elk (*Alces alces*)	7/1	21/1	–/–	–/–	–/–
Giant deer (*Megaceros giganteus*)	–/–	4/1	–/–	–/–	2/1
Reindeer (*Rangifer tarandus*)	606/13	998/7	5770/22	–/16	612/4
Bison (*Bison priscus*)	12/2	16/1	–/–	present	12/3
Wolf (*Canis lupus*)	4/1	–/–	3/1	–/–	–/–
Arctic fox (*Alopex lagopus*)	3/1	–/–	14/3	–/–	–/–
Bear (*Ursus arctos*)	–/–	–/–	–/–	–/–	1/1

SOURCES: Savich 1975, 50; Tatarinov 1977, 113; Chernysh 1987, 46; Borziyak 1993, 70–71, table 1; Borziac et al. 1997, 293, table 3.
[a]Number of identified specimens/minimum number of individuals

and/or networks covering thousands of square kilometers, and is also reminiscent of recent foraging peoples in the Arctic (e.g., McGhee 1997).

East European Plain: Southwest Region

The composition of large mammal assemblages in the southwest region differs from that of the central East European Plain. Assemblages in the Middle Dnestr Valley sites are dominated by reindeer and horse throughout the Last Glacial Maximum and Late Glacial (Chernysh 1959; Tatarinov 1977; Borziyak 1993). These two taxa typically account for roughly 80 to 90 percent of the identifiable large mammal remains. Mammoth is less common (typically 15 percent or less), and completely absent in the terminal Pleistocene layers. Other large mammals that are present in modest numbers include steppe bison and red deer (table 6.4).

The majority of the large mammal remains in the Middle Dnestr Valley sites seem to represent animals hunted by their inhabitants, although there is little supporting taphonomic data for this conclusion. The availability of wood in this region evidently eliminated the need for heavy consumption of bone fuel and collection of large bones for constructional material (Klein 1973, 53–54). However, mammoth remains may have been collected for raw material use. Although available information on skeletal parts is limited, it is apparent that a high proportion of the mammoth debris in these sites is represented by ivory fragments (Borziyak 1993, 69). Hunting may have been largely confined to reindeer, horse, and other smaller taxa. Plant foods could have played a more important role in the diet in comparison to the central and southern plain, and the sites in the southwest plain have consistently yielded grinding stones widely thought to have been used for preparation of seeds, roots, and other plant parts (Chernysh 1959; Borziyak 1993, 78–79; Borziac et al. 1997, 298).

East European Plain: Central Region

Faunal assemblages in the central plain sites are heavily dominated by mammoth and smaller fur-bearing mammals. Aside from the carnivores, mammoth often represents more than 95 percent of the large mammals at these sites; only isolated remains of reindeer, horse, bison, and other taxa are present at many of them (Vereshchagin and Kuz'mina 1977; Soffer 1985, 126–135, table 2.7). Taphonomic evidence that the mammoth bones were collected from natural occurrences has accumulated for many years (Gromov 1948; Pidoplichko 1969; Klein 1973, 52–54; Soffer 1993, 39). Virtually all portions of the skeleton are present, and occasionally sequences of bones in anatomical order are observed. Bones and tusks at many sites exhibit significant variations in the degree of weathering, and often display traces of carnivore damage while lacking tool cut marks. At

Table 6.5. Large Mammal Remains in Late Upper Paleolithic Sites of the East European Plain: Central Region

Taxon	Sites and level					
	Eliseevichi	Yudinovo	Mezin	Mezhirich	Kostenki I Layer 1	Gagarino
Woolly mammoth (Mammuthus primigenius)	–/60[a]	102/10	3979/116	3677/149	734/1	108/–
Woolly rhinoceros (Coelodonta antiquitatis)	–/–	–/–	17/3	–/–	–/–	22/–
Horse (Equus caballus)	1/1	6/2	659/61	2/2	27/1	–/–
Red deer (Cervus elaphus)	1/1	–/–	–/–	–/–	–/–	–/–
Giant deer (Megaceros giganteus)	–/–	–/–	1/1	–/–	–/–	–/–
Reindeer (Rangifer tarandus)	–/–	–/–	444/83	37/10	2/1	8/–
Musk ox (Ovibos moschatus)	–/–	–/–	188/17	–/–	–/–	–/–
Bison (Bison priscus)	–/–	6/2	19/5	17/6	–/–	2/–
Wolf (Canis lupus)	903/36	8/3	1004/59	79/12	370/9	–/–
Arctic fox (Alopex lagopus)	14654/287	109/25	1842/112	146/11	139/3	1663/–
Bear (Ursus arctos)	89/10	7/2	35/7	12/3	1/1	–/–

SOURCES: Shovkoplyas 1965a, 97; Vereshchagin and Kuz'mina 1977, 80–100; Tarasov 1979, 33, table 3; Kornietz et al. 1981, 115, table 4.
[a]Number of identified specimens/minimum number of individuals

Mezin (Desna River), analysis of collagen in mammoth bones revealed major variations suggesting collection of materials from various spatial and temporal contexts (Pidoplichko 1969, 91). At Gagarino (Middle Don River), large discrepancies in radiocarbon dates obtained on mammoth remains from the single occupation layer were attributed to the same phenomenon (Tarasov 1979, 149) (table 6.5).

Evidence that mammoths were hunted and consumed as food in the central plain is relatively scarce. Tool cut marks on bones are very rare, but have been reported from Novgorod-Severskii (Desna River), which lacks structural remains of mammoth bone (Boriskovskii 1953, 298). The site of Fastov, which is located along a small tributary of the Dnepr system on the eastern margin of the Volyn-Podolian Upland, presents a rare example of a short-term occupation (Shovkoplyas 1956; Boriskovskii and Praslov 1964, 31). The site contains several hundred bones of mammoth (80 percent) and horse (20 percent) associated with two former hearths and roughly 1,700 artifacts (chiefly lithic debris). An estimated nine juvenile and two adult mammoths are present. Although taphonomic data are lacking, Fastov appears to represent a small hunting camp where large mammalian prey were processed. A kill-butchery site might be present at Oktyabr'skoe on the southwest margin of the Central Russian Upland, which contains a small quantity of tools and lithic waste associated with the cut-marked bones of a single mammoth (Zamyatnin 1940, 100–101).

A few sites contain larger numbers of ungulate remains that may provide further insight into the large mammal component of the diet. At Mezin, hundreds of bones of horse, reindeer, and musk ox were recovered in addition to those of mammoth and carnivores (Shovkoplyas 1965a, 96–105). Significantly, Mezin is one of the few late Upper Paleolithic localities in the central plain to yield some wood charcoal, indicating that bone fuel may have been used less intensively here. Although the reindeer are probably inflated by collection of antlers for raw material, most of these remains appear to represent food debris and reflect a diet similar to that of the southwest region (without woodland species such as red deer). The absence of newborn and very young reindeer in the sample seems to indicate autumn or winter kills (Pidoplichko 1969, 83). In addition to Mezin, Novgorod-Severskii and Gontsy (Dnepr Basin) contained large proportions of reindeer remains, and Kostenki IV (Middle Don River) yielded several hundred horse bones and teeth (Boriskovskii 1953, 297–298; Pidoplichko 1969, 51; Vereshchagin and Kuz'mina 1977, 102).

Carnivores, small mammals, and other vertebrates also probably contributed to the meat component of the diet on the central plain. As noted earlier, large numbers of arctic fox, wolf, and hare remains have been found in these sites. Nearly complete skeletons of arctic fox and wolf have been recovered from some localities (e.g., Mezin, Kostenki XIII [Shovkoplyas 1965a, 99; Rogachev and Belyaeva 1982, 140]), suggesting that carcasses were skinned for their pelts and discarded. However, with the exception of paw bones, most arctic fox and wolf remains—as well as all hare remains—are disarticulated and may represent food debris. Cut marks were observed on arctic fox bones at Yudinovo (Burova 1999).

The remains of birds have been found at a number of sites on the central plain. The most commonly represented taxon is willow ptarmigan (*Lagopus lagopus*), but other game birds such as black grouse (*Lyrurus tetrix*), teal (*Anas* sp.), and goose (*Anser* sp.) are also present (Klein 1973, 57). These remains probably represent food debris, although some of them might have been accumulated by predatory birds that inhabited the sites—nesting in the ruins of collapsed mammoth-bone structures—after the departure of their human occupants (Pidoplichko 1969, 85).

The reconstruction of foraging strategy and settlement systems in the central plain has been complicated by uncertainties regarding the season(s) of occupation at the major sites (which may be especially difficult to determine because of the consumption of animal food debris as fuel). There is evidence of both cold and warm season occupation at the larger encampments. The absence of newborn and very young individuals among the reindeer remains at sites such as Mezin and Dobranichevka is interpreted as an indication of winter occupation (Klein 1973, 54–55). However, the numerous pits in these sites could not have been excavated into frozen loess, and probably were designed for warm-season use. The trapping of smaller fur-bearing mammals is also assumed to have been a winter activity (when pelts were in prime condition). Olga Soffer (1985, 425–427) has suggested that both winter and summer base camps were occupied in the main valleys, and that the latter were sometimes located on higher and better-drained surfaces (e.g., Pogon-Bugorok on the Desna River). No seasonality data are available for the rare examples of smaller short-term camps (e.g., Fastov), but these sites—which seem to be concentrated in the upper valleys and watersheds—may have been occupied during summer or fall hunting forays outside of the main valleys (Klein 1973, 58).

Southern Plain, Crimea, and Northern Caucasus

On the southern plain, Upper Paleolithic large mammal assemblages are generally dominated by steppe bison, which is rare in other parts of the East European Plain. Although some assemblages (e.g., Osokorovka, Anetovka II, Kamennaya Balka II) also contain other large mammals including horse, reindeer, and saiga (Leonova 1994), bison represented the primary food source in the region. At Amvrosievka (Donets Ridge), bison herds were evidently driven into a ravine, and the carcasses were processed at an adjacent occupation area. Thousands of bones were recovered from the ravine, and some of them exhibit cut marks. A variety of age groups are present, and males constitute roughly 60 percent of the sample (Krotova and Belan 1993). The site seems to have been used for multiple kill events that probably took place during the winter or spring. At Anetovka II (Southern Bug tributary), which appears to contain a butchery area, a similar pattern of bison procurement is evident (Bibikova and Starkin 1989).

Data from the Crimea and Northern Caucasus are comparatively limited. More diverse plant communities in these regions probably reduced dietary dependence on meat, especially relative to the central and southern plain. Saiga and red deer are particularly common in the rockshelters of the Crimea, and the latter reflects the presence of woodland (Vekilova 1971, 124–125). The only measurable sample of large mammals from the late Upper Paleolithic of the Northern Caucasus is reported from the rockshelter of Satanai, where horse represents 98 percent of the assemblage (Amirkhanov 1986, 12–13).

Late Upper Paleolithic Socioecology

During the late Upper Paleolithic, compelling archaeological evidence for organization along the lines of recent hunter-gatherer peoples emerges for the first time. Although such organization probably existed among the modern humans of the early Upper Paleolithic—and is widely perceived as having been instrumental in the settlement of areas like the East European Plain—the supporting evidence is more fragmentary (see chapter 5). By contrast, the late Upper Paleolithic yields occupation floor patterns with clear analogues to the structure of recent hunter-gatherer residential camps; these are most evident in the Epigravettian sites of the central plain. As in the case of earlier periods, the East European record provides an unparalleled opportunity to assess site structure and its implications for organization. As noted earlier, evidence for the long-distance transport of raw materials exceeds that of the preceding period and indicates movements and/or networks on a scale comparable to recent hunter-gatherers in very cold and/or arid environments.

In the southwest region of the East European Plain, large occupation floor areas have been mapped at late Upper Paleolithic sites in several areas. In the Middle Dnestr Valley, Layers 1–7 at Molodova V were mapped over an area of 860–970 square meters, and the uppermost level at Babin I was mapped over an area of 206 square meters (Chernysh 1959, 34–36; 1987). At Chutuleshty, which is located on a tributary of the Lower

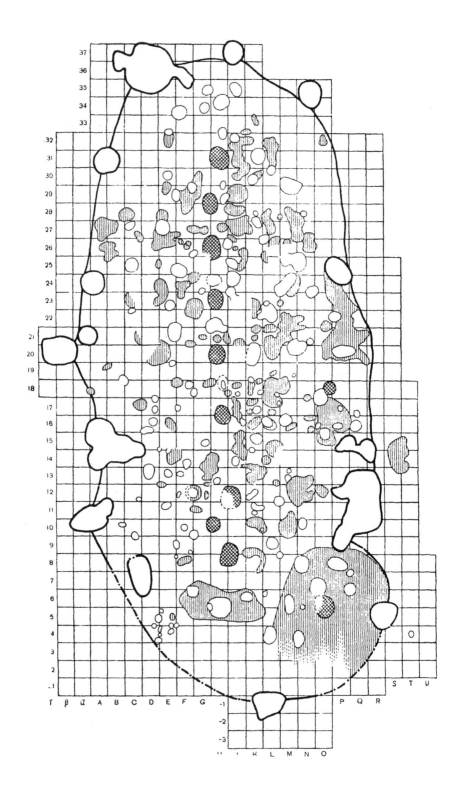

Dnestr River, an occupation area of 245 square meters was recorded (Ketraru 1965, 54–60). In the northern Volyn-Podolian Upland, a total of 1,626 square meters were mapped for Layer 1 at Kulichivka (Savich 1975, 47). These occupation floors reveal debris concentrations of varying size and composition, typically associated with former hearths. Many of them also contain probable traces of isolated former structures, most notably Layer 3 at Molodova V and Layer 1 at Kulichivka (Chernysh 1959, 99; Savich 1977). However, while most of these sites appear to represent long-term occupations, none of them exhibits the site structure of a recent hunter-gatherer residential camp, because of uncertainties concerning the identification of multiple dwelling units.

Late Upper Paleolithic sites in the central plain have been mapped on an even more impressive scale than those of the southwest region. Among the larger mapped occupation areas in the Dnepr-Desna Basin are Mezin (over 1,200 square meters), Dobranichevka (over 2,000 square meters), Pushkari I (518 square meters), and Mezhirich (over 500 square meters) (Shovkoplyas 1965a; Shovkoplyas et al. 1981, 106; Belyaeva 1997, 5; Soffer et al. 1997). At Avdeevo (west Central Russian Upland), more than 1,300 square meters have been exposed and mapped (Rogachev 1953; Velichko et al. 1981, "Avdeevo," 53). On the Middle Don River, large occupation areas have been mapped at Kostenki I, Layer 1 (over 1,600 square meters), Kostenki IV, Layers 1–2 (922 square meters), and Borshchevo I (392 square meters) (Praslov and Rogachev 1982).

Occupation floors at the largest sites exhibit at least two types of patterns that may be grouped according to time period. The first pattern is found in at least some of the major sites of the Eastern Gravettian techno-complex that are dated to the early Last Glacial Maximum (ca. 24,000–21,000 years BP). Occupation areas at Kostenki I and Avdeevo are characterized by large extended oval arrangements of deep pits surrounding a line of former hearths. At least some elements of this pattern appear to be present at other Eastern Gravettian sites of the central plain, including Khotylevo II and Pushkari I (Desna River), Gagarino and Kostenki IV (Middle Don River), and Zaraisk (Oka River tributary) (Tarasov 1979; Velichko et al. 1981, "Khotylevskie Stoyanki"; Praslov and Rogachev 1982; Amirkhanov 1997; Belyaeva 1997) (fig. 6.16).

Although the feature complex at Gagarino is widely interpreted as a small dwelling with associated storage pit (Tarasov 1979), the feature complexes at Kostenki I and Avdeevo appear to reflect occupations by a

Figure 6.16. *(opposite)* Arrangement of pits and hearths of the first feature complex ("longhouse") in Layer 1 at Kostenki I on the Middle Don River (after Efimenko 1958, 39, fig. 5).

relatively large number of people. Remarkably similar floor patterns are known among the Late Dorset sites of the Canadian Arctic (e.g., Schledermann 1978; McGhee 1997) that apparently represent open-air enclosures constructed during seasonal aggregations of people who were widely dispersed at other times of the year. The famous "longhouses" of the Eastern Gravettian probably reflect a similar pattern of periodic aggregation, which may have had both social and economic benefits to the assembled group. These events may have been scheduled to coincide with seasonal concentrations of large mammal prey (the remains of which were subsequently consumed in the hearths). Seasonality data are limited, but seem to indicate that these sites were occupied during the winter (e.g., Grigor'ev 1993, 63).

The second type of occupation floor pattern found at the larger sites in the central plain is associated with the Epigravettian (ca. 17,000–10,000 years BP). Many of these sites contain the remains of multiple structures of mammoth bone that measure several meters in diameter and presumably were occupied by small groups of people (perhaps 4–6 persons). At three localities in the Dnepr-Desna Basin, four such structures have been found on the same occupation level in each site (i.e., Dobranichevka, Yudinovo, and Mezhirich [Shovkoplyas 1972; Abramova 1993, 86–93; Soffer et al. 1997]). At other sites in the Dnepr-Desna Basin and on the Middle Don River, at least two or three former structures have been identified (i.e., Mezin, possibly Kirillovskaya, Gontsy, and Kostenki XI, Layer 1a [Shovkoplyas 1965a, 35–95; Pidoplichko 1969; Rogachev and Popov 1982, 120–124]). As noted earlier, individual structures are often associated with external pits, hearths, and debris concentrations (fig. 6.17).

The occupation floors that contain multiple former structures exhibit a site structure similar to that of recent hunter-gatherer residential camps. Such camps typically comprise groups of small dwelling units occupied by related nuclear or small extended families (e.g., Gould 1980; Binford 1983). However, this interpretation assumes that the dwellings—sometimes widely spaced apart—were inhabited simultaneously, which remains difficult to demonstrate. Unlike the Eastern Gravettian feature complexes, these former dwellings represent potentially independent units that might have been used at different times. Soffer et al. (1997, 57–58) argue that the presence of an interconnecting midden deposit indicates that at least two of the dwellings at Mezhirich were occupied simultaneously. Use of the dwellings by family units is supported by the diverse array of artifacts and debris recovered from their interiors, which reflect a variety of domestic activities (Pidoplichko 1969, 1976; Soffer 1985).

Several sites in the central plain appear to represent small short-term occupations related to a limited range of activities. Such special-function

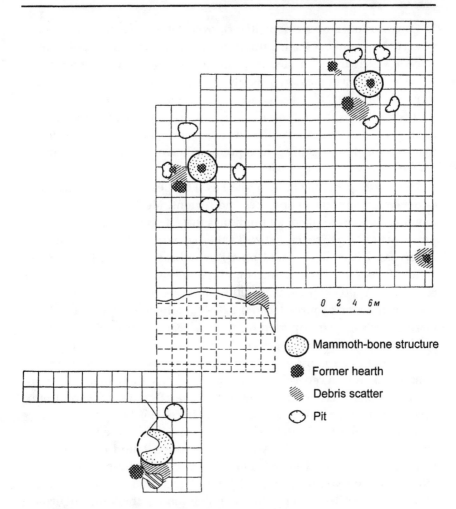

Figure 6.17. Arrangement of mammoth bone structures and associated hearths and debris scatters on the occupation floor of Dobranichevka on the Supoi River in the Dnepr Basin (after Shovkoplyas 1972, 178, fig. 1).

sites are also characteristic of recent hunter-gatherer settlement systems, but may provide little information on the size and organization of their occupants. These sites are likely underrepresented because of their comparatively limited archaeological visibility. The most carefully researched example is Fastov (described earlier), which seems to have been used as a temporary camp for the processing of mammoth and horse carcasses (Shovkoplyas 1956).

The southernmost areas of the East European Plain also provide some information on site structure, although long-term occupations with complex feature arrangements are relatively scarce in this region.

At Osokorovka (Lower Dnepr River), over 900 square meters were exposed of Horizon 3, revealing traces of up to five small dwellings—at least three of which are delineated by postmold arrangements (Kolosov 1964, 44). The structures measure roughly 3–4 meters in diameter and are associated with external debris concentrations and hearths; the overall pattern is similar to the organization at Mezhirich although deep pits are lacking. In contrast to the southwest and central regions, many sites in the southern plain appear likely to represent short-term occupations analogous to Fastov (Leonova 1994).

The late Upper Paleolithic of the East European Plain contains much evidence for contacts among sites and regions. Much of this evidence is found in the central plain, and indicates that the occupants of these sites were part of larger social networks that covered wide areas. Nonlocal materials are common in late Upper Paleolithic sites (Klein 1973, 88–89; Soffer 1985, 371–372). Flint was imported to Kostenki from a distance of roughly 150 kilometers to the southwest. Amber was brought to Gontsy and Mezin from a source located 220–225 kilometers away, and fossil marine shells were transported from areas near the Black Sea over 600–650 kilometers north to Eliseevichi, Yudinovo, and Timonovka. Such distances lie at the upper limit for maximum distances covered by recent hunter-gatherers in very cold or arid environments (Kelly 1995, 112–115), and trade might account for many of the materials transported over hundreds of kilometers from their sources (Soffer 1985, 442–444).

Stylistic similarities among artifacts and features also reflect wider social networks, and the most striking example is found in the Eastern Gravettian sites of the central plain, which are often distinguished as the "Kostenki-Avdeevo archaeological culture" (e.g., Praslov and Rogachev 1982; Grigor'ev 1993). The close parallels among the Venus figurines, tools, and feature arrangements in these sites indicate a high degree of social interaction over an area of several thousand square kilometers. Such patterns of stylistic similarities are less clear during the Epigravettian period on the central plain. One reason for this could be that the pattern of similarity among the Kostenki-Avdeevo sites—as well as the similarities with the Eastern Gravettian sites in Central Europe—is primarily the consequence of the initial dispersal of the population that created this culture. However, like so many other aspects of their loess-steppe adaptation, the geographic distribution of Eastern Gravettian art (and its implications for widely dispersed social networks) follows the pattern of prehistoric and recent hunter-gatherers in arctic environments.

CHAPTER 7

Retrospective

Eastern Europe provides a unique perspective on the later evolution of the genus *Homo*. On the west-east climate gradient of northern Eurasia, Eastern Europe is intermediate between the mild maritime setting of Western Europe and the extreme continental regime of northern Asia. During the oscillating glacial cycles of the Middle and Late Pleistocene, it became a margin of the hominid range, and several representatives of *Homo* alternately colonized and abandoned various regions between the Carpathian and Ural Mountains. This settlement history is tied closely to the evolution of adaptations to cold environments.

The occupation of Eastern Europe was a consequence of the emergence of a specialized northern variant of *Homo* in the late Middle Pleistocene (Hublin 1998). Prior to the appearance of the Neanderthals, settlement of the region was very limited and possibly confined to its southern periphery (although the effects of Middle Pleistocene glaciation and sea level change and generally lower visibility of open-air sites must be accounted for in this assessment [Praslov 1995]). The inability of Middle Pleistocene hominids to colonize northern Eurasia (above latitude 40 degrees North) outside Western and Central Europe probably reflects a lack of any of the morphological and behavioral adaptations to colder environments that distinguish recent hunter-gatherers in northern latitudes from those of equatorial regions.

The earlier inhabitants of Europe (*Homo heidelbergensis*) seem to have been differentiated from southern forms of *Homo* chiefly by geographic isolation and drift. In fact, *H. heidelbergensis* was present in both Africa and Europe for at least a hundred thousand years during the Middle Pleistocene (Rightmire 1998). The occupation of Western and Central Europe—primarily during the warmest interglacials of this period (OIS 13–9)—may have been simply an extension of the *Homo* range into higher latitudes under special circumstances. The latter may have included both the effect of warm interglacial climates on maritime areas of westernmost Eurasia, and new scavenging opportunities created by multiple carnivore extinctions [Turner 1992]).

There is little evidence for new morphological or behavioral adaptations among the earlier Middle Pleistocene hominids of Europe, and their niche may not have differed significantly from that of the hominids who initially dispersed across southern Eurasia during the Early Pleistocene

(*H. ergaster*) (e.g., Cachel and Harris 1998). Like recent hunter-gatherers in equatorial environments, they were most probably dependent on a diet high in plant foods. Although there is evidence of hominid consumption of meat (derived from large mammals) as early as the late Pliocene, it may have been obtained primarily through scavenging (e.g., Shipman 1986). Their foraging range seems to have been small, and they may have been unable to cope with widely dispersed (and seasonally unavailable) food resources. The use of controlled fire might represent a Middle Pleistocene adaptation to northern environments, but there is some evidence that this behavior may have evolved much earlier in southern latitudes (Gamble 1994). No significant changes in their lithic technology are evident (Schick and Toth 1993).

Throughout the Middle Pleistocene, Eastern Europe seems to have been something of a settlement vacuum for Eurasian hominids. It lay on the margins of *Homo* populations in Central Europe and the Near East, and a fragmentary archaeological record suggests that various groups occupied the peripheries at different times. People who made tools similar to the Central European industries seem to have been present during the earlier Middle Pleistocene (OIS 13–11), while groups from the Near East may have subsequently occupied the Northern Caucasus (OIS 7–6) (Doronichev 1992; Hidjrati et al. 1997).

In the late Middle Pleistocene (OIS 8–6), which was characterized by generally cooler climates, fully developed or classic Neanderthals evolved in Western Europe (*Homo neanderthalensis*). They are the only known specialized northern form of *Homo*, and they exhibited many of the adaptations to cold environments observed among recent hunter-gatherers in northern latitudes. The size and shape of their bodies conformed to the predictions of Bergmann's and Allen's rules for homiotherms in cold climates (e.g., Coon 1962; Trinkaus 1981), and their morphology has been characterized as "hyperpolar" on a scale of modern human populations (Holliday 1997). Stable isotope analyses of their bones indicate a heavy meat diet (Richards et al. 2000), and the character of faunal remains in their sites suggests that they were competent hunters of large mammals such as red deer and bison (e.g., Chase 1986; Gaudzinski 1996). Their lithic technology was more sophisticated than that of their predecessors, and was designed in part to produce more efficient hafted tools (Mellars 1996).

The adaptive shift represented by the Neanderthals is strikingly manifest in the temporal and spatial distribution of their sites. In contrast to their predecessors, they were clearly present in Western Europe under full glacial conditions at the end of the Middle Pleistocene (OIS 8 and OIS 6), and their sites are comparatively scarce in Western Europe during the Last

Interglacial climatic optimum (OIS 5e) (Gamble 1986). They also became the first hominids to settle widely in areas east of the Carpathians, including much of Eastern Europe and limited portions of southern Siberia. (They may have been present on the East European Plain during the final Middle Pleistocene interglacial [OIS 7] or earlier, but their sites do not become common until the beginning of the Late Pleistocene [OIS 5e] [Klein 1973; Praslov 1984a].) While their cold-adapted morphology could have been a prerequisite for colonization of these areas, their ability to procure significant amounts of meat through the hunting of large mammals seems equally likely to have been critical to occupation of habitat with lower primary productivity and reduced density of digestible plant foods.

As the Neanderthals spread east into the cooler and drier regions of northern Eurasia, they do not seem to have developed any major new adaptations to the latter. Their morphology and sites are remarkably uniform throughout their geographic range (although the morphology of the Siberian population remains largely unknown [Goebel 2000]). However, groups that expanded into more southern latitudes in the Near East exhibit less extreme morphological adaptations to cold climates (Trinkaus 1981). In Eastern Europe, the Neanderthals used Levallois core techniques somewhat less frequently than elsewhere, and they produced an unusually high proportion of bifacial tools, but their technology was fundamentally similar throughout their distribution (Klein 1969a; Praslov 1984b). The large mammals hunted by the East European Neanderthals contain a higher proportion of typical cold-steppe dwelling taxa (e.g., saiga, bison), but their diet and foraging strategy also seems to have been similar to that of Neanderthals in other parts of Eurasia (Hoffecker 1999a).

The East European record indicates that the Neanderthals were limited in the range of habitats that they could occupy. During cold phases, they apparently abandoned open periglacial loess-steppe areas of the East European Plain (Hoffecker 1987; Soffer 1989). In fact, widespread settlement of the central plain is firmly documented only for the climatic optimum of the Last Interglacial (OIS 5e) at sites such as Khotylevo I on the Middle Desna River (Zavernyaev 1978). The dating of sites on the central plain to subsequent periods (i.e., OIS 5d–3) is problematic. Neanderthal occupations are well represented in the southwest region of the plain—where conditions were somewhat milder—during the cooler interstadials of the Early Glacial (OIS 5a and 5c), but may disappear with the onset of full glacial climates at the start of the Pleniglacial (OIS 4) (Chernysh 1987).

When compared with recent hunter-gatherers in cold environments, the Neanderthal pattern differs significantly in two respects—at least one of which probably accounts for their inability to occupy periglacial loess-steppe habitat. Both the complexity of their technology and the range of their movements are more comparable to those of recent foraging peoples in temperate rather than in arctic/subarctic environments. Their simple technology (apparently without untended facilities or storage devices) may not have provided the needed efficiency of a foraging economy on the loess-steppe (Torrence 1983). Perhaps even more critical was the apparent lack of advanced cold protection technology in the form of tailored fur clothing and insulated shelters in areas where winter temperatures fell below –20 degrees C (Velichko 1984). The Neanderthals apparently worked hides and probably used them for cold protection (Praslov and Semenov 1969), but without tailored clothing and artificial shelters they would have been unable to forage and perform other economic tasks in very low temperatures.

The foraging range of the Neanderthals—especially in the context of cooler environments—probably exceeded that of their predecessors (Roebroeks et al. 1988; Mellars 1996). However, their movements (inferred from transport distances of raw materials from their sources) seem to have been on a lower order of magnitude relative to recent hunter-gatherers subsisting primarily on meat (Kelly 1995). While the mobility demands of the periglacial loess-steppe seem likely to have been lower than those of the modern tundra (due to higher productivity and habitat diversity), the enormous transport distances of raw materials in sites occupied by modern humans in this environment suggest movements and networks on a scale similar to that of recent hunter-gatherers of the High Arctic. If the Neanderthals lacked the ability to forage across wide areas and/or establish large social networks over such areas, this may have been a further constraint to sustained settlement of the central East European Plain (Gamble 1986).

The emergence of the Neanderthals in the late Middle Pleistocene subdivided representatives of *Homo* in western Eurasia and Africa into northern and southern forms. (No comparable subdivision evolved in Eastern Asia, apparently because no opportunities existed for geographic isolation of a large population in a northern region of milder and alternating climates such as Western Europe.) The interaction between the northern and southern groups of *Homo* in Europe and Africa—much of which seems to have taken place in the Near East—is perhaps one of the most important aspects of later human evolution. Some paleoanthropologists believe that there was substantial gene flow between the two groups during the Late Pleistocene (e.g., Wolpoff 1999). Given their apparent degree of morphological and genetic divergence—and similar ecological niches—a competitive relationship is probable.

At the beginning of the Pleniglacial (OIS 4), Neanderthals seem to have extended their range southward, and perhaps replaced an existing early modern (*Homo sapiens*) population in the Near East (Bar-Yosef 1998). This move may have been triggered in part by the more or less simultaneous abandonment of parts of Eastern Europe described earlier. In any case, Neanderthals presumably enjoyed some competitive advantage in the Near East under cooler climatic conditions. Modern humans were again present in the Near East roughly 50,000 years BP (earlier part of OIS 3) and apparently replaced the local Neanderthal population.

The global dispersal of modern humans after 50,000 years BP is arguably tied to the appearance of symbols and symbolic language, which seem to have materialized very suddenly among the *Homo sapiens* population in Africa and/or the Near East at this time. It has been suggested that neurological changes—triggered by a random mutation—were the likely immediate source of this development (Klein 1999, 514–517), but its wider evolutionary context could have been competitive interactions with the Neanderthals. Symbols had an enormous impact on human ecology, because they seem to underlie both the technological ability and organizational structures of "behaviorally modern humans" (e.g., Mithen 1996; Gamble 1999).

Eastern Europe may provide the most suitable setting in Eurasia to observe the role of symbol-based technology and organization in the dispersal of modern humans. The latter apparently occupied the East European Plain during the Middle Pleniglacial (OIS 3) prior to 40,000 years BP. The earliest modern human sites are found at Kostenki on the Middle Don River in the central plain (Praslov and Rogachev 1982; Sinitsyn et al. 1997). These sites are at least as old as any modern human sites in Europe (Mellars 1996). Because the central plain seems to have been abandoned by the Neanderthals by the beginning of OIS 4, it may have provided an open niche for a modern human population entering Europe (Hoffecker 1999a). Occupation of areas such as the Crimea and Northern Caucasus (where Neanderthals were still present until ca. 30,000 years BP) seems to have occurred later (Marks and Chabai 1998; Golovanova et al. 1999).

Derived from southern latitudes, modern humans were morphologically adapted to tropical and subtropical environments, and particularly unsuited for Pleniglacial climates in Eastern Europe (Trinkaus 1981). But in all other respects, the people who colonized northern Eurasia during OIS 3 followed the pattern of recent hunter-gatherers in subarctic and arctic environments. The early Upper Paleolithic archaeological record of the central plain contains evidence of complex and innovative technology in the form of tailored clothing and heated artificial shelters, as well as untended facilities for small-mammal trapping, and novel devices for tool production such as rotary drills. It also reveals impressive increases in the long-distance transport of materials, indicating the existence of wider movements and networks that may have been critical to foraging in periglacial loess-steppe environments (Boriskovskii 1963; Klein 1969b; Praslov and Rogachev 1982).

The archaeological record of the Middle Pleniglacial (OIS 3) in Eastern Europe is complex, and there remain many uncertainties about the process of the transition to modern humans in this region. As in other parts of Europe, Upper Paleolithic artifact assemblages containing many archaic elements that are typical of Neanderthal industries are present (e.g., Strelets Culture, Brynzeny Culture) (Rogachev and Anikovich 1984). It is not clear whether modern humans produced these assemblages, and what relationship they bear to other Upper Paleolithic industries that were indisputably or most probably made by modern humans (e.g., Gorodtsov Culture, Spitsyn Culture, Molodova Culture). One possibility is that Neanderthals made at least some of these artifacts (e.g., possibly Brynzeny) and that they are analogous to West European industries such as the Chatelperronian (Allsworth-Jones 1986). An alternative possibility is that two or more dissimilar populations of modern humans may have entered Eastern Europe during OIS 3 from different routes (e.g., Balkans and Caucasus) (Cohen and Stepanchuk 1999).

The success of modern human technology and organization is most dramatically apparent after the beginning of the Last Glacial Maximum (OIS 2), when modern humans occupied the central East European Plain under full glacial conditions. They created a remarkable adaptation to this bizarre environment that employed ingenious technology and complex organizational structures. During the early phase of the Last Glacial Maximum (ca. 25,000–22,000 years BP), this adaptation took the form of the widespread but highly distinctive "Eastern Gravettian Culture" (also present in loess-steppe areas of Central Europe). Settlement was reduced or abandoned in some areas during the maximum cold peak (21,000–18,000 years BP), and later sites throughout Eastern Europe are termed Epigravettian (17,000–10,000 years BP).

Both the Eastern Gravettian and Epigravettian sites of the East European Plain reveal a pattern of expanding innovation and complexity in technology. The absence of wood in many areas compelled heavy use of bone, antler, and ivory, and rich inventories of nonstone artifacts (which best illustrate this technology) contain examples of microblade-inset spear points, a probable needle case, and possible components of a small-mammal trap (Pidoplichko 1976; Tarasov 1979). Deep pits were excavated and used for cold storage of food and/or fuel. Semisubterranean houses were constructed at some sites, while artificial shelters of mammoth bone (heated with bone fuel) were erected at many of the Epigravettian occupations (Klein 1973; Rogachev and Anikovich 1984; Soffer 1985; Kozlowski 1986; Grigor'ev 1993).

The large open-air sites of the East European Plain contain complex arrangements of features and debris, and a long tradition of broad horizontal excavation of these floors has yielded a an unparalleled mass of information on Pleistocene site structure. Eastern Gravettian occupations include linear feature complexes (once interpreted as the remains of longhouses) that are reminiscent of the aggregation sites of late prehistoric peoples in the Arctic (e.g., McGhee 1997), and probably reflect a similar pattern of dispersal and aggregation of social groups across wide areas (Efimenko 1958; Klein 1969b, 1973). Among the younger Epigravettian sites are arrangements of multiple dwellings and associated activity areas that appear similar to residential camps—typically part of a larger tribal network—among many recent hunter-gatherers (Pidoplichko 1969; Soffer 1985). Equally important are the materials (e.g., shells, flint) transported over hundreds of kilometers from their sources, and the appearance of similar art forms (e.g., "Venus figurines") across immense areas, both of which suggest long-distance movements and/or cooperative networks as among late prehistoric and recent hunter-gatherers in arctic landscapes (Kelly 1995; McGhee 1997).

The combined forces of technology and organization unleashed by the evolution of symbols 50,000 years ago had produced complex societies and simple industry within a few millennia after the end of the Pleistocene. Technological innovation and organizational change seem to have continually stimulated each other, accelerating the pace of development within and among modern human societies throughout the world, although it is apparent that climate oscillations remained a periodic stimulus to change. Our lives are dominated by the consequences of events that took place 50,000 years ago.

Bibliography

Abramova, Z. A. 1967. "Paleolithic Art in the U.S.S.R." *Arctic Anthropology* 4 (2): 1–179.
———. 1993. "Two Examples of Terminal Paleolithic Adaptations." In *From Kostenki to Clovis*, edited by O. Soffer and N. D. Praslov, 85–100. New York: Plenum Press.
Adams, R. M. 1996. *Paths of Fire: An Anthropologist's Inquiry into Western Technology*. Princeton: Princeton University Press.
Alekseev, V. P. 1978. *Paleoantropologiya Zemnogo Shara i Formirovanie Chelovecheskikh Ras: Paleolit*. Moscow: Nauka.
Alekseeva, L. I. 1987. "Teriofauna Mnogosloinoi Stoyanki Molodova V." In *Mnogosloinaya Paleoliticheskaya Stoyanka Molodova V: Lyudi Kamennogo Veka i Okruzhayushchaya Sreda*, edited by I. K. Ivanova and S. M. Tseitlin, 153–162. Moscow: Nauka.
———. 1990. *Teriofauna Verkhnego Pleistotsena Vostochnoi Evropy (Krupnye Mlekopitayushchie)*. Moscow: Nauka.
Allen, H. 1990. "Environmental History in Southwestern New South Wales During the Late Pleistocene." In *The World at 18 000 BP*. Vol. 2: *Low Latitudes*, edited by C. Gamble and O. Soffer, 296–321. London: Unwin Hyman.
Allsworth-Jones, P. 1986. *The Szeletian and the Transition from Middle to Upper Paleolithic in Central Europe*. Oxford, U.K.: Clarendon Press.
———. 1990. "The Szeletian and the Stratigraphic Succession in Central Europe and Adjacent Areas: Main Trends, Recent Results, and Problems for Resolution." In *The Emergence of Modern Humans*, edited by P. Mellars, 160–242. Edinburgh: Edinburgh University Press.
Amirkhanov, Kh. A. 1986. *Verkhnii Paleolit Prikuban'ya*. Moscow: Nauka.
———. 1997. "Problemy Stratigrafii i Khronologii Kul'turnykh Otlozhenii Zaraiskoi Stoyanki." *Sovetskaya Arkheologiya* 4: 5–16.
Amirkhanov, Kh. A., M. V. Anikovich, and I. A. Borziyak. 1980. "K Probleme Perekhoda ot Must'e k Verkhnemu Paleolitu na Territorii Russkoi Ravniny i Kavkaza." *Sovetskaya Arkheologiya* 2: 5–22.
Anderson-Gerfaud, P. 1990. "Aspects of Behaviour in the Middle Palaeolithic: Functional Analysis of Stone Tools from Southwest France." In *The Emergence of Modern Humans*, edited by P. Mellars, 389–418. Edinburgh: Edinburgh University Press.
Anikovich, M. V. 1993. "O Znachenii Kostenkovsko-Borshchevskogo Raiona v Sovremennom Paleolitovedenii." *Peterburgskii Arkheologicheskii Vestnik* 3: 3–19.
Anisyutkin, N. K. 1969. "Must'erskaya Stoyanka Stinka na Srednem Dnestre (Predvaritel'noe Soobshchenie)." *Arkheologicheskii Sbornik* 11: 5–17.
———. 1981a. "Arkheologicheskoe Izuchenie Must'erskoi Stoyanki Ketrosy." In *Ketrosy: Must'erskaya Stoyanka na Srednem Dnestre*, edited by N. D. Praslov, 7–53. Moscow: Nauka.

———. 1981b. "Stratifitsirovannye Nakhodki Domust'erskoi Epokhi v Chernovitskoi Oblasti." *Kratkie Soobshchenie Instituta Arkheologii* 165: 55–58.

———. 1987. "De Nouvelles Données sur le Paléolithique Ancien de la Moldavie." *L'Anthropologie* 91: 69–74.

Anisyutkin, N. K., I. A. Borziyak, and N. A. Ketraru. 1986. *Pervobytnyi Chelovek v Grotakh Trinka I–III*. Kishinev: Shtiinsta.

Anisyutkin, N. K., and G. M. Levkovskaya. 1992. "Mestonakhozhdeniya Rannego Paleolita v Raione g. Dubossary i Novye Dannye o Paleogeografii Srednego Pleistotsena Nizhnego Podnestrov'ya." In *Rannepaleoliticheskie Kompleksy Evrazii*, edited by A. P. Derevyanko and V. T. Petrin, 82–92.Novosibirsk: Nauka.

Archibold, O. W. 1995. *Ecology of World Vegetation*. London: Chapman and Hall.

Arsuaga, J. L., I. Martinez, A. Gracia, and C. Lorenzo.1997. "The Sima de los Huesos Crania (Sierra de Atapuerca, Spain): A Comparative Study." *Journal of Human Evolution* 33: 219–281.

Ascenzi, A., I. Biddittu, P. F. Cassoli, A. G. Segre, and E. Segre-Naldini. 1996. "A Calvarium of Late *Homo erectus* from Ceprano, Italy." *Journal of Human Evolution* 31: 409–423.

Autlev, P. U. 1964. "Gubskaya Paleoliticheskaya Stoyanka." *Sovetskaya Arkheologiya* 4: 172–176.

Avernianov, A. O. 1998. "Late Pleistocene Hares (*Lepus*) of the Russian Plain." In *Quaternary Paleozoology in the Northern Hemisphere*, edited by J. J. Saunders, B. W. Styles, and G. F. Baryshnikov, 41–68. Illinois State Museum Scientific Papers 27. Springfield: Illinois State Museum.

Bader, O. N. 1965. *Kapovaya Peshchera: Paleoliticheskaya Zhivopis'*. Moscow: Nauka.

———. 1978. *Sungir' Verkhnepaleoliticheskaya Stoyanka*. Moscow: Nauka.

Bahn, P. G., ed. 1996. *The Cambridge Illustrated History of Archaeology*. Cambridge, U.K.: Cambridge University Press.

Bailey, H. P. 1960. "A Method of Determining the Warmth and Temperateness of Climate." *Geografiska Annaler* 42 (1): 1–16.

Bar-Yosef, O. 1989. "Upper Pleistocene Cultural Stratigraphy in Southwest Asia." In *The Emergence of Modern Humans*, edited by E. Trinkaus, 154–180. Cambridge, U.K.: Cambridge University Press.

———. 1998. "The Chronology of the Middle Paleolithic of the Levant." In *Neandertals and Modern Humans in Western Asia*, edited by T. Akazawa, K. Aoki, and O. Bar-Yosef, 39–56. New York: Plenum Press.

Baryshnikov, G. F. 1993. "Krupnye Mlekopitayushchie Ashel'skoi Stoyanki v Peshchere Treugol'naya na Severnom Kavkaze." *Trudy Zoologicheskogo Instituta AN SSSR* 249: 3–47.

Baryshnikov, G., and J. F. Hoffecker. 1994. "Mousterian Hunters of the NW Caucasus: Preliminary Results of Recent Investigations." *Journal of Field Archaeology* 21: 1–14.

Baryshnikov, G., J. F. Hoffecker, and R. L. Burgess. 1996. "Zooarchaeology and Palaeontology of Mezmaiskaya Cave, Northwestern Caucasus." *Journal of Archaeological Science* 23: 313–335.

Baryshnikov, G. F., A. K. Kasparov, and A. N. Tikhonov. 1990. "Saiga Paleolita Kryma." *Trudy Zoologicheskogo Instituta AN SSSR* 212: 3–48.

Belyaeva, E. V. 1992. "Novye Issledovaniya Monasheskoi Peshchery na Gubse." In *Voprosy Arkheologii Adygei*, edited by D. Kh. Mekulov, 182–193. Maikop: ANIIEYaLI.

Belyaeva, V. I. 1997. "Issledovaniya Novogo Uchastka Poseleniya na Paleoliticheskoi Stoyanke Pushkari I." In *Pushkarevskii Sbornik*, edited by V. I. Belyaeva, 5–18. Saint Petersburg: Obrazovanie Kul'tura.

Beregovaya, N. A. 1960. *Paleoliticheskie Mestonakhozhdeniya SSSR*. Materialy i Issledovaniya po Arkheologii SSSR 81.

Beyries, S. 1988. "Functional Variability of Lithic Sets in the Middle Paleolithic." In *Upper Pleistocene Prehistory of Western Eurasia*, edited by H. L. Dibble and A. Montet-White, 213–224. Philadelphia: University of Pennsylvania Museum.

Bibikova, V. I., and A. V. Starkin.1989. "Teriokompleks Pozdnepaleoliticheskogo Poseleniya Anetovka II." In *Chetvertichnyi Period: Paleontologiya i Arkheologiya*, edited by A. L. Yanshin, 8–16. Kishinev: Shtiintsa.

Binford, L. R. 1980. "Willow Smoke and Dogs' Tails: Hunter-Gatherer Settlement Systems and Archaeological Site Formation." *American Antiquity* 45 (1): 4–20.

———. 1981. *Bones: Ancient Men and Modern Myths*. New York: Academic Press.

———. 1983. *In Pursuit of the Past*. New York: Thames and Hudson.

———. 1984. *Faunal Remains from Klasies River Mouth*. Orlando, Fla.: Academic Press.

———. 1985. "Human Ancestors: Changing Views of Their Behavior." *Journal of Anthropological Archaeology* 4: 292–327.

———. 1990. "Mobility, Housing, and Environment: A Comparative Study." *Journal of Anthropological Research* 46: 119–152.

———. 1993. "Bones for Stones: Considerations of Analogues for Features Found on the Central Russian Plain." In *From Kostenki to Clovis*, edited by O. Soffer and N. D. Praslov, 101–124. New York: Plenum Press.

Bocherens, H., D. Billiou, A. Mariotti, M. Patou-Mathis, M. Otte, D. Bonjean, and M. Toussaint. 1999. "Palaeoenvironmental and Palaeodietary Implications of Isotopic Biogeochemistry of Last Interglacial Neanderthal and Mammal Bones in Scladina Cave (Belgium)." *Journal of Archaeological Science* 26: 599–607.

Boëda, E. 1993. *Le Concept Levallois: Variabilité des Méthodes*. Paris: CNRS.

Bolikhovskaya, N. S., and G. A. Pashkevich. 1982. "Dinamika Rastitel'nosti v Okrestnostyakh Stoyanki Molodova i v Pozdnem Pleistotsene." In *Molodova I: Unikal'noe Must'erskoe Poselenie na Srednem Dnestre*, edited by G. I. Goretskii and I. K. Ivanova, 120–144. Moscow: Nauka.

Bonch-Osmolovskii, G. A. 1934. "Résultats de l'Étude du Paléolithique du Crimée." *Transactions of the II International Conference of the Association on the Study of the Quaternary Period in Europe* 5: 113–183.

———. 1940. *Paleolit Kryma*, vol. 1. Moscow: AN SSSR.

———. 1941. "Kist' Iskopaemogo Cheloveka iz Grota Kiik-Koba." *Paleolit Kryma*, vol. 2. Moscow: AN SSSR.

———. 1954. "Skelet Stopy i Goleni Iskopaemogo Cheloveka iz Grota Kiik-Koba." *Paleolit Kryma*, vol. 3. Moscow: AN SSSR.

Bordaz, J. 1970. *Tools of the Old and New Stone Age*. Garden City: The Natural History Press.

Bordes, F. 1961. *Typologie du Paléolithique Ancien et Moyen*. Bordeaux: Delmas.

Boriskovskii, P. I. 1932. "K Voprosu o Stadial'nosti v Razvitii Verkhnego Paleolita." *Izvestiya GAIMK* 14 (4): 9–40.

―――. 1953. *Paleolit Ukrainy*. Materialy i Issledovaniya po Arkheologii SSSR 40.

―――. 1963. *Ocherki po Paleolitu Basseina Dona*. Materialy i Issledovaniya po Arkheologii SSSR 121.

Boriskovskii, P. I., and N. D. Praslov. 1964. "Paleolit Basseina Dnepra i Priazov'ya." *Svod Arkheologicheskikh Istochnikov* A 1-5.

Boriskovskii, P. I., N. D. Praslov, and M. V. Anikovich. 1982. "Kostenki 17." In *Paleolit Kostenkovsko-Borshchevskogo Raiona na Donu 1879–1979*, edited by N. D. Praslov and A. N. Rogachev, 181–186. Leningrad: Nauka.

Borziac, I. A., P. Allsworth-Jones, C. French, S. I. Medyanik, W. J. Rink, and H. K. Lee, 1997. "The Upper Palaeolithic Site of Ciuntu on the Middle Pruth, Moldova: A Multidisciplinary Study and Reinterpretation." *Proceedings of the Prehistoric Society* 63: 285–301.

Borziyak, I. A. 1993. "Subsistence Practices of Late Paleolithic Groups along the Dnestr River and Its Tributaries." In *From Kostenki to Clovis*, edited by O. Soffer and N. D. Praslov, 67–84. New York: Plenum Press.

Borziyak, I. A., and N. A. Ketraru. 1976. "Issledovaniya Paleoliticheskoi Stoyanki v Grote Chuntu." In *Arkheologicheskie Otkrytiya 1975 Goda*, 468–469. Moscow: Nauka.

Borziyak, I. A., and S. I. Kovalenko. 1989. "Nekotorye Dannye o Mnogosloinoi Paleoliticheskoi Stoyanke Kosoutsy na Srednem Dnestre." In *Chetvertichnyi Period: Paleontologiya i Arkheologiya*, 201–218. Kishinev: Shtiintsa.

Boserup, E. 1981. *Population and Technological Change*. Chicago: University of Chicago Press.

Bosinski, G. 1967. *Die Mittelpaläolithischen Funde im Westlichen Mitteleuropa*. Cologne: Herman Bohlau.

Bowen, D. Q. 1978. *Quaternary Geology: A Stratigraphic Framework for Multidisciplinary Study*. Oxford, U.K.: Pergamon Press.

Brace, C. L. 1964. "The Fate of the 'Classic' Neanderthals: A Consideration of Hominid Catastrophism." *Current Anthropology* 5: 3–43.

Bradley, B. A., M. Anikovich, and E. Giria. 1995. "Early Upper Paleolithic in the Russian Plain: Streletskayan Flaked Stone Artifacts and Technology." *Antiquity* 69: 989–998.

Bradley, R. S. 1985. *Quaternary Paleoclimatology: Methods of Paleoclimatic Reconstruction*. Boston: Allen and Unwin.

Brain, C. K. 1981. *The Hunters or the Hunted? An Introduction to African Cave Taphonomy*. Chicago: University of Chicago Press.

Bräuer, G. 1984. "A Craniological Approach to to the Origin of Anatomically Modern Homo sapiens in Africa and Implications for the Appearance of Modern Europeans." In *The Origins of Modern Humans*, edited by F. H. Smith and F. Spencer, 327–410. New York: Alan R. Liss.

―――. 1989. "The Evolution of Modern Humans: A Comparison of the African and Non-African Evidence." In *The Human Revolution*, edited by P. Mellars and C. Stringer, 123–154. Princeton: Princeton University Press.

Bud'ko, V. D. 1964. "O Zhilishchakh Berdyzhskoi Paleoliticheskoi Stoyanki." *Kratkie Soobshcheniya Instituta Arkheologii* 101: 31–34.

Bud'ko, V. D., L. N. Voznyachuk, and V. I. Kochetkov. 1970. "Nekotorye Rezul'tati Raskopki Berdyzhskoi Stoyanki." In *Arkheologicheskie Otkrytiya 1969 Goda*, 295–296. Moscow: Nauka.

Bud'ko, V. D., L. N. Voznyachuk, and E. G. Laechits. 1971. "Paleoliticheskaya Stoyanka Berdyzh." In *Arkheologicheskie Otkrytiya 1970 Goda*, 303–304. Moscow: Nauka.

Burke, A. 1999a. "Butchering and Scavenging at the Middle Paleolithic Site of Starosele." In *The Middle Paleolithic of Western Crimea*. Vol. 2, edited by V. P. Chabai and K. Monigal, 1–27. ERAUL 87.

———. 1999b. "Kabazi V: Faunal Exploitation at a Middle Paleolithic Rockshelter in Western Crimea." In *The Middle Paleolithic of Western Crimea*. Vol. 2, Edited by V. P. Chabai and K. Monigal, 29–39. ERAUL 87.

———. 2000. "The View from Starosele: Faunal Exploitation at a Middle Palaeolithic Site in Western Crimea." *International Journal of Osteoarchaeology* 10: 325–335.

Burova, N. D. 1999. "Metodika Tafonomicheskogo Issledovaniya Kostnogo Skopleniya Krupnykh Mlekopitayushchikh na Verkhnepaleoliticheskom Poselenii Yudinovo." Paper presented at *Osobennosti Razvitiya Verkhnego Paleolita Vostochnoi Evropy*, Saint Petersburg, Russia, 15–19 November 1999.

Butzer, K. W. 1971. *Environment and Archaeology: An Ecological Approach to Prehistory*. 2nd ed. Chicago: Aldine.

———. 1982. *Archaeology as Human Ecology*. Cambridge, U.K.: Cambridge University Press.

Cachel, S., and J.W.K. Harris. 1998. "The Lifeways of *Homo erectus* Inferred from Archaeology and Evolutionary Ecology: A Perspective from East Africa." In *Early Human Behaviour in Global Context: The Rise and Diversity of the Lower Palaeolithic Record*, edited by M. D. Petraglia and R. Korisettar, 108–132. London: Routledge.

Carbonell, E., J. M. Bermudez de Castro, J. L. Arsuaga, J. C. Diez, A. Rosas, G. Cuenca-Bescos, R. Sala, M. Mosquera, X. P. Rodriguez. 1995. "Lower Pleistocene Hominids and Artifacts from Atapuerca-TD6 (Spain)." *Science* 269: 826–832.

Carbonell, E., and Z. Castro-Curel. 1992. "Palaeolithic Wooden Artifacts from the Abric Romani (Capellades, Barcelona, Spain)." *Journal of Archaeological Science* 19: 707–719.

Carciumaru, M. 1980. *Mediul Geografic in Pleistocenul Superior si Culture Paleolitice din Romania*. Bucharest: Editura Academiei Republicii Socialiste Romania.

Castro-Curel, Z., and E. Carbonell. 1995. "Wood Pseudomorphs from Level I at Abric Romani, Barcelona, Spain." *Journal of Field Archaeology* 22: 376–384.

Chabai, V. P. 1998. "Kabazi II: The Western Crimean Mousterian Assemblages of Unit II, Levels II/7–II/8C." In *The Middle Paleolithic of Western Crimea*. Vol. 1, edited by A. E. Marks and V. P. Chabai, 201–252. ERAUL 84.

Chabai, V. P., and Yu. E. Demidenko. 1998. "The Classification of Flint Artifacts." In *The Middle Paleolithic of Western Crimea*. Vol. 1, edited by A. E. Marks and V. P. Chabai, 31–51. ERAUL 84.

Chabai, V. P., and C. R. Ferring. 1998. "Kabazi II: Introduction." In *The Middle Paleolithic of Western Crimea*. Vol. 1, edited by A. E. Marks and V. P. Chabai, 167–200. ERAUL 84.

Chabai, V. P., and A. E. Marks. 1998. "Preliminary Synthesis: Middle Paleolithic

Assemblage Variability in Western Crimea." In *The Middle Paleolithic of Western Crimea*. Vol. 1, edited by A. E. Marks and V. P. Chabai, 355–367. ERAUL 84.

Chabai, V. P., A. E. Marks, and A. I. Yevtushenko. 1995. "Views of the Crimean Middle Paleolithic: Past and Present." *Préhistoire Européene* 7: 59–79.

Chase, P. G. 1986. *The Hunters of Combe Grenal: Approaches to Middle Paleolithic Subsistence in Europe*. Oxford, U.K.: British Archaeological Reports International Series S-286.

Chase, P. G., and H. L. Dibble. 1987. "Middle Paleolithic Symbolism: A Review of Current Evidence and Interpretations." *Journal of Anthropological Archaeology* 6: 263–296.

Chepalyga, A. L. 1984. "Inland Sea Basins." In *Late Quaternary Environments of the Soviet Union*, edited by A. A. Velichko, 229–247. Minneapolis: University of Minnesota Press.

Chernysh, A. P. 1959. Pozdnii Paleolit Srednego Pridnestrov'ya. *Trudy Komissii po Izucheniyu Chetvertichnogo Perioda* 15: 5–214.

———. 1961. *Paleolitichna Stoyanka Molodove V*. Kiev: AN USSR.

———. 1965. Rannii i Srednii Paleolit Pridnestrov'ya. *Trudy Komissii po Izucheniyu Chetvertichnogo Perioda* 25.

———. 1968. "Paleoliticheskaya Stoyanka Ataki I v Pridnestrov'e." *Byulleten' Komissii po Izucheniyu Chetvertichnogo Perioda* 35: 102–112.

———. 1971. "Issledovaniya Stoyanki Oselivka I v 1966–1967 gg." *Kratkie Soobshcheniya Instituta Arkheologii* 126: 68–77.

———. 1973. *Paleolit i Mezolit Pridnestrov'ya*. Moscow: Nauka.

———. 1977. "Mnogosloinaya Paleoliticheskaya Stoyanka Korman' IV i Ee Mesto v Paleolite." In *Mnogosloinaya Paleoliticheskaya Stoyanka Korman' IV na Srednem Dnestre*, edited by G. I. Goretskii and S. M. Tseitlin, 7–77. Moscow: Nauka.

———. 1982. "Mnogosloinaya Paleoliticheskaya Stoyanka Molodova I." In *Molodova I: Unikal'noe Must'erskoe Poselenie na Srednem Dnestre*, edited by G. I. Goretskii and I. K. Ivanova, 6–102. Moscow: Nauka.

———. 1987. "Etalonnaya Mnogosloinaya Stoyanka Molodova V. Arkheologiya." In *Mnogosloinaya Paleoliticheskaya Stoyanka Molodova V: Lyudi Kamennogo Veka i Okruzhayushchaya Sreda*, edited by I. K. Ivanova and S. M. Tseitlin, 7–93. Moscow: Nauka.

Chiguryaeva, A. A., and N. Ya. Khvalina. 1961. "O Kharaktere Rastitel'nosti v Raione Stalingrada v Epokhu Srednego Paleolita." *Kratkie Soobshcheniya Instituta Arkheologii* 82: 37–41.

Close, A., and F. Wendorf. 1990. "North Africa at 18 000 BP." In *The World at 18 000 BP*, vol. 2: *Low Latitudes*, edited by C. Gamble and O. Soffer, 41–57. London: Unwin Hyman.

Cohen, V. Yu., and V. N. Stepanchuk. 1999. "Late Middle and Early Upper Paleolithic Evidence from the East European Plain and Caucasus: A New Look at Variability, Interactions, and Transitions." *Journal of World Prehistory* 13 (3): 265–319.

Commont, V. 1909. "L'Industrie Mousterienne dans la Région du Nord de la France." In *Compte Rendu de la 5ième Session du Congrès Préhistorique de France*, 115–157. Paris: Société Préhistorique Française.

Coon, C. S. 1962. *The Origin of Races*. New York: Alfred A. Knopf.
———. 1971. *The Hunting Peoples*. New York: Little, Brown and Co.
Danilova, E. I. 1979a. "Zatylochnaya Kost' Neandertal'tsa iz Transhei Zaskal'naya v vozle Ak-Kaya." In *Issledovanie Paleolita v Krymu*, edited by Yu. G. Kolosov, 76–84. Kiev: Naukova Dumka.
———. 1979b. "Fragment Pyastnoi Kosti Neandertal'tsa." In *Issledovanie Paleolita v Krymu*, edited by Yu. G. Kolosov, 84–85. Kiev: Naukova Dumka.
Darwin, C. 1859. *On the Origin of Species*. London: John Murray.
David, A. I. 1981. "Ostatki Mlekopitayushchikh iz Raskopok Paleoliticheskoi Stoyanki Ketrosy." In *Ketrosy: Must'erskaya Stoyanka na Srednem Dnestre*, edited by N. D. Praslov, 135–142. Moscow: Nauka.
David, A. I., and N. A. Ketraru. 1970. "Fauna Mlekopitayushchikh Paleolita Moldavii." In *Fauna Kainozoya Moldavii*, 3–53. Kishinev: AN MSSR.
Debets, G. F. 1955. "Paleoantropologicheskie Nakhodki v Kostenkakh." *Sovetskaya Etnografiya* 1: 43–53.
———. 1967. "Skelet Pozdnepaleoliticheskogo Cheloveka iz Pogrebeniya na Sungirskoi Stoyanke." *Sovetskaya Arkheologiya* 3: 160–164.
Deetz, J. 1967. *Invitation to Archaeology*. Garden City, N.Y.: The Natural History Press.
Demidenko, Yu. E. 1999. "Problema Identifikatsii Orin'yaka v Vostochnoi Evrope i Atributsiya Orin'yaka Syureni I (Krym) i Kostenok I (Srednii Don)." Paper presented at *Osobennosti Razvitiya Verkhnego Paleolita Vostochnoi Evropy*, Saint Petersburg, Russia, 15–19 November 1999.
Dennell, R. 1983. *European Economic Prehistory: A New Approach*. London: Academic Press.
D'Errico, F., and P. Villa. 1997. "Holes and Grooves: The Contribution of Microscopy and Taphonomy to the Problem of Art Origins." *Journal of Human Evolution* 33: 1–31.
Dibble, H. L. 1987. "The Interpretation of Middle Paleolithic Scraper Morphology." *American Antiquity* 52: 109–117.
Doronichev, V. B. 1992. "Ranneashel'skaya Stoyanka v Treugol'noi Peshchere." In *Voprosy Arkheologii Adygei*, edited by D. Kh. Mekulov, 102–134. Maikop: ANIIEYaLI.
Efimenko, P. P. 1938. *Pervobytnoe Obshchestvo*. 2nd ed. Leningrad: USSR Academy of Sciences.
———. 1958. *Kostenki I*. Moscow: Akademiya Nauk SSSR.
Eldredge, N. 1985. *Time Frames: The Evolution of Punctuated Equilibria*. Princeton: Princeton University Press.
Elias, S. A. 1994. *Quaternary Insects and Their Environments*. Washington, D.C.: Smithsonian Institution Press.
Engels, F. 1942. (1884). *The Origin of the Family, Private Property, and the State*. New York: International Publishers.
Enloe, J. G., F. David, and G. Baryshnikov. 2000. "Hyenas and Hunters: Zooarchaeological Investigations at Prolom II Cave, Crimea." *International Journal of Osteoarchaeology* 10: 310–324.
Fagan, B. 1990. *The Journey from Eden: The Peopling of Our World*. London: Thames and Hudson.

Farizy, C., and F. David. 1992. "Subsistence and Behavioral Patterns of Some Middle Paleolithic Local Groups." In *The Middle Paleolithic: Adaptation, Behavior, and Variability*, edited by H. L. Dibble and P. A. Mellars, 87–96. Philadelphia: University of Pennsylvania.

Faustova, M. A. 1984. "Late Pleistocene Glaciation of the European USSR." In *Late Quaternary Environments of the Soviet Union*, edited by A. A. Velichko, 3–12. Minneapolis: University of Minnesota Press.

Féblot-Augustins, J. 1999. "Raw Material Transport Patterns and Settlement Systems in the European Lower and Middle Palaeolithic: Continuity, Change, and Variability." In *The Middle Palaeolithic Occupation of Europe*, edited by W. Roebroeks and C. Gamble, 193–214. Leiden: University of Leiden.

Ferring, C. R. 1998. "The Geologic Setting of Mousterian Sites in Western Crimea." In *The Middle Paleolithic of Western Crimea*. Vol. 1, edited by A. E. Marks and V. P. Chabai, 17–30. ERAUL 84.

Foley, R. A. 1987. *Another Unique Species: Patterns in Human Evolutionary Ecology*. New York: John Wiley and Sons.

———. 1989. "The Evolution of Hominid Social Behaviour." In *Comparative Socioecology: The Behavioural Ecology of Humans and Other Mammals*, edited by V. Standen and R. A. Foley, 473–494. Oxford, U.K.: Blackwell Scientific Publications.

Formozov, A. A. 1958. *Peshchernaya Stoyanka Starosel'e i Ee Mesto v Paleolite*. Materialy i Issledovaniya po Arkheologii SSSR 71.

———. 1959. "Must'erskaya Stoyanka Kabazi v Krymu." *Sovetskaya Arkheologiya* 29–30: 143–158.

———. 1965. *Kamennyi Vek i Eneolit Prikuban'ya*. Moscow: Nauka.

———. 1971. "Kamennomostskaya Peshchera—Mnogosloinaya Stoyanka v Prikuban'e." *Materialy i Issledovaniya po Arkheologii SSSR* 173: 100–116.

Franciscus, R. G. 1999. "Neandertal Nasal Structures and Upper Respiratory Tract 'Specialization'." *Proceedings of the National Academy of Sciences* 96: 1805–1809.

Frayer, D. W. 1978. *Evolution of the Dentition in Upper Paleolithic and Mesolithic Europe*. University of Kansas Publications in Anthropology 10.

Frenzel, B. 1968. "The Pleistocene Vegetation of Northern Eurasia." *Science* 161: 637–649.

———. 1973. *Climatic Fluctuations of the Ice Age*. Translated by A.E.M. Nairn. Cleveland: Case Western Reserve University.

Frison, G. C., ed. 1974. *The Casper Site*. New York: Academic Press.

———. 1978. *Prehistoric Hunters of the High Plains*. New York: Academic Press.

Gabunia, L., et al. 2000. "Earliest Pleistocene Hominid Cranial Remains from Dmanisi, Republic of Georgia: Taxonomy, Geological Setting, and Age." *Science* 288: 1219–1025.

Gambier, D. 1989. "Fossil Hominids from the Early Upper Palaeolithic (Aurignacian) of France." In *The Human Revolution*, edited by P. Mellars and C. Stringer., 194–211. Princeton: Princeton University Press.

Gamble, C. 1982. "Interaction and Alliance in Palaeolithic Society." *Man* 17: 92–107.

———. 1986. *The Palaeolithic Settlement of Europe*. Cambridge, U.K.: Cambridge University Press.

———. 1994. *Timewalkers: The Prehistory of Global Colonization.* Cambridge, Mass.: Harvard University Press.

———. 1995. "The Earliest Occupation of Europe: The Environmental Background." In *The Earliest Occupation of Europe,* edited by W. Roebroeks and T. van Kolfschoten, 279–295. Leiden: University of Leiden.

———. 1999. *The Palaeolithic Societies of Europe.* Cambridge, U.K.: Cambridge University Press.

Gamble, C., and O. Soffer. eds. 1990. *The World at 18 000 BP.* Vol. 2, *Low Latitudes.* London: Unwin Hyman.

Gargett, R. H. 1989. "Grave Shortcomings: The Evidence for Neandertal Burial." *Current Anthropology* 30: 157–190.

Gaudzinski, S. 1996. "On Bovid Assemblages and Their Consequences for the Knowledge of Subsistence Patterns in the Middle Palaeolithic." *Proceedings of the Prehistoric Society* 62: 19–39.

Gerasimova, M. M. 1982. "Paleoantropologicheskie Nakhodki." In *Paleolit Kostenkovsko-Borshchevskogo Raiona na Donu 1879–1979,* edited by N. D. Praslov and A. N. Rogachev, 245–257. Leningrad: Nauka.

Gladilin, V. N. 1976. *Problemy Rannego Paleolita Vostochnoi Evropy.* Kiev: Naukova Dumka.

Gladilin, V. N., and V. I. Sitlivyi. 1990. *Ashel' Tsentral'noi Evropy.* Kiev: Naukova Dumka.

Gladkikh, M. I. 1981. "Issledovaniya v Mezhirichakh." In *Arkheologicheskie Otkrytiya 1980 Goda,* 239. Moscow: Nauka.

Gladkikh, M. I., N. L. Kornietz, and O. Soffer. 1984. "Mammoth-Bone Dwellings on the Russian Plain." *Scientific American* 251 (5): 164–175.

Goebel, T. 2000. "Pleistocene Human Colonization of Siberia and Peopling of the Americas: An Ecological Approach." *Evolutionary Anthropology* 8: 208–227.

Golovanova, L. V., J. F. Hoffecker, V. M. Kharitonov, and G. P. Romanova. 1999. "Mezmaiskaya Cave: A Neanderthal Occupation in the Northern Caucasus." *Current Anthropology* 41: 77–86.

Golovanova, L. V., G. M. Levkovskaya, and G. F. Baryshnikov. 1990. "Le Nouveau Site Moustérien en Grotte de Matouzka, Caucase Septentrional (Résultats des Fouilles de 1985–1987)." *L'Anthropologie* 94: 739–762.

Gorodtsov, V. A. 1935. "Timonovskaya Paleoliticheskaya Stoyanka." *Trudy Instituta Antropologii, Etnografii, i Arkheologii* 3.

———. 1941. "Rezul'tati Issledovaniya Il'skoi Paleoliticheskoi Stoyanki (Predvaritel'noe Soobshchenie)." *Materialy i Issledovaniya po Arkheologii SSSR* 2: 7–25.

Gould, R. A. 1980. *Living Archaeology.* Cambridge, U.K.: Cambridge University Press.

Gould, S. J., and R. C. Lewontin. 1979. "The Spandrels of San Marco and the Panglossian Paradigm: A Critique of the Adaptationist Programme." *Proceedings of the Royal Society of London* 205: 581–598.

Grichuk, V. P. 1972. "Osnovye Etapy Istorii Rastitel'nosti Yugo-Zapada Russkoi Ravniny v Pozdnem Pleistotsene." In *Palinologiya Pleistotsena,* edited by V. P. Grichuk, 9–53. Moscow: Nauka.

———. 1982. "Rastitel'nost' Evropy v Pozdnem Pleistotsene." In *Paleogeografiya*

Evropy za Poslednie Sto Tysyach Let, edited by I. P. Gerasimov and A. A. Velichko, 42–109. Moscow: Nauka.

———. 1984. "Late Pleistocene Vegetation History." In *Late Quaternary Environments of the Soviet Union*, edited by A. A. Velichko.,155–178. Minneapolis: University of Minnesota Press.

Grigor'ev, G. P. 1963. "Selet i Kostenkovsko-Streletskaya Kul'tura." *Sovetskaya Arkheologiya* 3–11.

———. 1967. "A New Reconstruction of the Above-Ground Dwelling of Kostenki." *Current Anthropology* 8: 344–349.

———. 1968. *Nachalo Verkhnego Paleolita i Proiskhozhdenie Homo sapiens*. Leningrad: Nauka.

———. 1970. "Verkhnii Paleolit." In *Kamennyi Vek na Territorii SSSR*, edited by A. A. Formozov, 43–63. Moscow: Nauka.

———. 1993. "The Kostenki-Avdeevo Archaeological Culture and the Willendorf-Pavlov-Kostenki-Avdeevo Cultural Unity." In *From Kostenki to Clovis*, edited by O. Soffer and N. D. Praslov, 51–65. New York: Plenum Press.

Grigor'eva, G. V. 1983. "Korpatch, Un Gisement Stratifié du Paléolithique Supérieur en Moldavie." *L'Anthropologie* 87: 215–220.

Gromov, V. I. 1948. *Paleontologicheskoe i Arkheologicheskoe Obosnovanie Stratigrafii Kontinental'nykh Otlozhenii Chetvertichnogo Perioda na Territorii SSSR (Mlekopitayushchie, Paleolit)*. Trudy Instituta Geograficheskikh Nauk AN SSSR 64, Geologicheskaya Seriya 17.

———. 1966a. "Fauna iz Stoyanki Sungir'." In *Verkhnepaleoliticheskaya Stoyanka Sungir'*, edited by V. N. Sukachev, V. I. Gromov, and O. N. Bader, 74–78. Moscow: Nauka.

———. 1966b. "Paleogeografiya i Geologicheskii Vozrast Stoyanki Sungir'." In *Verkhnepaleoliticheskaya Stoyanka Sungir'*, edited by V. N. Sukachev, V. I. Gromov, and O. N. Bader, 102–114. Moscow: Nauka.

Grosswald, M. G. 1980. "Late Weichselian Ice Sheet of Northern Eurasia." *Quaternary Research* 13: 1–32.

Guslitzer, B. I., and P. Yu. Pavlov. 1993. "Man and Nature in Northeastern Europe in the Middle and Late Pleistocene." In *From Kostenki to Clovis*, edited by O. Soffer and N. D. Praslov, 175–188. New York: Plenum Press.

Guthrie, R. D. 1982. "Mammals of the Mammoth Steppe as Paleoenvironmental Indicators." In *Paleoecology of Beringia*, edited by D. M. Hopkins, J. V. Matthews, C. E. Schweger, and S. B. Young, 307–326. New York: Academic Press.

———. 1990. *Frozen Fauna of the Mammoth Steppe: The Story of Blue Babe*. Chicago: University of Chicago Press.

Gvozdover, M. D. 1947. "Paleoliticheskaya Stoyanka Bugorok." *Kratkie Soobshcheniya Instituta Istorii Material'noi Kul'tury* 15: 92–97.

———. 1953. "Obrabotka Kosti i Kostyanye Izdeliya Avdeevskoi Stoyanki." *Materialy i Issledovaniya po Arkheologii SSSR* 39: 192–226.

———. 1964. "Pozdnepaleoliticheskie Pamyatniki Nizhnego Dona." *Svod Arkheologicheskikh Istochnikov* A 1-5: 37–41.

———. 1989. "The Typology of Female Figurines of the Kostenki Paleolithic Culture." *Soviet Anthropology and Archaeology* 27 (4): 32–94.

Gvozdover, M. D., and G. P. Grigor'ev. 1978. "Ocherednoi God Raboty na Avdeevskoi Paleoliticheskoi Stoyanke bliz Kurska." In *Arkheologicheskie Otkrytiya 1977 Goda*, 54–55. Moscow : Nauka.

Hahn, J. 1972. "Aurignacian Signs, Pendants, and Art Objects in Central and Eastern Europe." *World Archaeology* 3: 252–266.

Hamilton, W. D. 1964. "The Genetical Evolution of Social Behaviour." *Journal of Theoretical Biology* 7: 1–32.

Hapgood, P. J. 1989. "The Origin of Anatomically Modern Humans in Australasia." In *The Human Revolution*, edited by P. Mellars and C. Stringer, 245–273. Princeton: Princeton University Press.

Harrison, G. A., J. S. Weiner, J. M. Tanner, and N.A. Barnicot. 1977. *Human Biology*. 2nd ed. Oxford, U.K.: Oxford University Press.

Hedges, R. E., R. A. Housley, P. B. Pettitt, C. Bronk Ramsey, and G. J. Van Klinken. 1996. "Radiocarbon Dates from the Oxford AMS System: Archaeometry Datelist 21." *Archaeometry* 38: 181–207.

Henry, D. O. 1995. *Prehistoric Cultural Ecology and Evolution: Insights from Southern Jordan*. New York: Plenum Press.

Hibbert, D. 1982. "History of the Steppe-Tundra Concept." In *Paleoecology of Beringia*, edited by D. M. Hopkins, J. V. Matthews, C. E. Schweger, and S. B. Young, 153–156. New York: Academic Press.

Hidjrati, N. I. 1987. "K Probleme Interpretatsii Nizhnepaleoliticheskikh Otlozhenii Peshchery Lasok (Myshtulagty Lagat) v Severnoi Osetii." In *Problemy Interpretatsii Arkheologicheskikh Istochnikov*, edited by T. B. Turgiev, 141–154. Ordzhonikidze: North Osetia State University.

Hidjrati, N. I., L. R. Kimball, T. Koetje, and A. Deino, 1997. "Weasel Cave: 15 Years of Archaeological Investigations in the Middle Palaeolithic of the North-Central Caucasus, North Osetia, Russia." Paper presented at the 62nd annual meeting of the Society for American Archaeology, Nashville, Tennessee.

Hoffecker, J. F. 1987. "Upper Pleistocene Loess Stratigraphy and Paleolithic Site Chronology on the Russian Plain. *Geoarchaeology: An International Journal* 2: 259–284.

———. 1988. "Early Upper Paleolithic Sites of the European USSR." In *The Early Upper Paleolithic*, edited by J.F. Hoffecker and C.A. Wolf, 237–272. BAR International Series 437. Oxford, U.K.: British Archaeological Reports.

———. 1999a. "Neanderthals and Modern Humans in Eastern Europe." *Evolutionary Anthropology* 7: 129–141.

———. 1999b. "Treugol'naya Cave: A Middle Pleistocene Faunal Assemblage from Eastern Europe." Unpublished report to the L.S.B. Leakey Foundation, December 1999.

Hoffecker, J. F., and G. Baryshnikov. 1998. "Neanderthal Ecology in the Northwestern Caucasus: Faunal Remains from the Borisovskoe Gorge Sites." In *Quaternary Paleozoology in the Northern Hemisphere*, edited by J. J. Saunders, B. W. Styles, and G. Baryshnikov, 187–211. Illinois State Museum Scientific Papers, vol. 27. Springfield: Illinois State Museum.

Hoffecker, J. F., G. F. Baryshnikov, and O. R. Potapova. 1991. "Vertebrate Remains from the Mousterian Site of Il'skaya I (Northern Caucasus, USSR): New Analysis and Interpretation." *Journal of Archaeological Science* 18: 113–147.

Hoffecker, J. F., and N. Cleghorn. 2000. "Mousterian Hunting Patterns in the Northwestern Caucasus and the Ecology of the Neanderthals." *International Journal of Osteoarchaeology* 10: 368–378.

Holliday, T. W. 1997. "Postcranial Evidence of Cold Adaptation in European Neandertals." *American Journal of Physical Anthropology* 104: 245–258.

———. 1999. "Brachial and Crural Indices of European Late Upper Paleolithic and Mesolithic Humans." *Journal of Human Evolution* 36: 549–566.

Holliday, V. T. 1987. "A Reexamination of Late-Pleistocene Boreal Forest Reconstructions for the Southern Plains." *Quaternary Research* 28: 238–244.

Holloway, R. L. 1985. "The Poor Brain of *Homo sapiens neanderthalensis*: See What You Please . . ." In *Ancestors: The Hard Evidence*, edited by E. Delson, 319–324. New York: Alan R. Liss, Inc.

Howe, J. E. 1976. "Pre-Agricultural Society in Soviet Theory and Method." *Arctic Anthropology* 13: 84–115.

Howell, F. C. 1952. "Pleistocene Glacial Ecology and the Evolution of 'Classic Neandertal' Man." *Southwestern Journal of Anthropology* 8: 377–410.

———. 1957. "The Evolutionary Significance of Variations and Varieties of 'Neanderthal' Man" *Quarterly Review of Biology* 32: 330–347.

Hublin, J.-J. 1998. "Climatic Changes, Paleogeography, and the Evolution of the Neandertals." In *Neandertals and Modern Humans in Western Asia*, edited by T. Akazawa, K. Aoki, and O. Bar-Yosef, 295–310. New York: Plenum Press.

Ingman, M., H. Kaessmann, S. Pääbo, and U. Gyllensten. 2000. "Mitochondrial Genome Variation and the Origin of Modern Humans." *Nature* 408: 708–713.

Ivanova, I. K. 1959. "Geologicheskie Usloviya Nakhozhdeniya Paleoliticheskikh Stoyanok Srednego Pridnestrov'ya." *Trudy Komissii po Izucheniyu Chetvertichnogo Perioda* 15: 215–278.

———. 1968. "Geologicheskoe Stroenie Raiona Paleoliticheskoi Stoyanki Ataki I v Srednem Dnestre." *Byulleten' Komissii po Izucheniyu Chetvertichnogo Perioda* 35: 113–119.

———. 1969. "Geologicheskoe Stroenie Doliny r. Dnestr v Raione Must'erskogo Mestonakhozhdeniya Stinka." *Byulleten' Komissii po Izucheniyu Chetvertichnogo Perioda* 36: 129–136.

———. 1977. "Geologiya i Paleogeografiya Stoyanki Korman' IV na Obshchem Fone Geologicheskoi Istorii Kamennogo Veka Srednego Pridnestrov'ya." In *Mnogosloinaya Paleoliticheskaya Stoyanka Korman' IV na Srednem Dnestre*, edited by G. I. Goretskii and S. M. Tseitlin, 126–181. Moscow: Nauka.

———. 1981. "Bassein Srednego Dnestra." In *Arkheologiya i Paleogeografiya Pozdnego Paleolita Russkoi Ravniny*, edited by A. A. Velichko, 120–131. Moscow: Nauka.

———. 1982. "Geologiya i Paleogeografiya Must'erskogo Poseleniya Molodova I." In *Molodova I: Unikal'noe Must'erskoe Poselenie na Srednem Dnestre*, edited by G. I. Goretskii and I. K. Ivanova, 188–235. Moscow: Nauka.

———. 1987. "Paleogeografiya i Paleoekologiya Sredy Obitaniya Lyudei Kamennogo Veka na Srednem Dnestre. Stoyanka Molodova V." In *Mnogosloinaya Paleoliticheskaya Stoyanka Molodova V: Lyudi Kamennogo Veka i Okruzhayushchaya Sreda*, edited by I. K. Ivanova and S. M. Tseitlin, 94–123. Moscow: Nauka.

Ivanova, I. K., N. S. Bolikhovskaya, N. V. Rengarten. 1981. "Geologicheskii Vozrast i Prirodnaya Obstanovka Must'erskoi Stoyanki Ketrosy." In *Ketrosy: Must'erskaya Stoyanka na Srednem Dnestre*, edited by N. D. Praslov, 152–161. Moscow: Nauka.

Ivanova, I. K., and N. V. Rengarten. 1975. "Materialy k Geologii i Paleogeografii Paleoliticheskoi Stoyanki Kulichivka." *Byulleten' Komissii po Izucheniyu Chetvertichnogo Perioda* 44: 52–68.

Jones, P. R. 1980. "Experimental Butchery with Modern Stone Tools and Its Relevance for Palaeolithic Archaeology." *World Archaeology* 12: 153–175.

Kalechits, E. G. 1984. *Pervonachal'noe Zaselenie Territorii Belorussii*. Minsk: Nauka i Tekhnika.

Kanivets, V. I. 1976. *Paleolit Krainego Severo-Vostoka Evropy*. Moscow: Nauka.

Keeley, L. H. 1993. "Microwear Analysis of Lithics." In *The Lower Paleolithic Site at Hoxne, England*, edited by R. Singer, B. G. Gladfelter, and J. J. Wymer, 129–138. Chicago: University of Chicago Press.

Kelly, R. L. 1995. *The Foraging Spectrum: Diversity in Hunter-Gatherer Lifeways*. Washington, D.C.: Smithsonian Institution Press.

Ketraru, N. A. 1965. "Paleoliticheskaya Stoyanka Chutuleshty I." *Izvestiya AN MSSR* 12: 53–61.

———. 1969. "O Paleoliticheskoi Stoyanke v Peshchere u c. Buteshty." *Trudy GIKM MSSR* 2.

———. 1973. *Pamyatniki Epokh Paleolita i Neolita*. Kishinev: Shtiintsa.

Khrisanfova, E. N. 1965. "Bedrennaya Kost' Paleoantropa iz Romankova." *Voprosy Antropologii* 20: 80–89.

Klein, R. G. 1965. "The Middle Palaeolithic of the Crimea." *Arctic Anthropology* 3: 34–68.

———. 1966. "Chellean and Acheulean on the Territory of the Soviet Union; A Critical Review of the Evidence as Presented in the Literature." *American Anthropologist* 68 (2): 1–45.

———. 1969a. "The Mousterian of European Russia." *Proceedings of the Prehistoric Society* 25: 77–111.

———. 1969b. *Man and Culture in the Late Pleistocene: A Case Study*. San Francisco: Chandler.

———. 1973. *Ice-Age Hunters of the Ukraine*. Chicago: University of Chicago Press.

———. 1982. "Age (Mortality) Profiles as a Means of Distinguishing Hunted Species from Scavenged Ones in Stone Age Archaeological Sites." *Paleobiology* 8: 151–158.

———. 1994. "Southern Africa before the Iron Age." In *Integrative Paths to the Past: Paleoanthropological Advances in Honor of F. Clark Howell*, edited by R. S. Corruccini and R. L. Ciochon, 471–519. Englewood Cliffs, N.J.: Prentice Hall.

———. 1999. *The Human Career*. 2nd ed. Chicago: University of Chicago Press.

———. 2000. "Archeology and the Evolution of Human Behavior." *Evolutionary Anthropology* 9: 17–36.

Klein, R. G., I. K. Ivanova, and G. F. Debets. 1971. "U.S.S.R." In *Catalogue of Fossil Hominids*. Vol. 2, *Europe*, edited by K. P. Oakley, B. Campbell, and T. Mollseon, 311–335. London: British Museum.

Kolosov, Yu. G. 1964. "Nekotorye Pozdnepaleoliticheskie Stoyanki Porozhistoi Chasti Dnepra (Osokorovka, Dubovaya Balka, Yamburg)." *Svod Arkheologicheskikh Istochnikov* A 1-5: 42–49.

———. 1972. *Shaitan-Koba—Must'erska Stoyanka Krimu.* Kiev: Naukova Dumka.

———. 1983. *Must'erskie Stoyanki Raiona Belogorska.* Kiev: Naukova Dumka.

Kolosov, Yu. G., V. M. Kharitonov, and V. P. Yakimov. 1975. "Palaeoanthropic Specimens from the Site Zaskal'naya VI in the Crimea." In *Paleoanthropology, Morphology, and Palaeoecology*, edited by R. Tuttle, 419–428. The Hague: Mouton.

Kolosov, Yu. G., V. N. Stepanchuk, and V. P. Chabai. 1993. *Rannii Paleolit Kryma.* Kiev: Naukova Dumka.

Konyukova, L. G., V. V. Orlova, and Ts. A. Shver. 1971. *Klimaticheskie Kharakteristiki SSSR po Mesyatsam.* Leningrad: Gidrometeoizdat.

Korniets, N. L., M. I. Gladkikh, A. A. Velichko, G. V. Antonova, Yu. N. Gribchenko, E. M. Zelikson, E. I. Kurenkova, T. Kh. Khalcheva, and A. L. Chepalyga. 1981. "Mezhirich." In *Arkheologiya i Paleogeografiya Pozdnego Paleolita Russkoi Ravniny*, edited by I. P. Gerasimov, 106–119. Moscow: Nauka.

Kozlowski, J. K. 1986. "The Gravettian in Central and Eastern Europe." In *Advances in World Archaeology.* Vol. 5, edited by F. Wendorf and A. E. Close, 131–200. Orlando, Fla.: Academic Press.

———. 1988. "Transition from the Middle to the Early Upper Paleolithic in Central Europe and the Balkans." In *The Early Upper Paleolithic*, edited by J. F. Hoffecker and C. A. Wolf, 193–235. BAR International Series 437. Oxford, U.K.: British Archaeological Reports.

Krebs, C. J. 1978. *Ecology: The Experimental Analysis of Distribution and Abundance.* New York: Harper and Row.

Kretzoi, M., and V. Dobosi. eds. 1990. *Vertésszöllös: Man, Site, and Culture.* Budapest: Akademiai Kiado.

Krings, M., A. Stone, R. W. Schmitz, H. Krainitzki, M. Stoneking, and S. Pääbo. 1997. "Neanderthal DNA Sequences and the Origin of Modern Humans." *Cell* 90: 19–30.

Krotova, A. A., and N. G. Belan. 1993. "Amvrosievka: A Unique Upper Paleolithic Site in Eastern Europe." In *From Kostenki to Clovis*, edited by O. Soffer and N. D. Praslov, 125–142. New York: Plenum Press.

Ksenzov, V. P. 1988. *Paleolit i Mesolit Belorusskogo Podneprov'ya.* Minsk: Nauka i Tekhnika.

Kukharchuk, Yu. V. 1995. "Must'erskie Kompleksy Zhitomirskoi Stoyanki i Ikh Sootnoshenie c Industriei Rikhty." *Arkheologicheskii Al'manakh* 4: 53–74.

Kukla, G. J. 1975. "Loess Stratigraphy of Central Europe." In *After the Australopithecines*, edited by K. W. Butzer and G. Ll. Isaac, 99–188. The Hague: Mouton.

Lahr, M. M., and R. Foley. 1994. "Multiple Dispersals and Modern Human Origins." *Evolutionary Anthropology* 3: 48–60.

Laitman, J. T., R. C. Heimbuch, and C. S. Crelin. 1979. "The Basicranium of Fossil Hominids as an Indicator of Their Upper Respiratory Systems." *American Journal of Physical Anthropology* 51: 15–34.

Laitman, J. T., J. S. Reidenberg, S. Marquez, and P. J. Gannon. 1996. "What the

Nose Knows: New Understandings of Neanderthal Upper Respiratory Tract Specializations." *Proceedings of the National Academy of Sciences* 93: 10543–10545.

Laville, H., J.-P. Rigaud, and J. Sackett. 1980. *Rock Shelters of the Perigord*. New York: Academic Press.

Lazukov, G. I. 1957. "Geologiya Stoyanok Kostenkovsko-Borshevkogo Raiona." *Materialy i Issledovaniya po Arkheologiya SSSR* 59: 135–173.

———. 1982. "Kharakteristika Chetvertichnykh Otlozhenii Raiona." In *Paleolit Kostenkovsko-Borshchevskogo Raiona na Donu 1879–1979*, edited by N. D. Praslov and A. N. Rogachev, 13–37. Leningrad: Nauka.

Lee, R. B. 1972. "!Kung Spatial Organization: An Ecological and Historical Perspective." *Human Ecology* 1: 125–147.

Leith, H. 1975. "Primary Production of the Major Vegetation Units of the World." In *Primary Productivity of the Biosphere*, edited by H. Leith and R. H. Whittaker, 203–215. New York: Springer-Verlag.

Leonova, N. B. 1993. "Criteria for Estimating the Duration of Occupation at Paleolithic Sites." In *From Kostenki to Clovis*, edited by O. Soffer and N. D. Praslov, 149–157. New York: Plenum Press.

———. 1994. "The Upper Paleolithic of the Russian Steppe Zone." *Journal of World Prehistory* 8: 169–210.

Levkovskaya, G. M. 1981. "Palinologicheskaya Kharakteristika Must'erskogo Kul'turnogo Sloya Stoyanki Ketrosy." In *Ketrosy: Must'erskaya Stoyanka na Srednem Dnestre*, edited by N. D. Praslov, 125–135. Moscow: Nauka.

———. 1994. "Palinologicheskaya Kharakteristika Otlozhenii Barakaevskoi Peshchery." In *Neandertaltsy Gupsskogo Ushchel'ya na Severnom Kavkaze*, edited by V. P. Lyubin, 77–82. Maikop: Meoty.

Lieberman, P., J. T. Laitman, J. S. Reidenberg, and P. J. Gannon. 1992. "The Anatomy, Physiology, Acoustics, and Perception of Speech: Essential Elements in the Analysis of the Evolution of Human Speech." *Journal of Human Evolution* 23: 447–467.

Lindly, J. M., and G. A. Clark. 1990. "Symbolism and Modern Human Origins." *Current Anthropology* 31 (3): 233–261.

Lowe, J. J., and M.J.C.Walker. 1997. *Reconstructing Quaternary Environments*. 2nd ed. London: Longman.

Lozek, V. 1967. "Climatic Zones of Czechoslovakia during the Quaternary." In *Quaternary Paleoecology*, edited by E. J. Cushing and H. E. Wright, 381–392. New Haven: Yale University Press.

———. 1968. "The Loess Environments in Central Europe." In *Loess and Related Eolian Deposits of the World*, edited by C. B. Schultz and J. C. Frye, 67–80. Lincoln: University of Nebraska Press.

Lydolph, P. E. 1977. *Geography of the U.S.S.R.* 3rd ed. New York: John Wiley and Sons.

Lyubin, V. P. 1977. *Must'erskie Kul'tury Kavkaza*. Leningrad: Nauka.

———, ed. 1994. *Neandertal'tsy Gupsskogo Ushchel'ya*. Maikop: Meoty.

———. 1998. *Ashel'skaya Epokha na Kavkaze*. Saint Petersburg: Russian Academy of Sciences.

Lyubin, V. P., and P. U. Autlev. 1994. "Kamennaya Industriya." In *Neandertal'tsy Gupsskogo Ushchel'ya*, 99–141. Maikop: Meoty.

Lyubin, V. P., P. U. Autlev, A. A. Zubov, G. P. Romanova, and V. M. Kharitonov. 1986. "Otkrytie Skeletnykh Ostatkov Paleoantropa na Barakaevskoi Stoyanke (Zapadnyi Kavkaz)." *Voprosy Antropologii* 77: 60–70.

MacArthur, R. 1972. *Geographical Ecology*. New York: Harper and Row.

Markova, A. K. 1984. "Late Pleistocene Mammal Fauna of the Russian Plain." In *Late Quaternary Environments of the Soviet Union*, edited by A. A. Velichko, 209–218. Minneapolis: University of Minnesota Press.

Marks, A. E. 1998. "A New Middle to Upper Paleolithic 'Transitional' Assemblage from Buran Kaya III, Level C: A Preliminary Report." In *Préhistoire d'Anatolie—Genese de Deux Mondes*, edited by M. Otte, 353–366. Liege: ERAUL.

Marks, A. E., and V. P. Chabai. eds. 1998. *The Middle Paleolithic of Western Crimea*, vol. 1. ERAUL 84.

Marks, A. E., Yu. E. Demidenko, K. Monigal, V. I. Usik, and C. R. Ferring. 1998. "Starosele: The 1993-95 Excavations." In *The Middle Paleolithic of Western Crimea*. Vol. 1, edited by A. E. Marks and V. P. Chabai, 67–99. ERAUL 84.

Marks, A. E., Yu. E. Demidenko, K. Monigal, V. I. Usik, C. R. Ferring, A. Burke, J. Rink, and C. McKinney, 1997. "Starosele and the Starosele Child: New Excavations, New Results." *Current Anthropology* 38: 112–123.

Marks, A. E., and K. Monigal. 1998. "Starosele 1993–1995: The Lithic Artifacts." In *The Middle Paleolithic of Western Crimea*. Vol. 1, edited by A. E. Marks and V. P. Chabai, 117–165. ERAUL 84.

Marshack, A. 1979. "Upper Paleolithic Symbol Systems of the Russian Plain: Cognitive and Comparative Analysis." *Current Anthropology* 20: 271–311.

Maynard Smith, J. 1982. *Evolution and the Theory of Games*. Cambridge, U.K.: Cambridge University Press.

Mayr, E. 1970. *Populations, Species, and Evolution*. Cambridge, Mass.: Belknap Press.

McBrearty, S., and A. S. Brooks. 2000. "The Revolution that Wasn't: A New Interpretation of the Origin of Modern Human Behavior." *Journal of Human Evolution* 39 (5): 453–563.

McGhee, R. 1997. *Ancient People of the Arctic*. Vancouver: UBC Press.

McKinney, C. R. 1998. "Uranium Series Dating of Enamel, Dentine, and Bone from Kabazi II, Starosele, Kabazi V, and Gabo." In *The Middle Paleolithic of Western Crimea*. Vol. 1, edited by A. E. Marks and V. P. Chabai, 341–353. ERAUL 84.

Mellars, P. 1996. *The Neanderthal Legacy: An Archaeological Perspective from Western Europe*. Princeton: Princeton University Press.

Mellars, P., and C. Stringer. eds. 1989. *The Human Revolution*. Princeton: Princeton University Press.

Mertens, S. B. 1994. "The Middle Paleolithic of Romania." *Current Anthropology* 37: 515–521.

Mesyats, V. A. 1962. "Zhitomirskaya Rannepaleoliticheskaya Stoyanka (Predvaritel'noe Soobshchenie)." *Kratkie Soobshcheniya Instituta Arkheologii* 12: 53–56.

Miller, M. 1956. *Archaeology in the U.S.S.R*. New York: Frederick A. Praeger.

Mithen, S. 1994. "From Domain Specific to Generalized Intelligence: A Cognitive

Interpretation of the Middle/Upper Paleolithic Transition." In *The Ancient Mind: Elements of Cognitive Archaeology*, edited by C. Renfrew and E.B.W. Zubrow, 29–36. London: Cambridge University Press.

———. 1996. *The Prehistory of the Mind: The Cognitve Origins of Art, Religion, and Science*. London: Thames and Hudson.

Morgan, L. H. 1878. *Ancient Society*. Chicago: Charles Kerr.

Motuz, V. M. 1967. "Chetvertichnye Mollyuski iz Khotylevskogo Nizhnepaleoliticheskogo Mestonakhozhdeniya Bryanskoi Oblasti." *Byulleten' Komissii po Izucheniyu Chetvertichnogo Perioda* 33: 150–154.

Movius, H. L. 1950. "A Wooden Spear of Third Interglacial Age from Lower Saxony." *Southwestern Journal of Anthropology* 6: 139–142.

Mücher, H. J. 1986. "Aspects of Loess and Loess-Derived Slope Deposits: An Experimental and Micromorphological Approach." *Nederlandse Geografische Studies* 23.

Murdoch, J. 1892. *Ethnological Results of the Point Barrow Expedition*. Washington, D.C.: Smithsonian Institution Press.

Nakel'skii, S. K., and N. N. Karlov. 1965. "O Geologicheskom Vozraste i Znachenii Ostatkov Iskopaemogo Paleoliticheskogo Cheloveka, Naidennykh v Srednem Pridneprov'e." *Voprosy Antropologii* 20: 75–79.

Nekhoroshev, P. E. 1999. *Tekhnologicheskii Metod Izucheniya Pervichnogo Rasshchepleniya Kamnya Srednego Paleolita*. Saint Petersburg: Russian Academy of Sciences.

Nesmeyanov, S. A. 1989. "Geologiya i Paleogeografiya v Paleoekologicheskikh Rekonstruktsiyakh." In *Metodicheskie Problemy Rekonstruktsii v Arkheologii i Paleoekologii*, 225–260. Novosibirsk: Nauka.

———. 1999. *Geomorfologicheskie Aspekty Paleoekologii Gornogo Paleolita*. Moscow: Nauchnyi Mir.

New Oxford Atlas. 1978. Rev. ed. Oxford, U.K.: Oxford University Press.

Oakley, K. P. 1949. *Man the Tool-Maker*. Chicago: University of Chicago Press.

Odum, E. P. 1975. *Ecology*. 2nd ed. New York: Holt, Rinehart and Winston.

Oswalt, W. H. 1976. *An Anthropological Analysis of Food-Getting Technology*. New York: John Wiley and Sons.

Oswalt, W. H. 1987. "Technological Complexity: The Polar Eskimos and the Tareumiut." *Arctic Anthropology* 24 (2): 82–98.

Otte, M., P. Noiret, S. Tatartsev, and I. Lopez Bayon. 1996. "L'Aurignacien de Siuren I (Crimée): Fouilles 1994 et 1995." *XIII Congres d'UISPP* 6: 123–137.

Ovchinnikov, I. V., A. Gotherstrom, G. P. Romanova, V. M. Kharitonov, K. Lidon, and W. Goodwin. 2000. "Molecular Analysis of Neanderthal DNA from the Northern Caucasus." *Nature* 404: 490–493.

Pacey, A. 1990. *Technology in World Civilization*. Cambridge, Mass.: MIT Press.

Parfitt, S. A., and M. B. Roberts. 1999. "Human Modification of Faunal Remains." In *Boxgrove: A Middle Pleistocene Hominid Site at Eartham Quarry, Boxgrove, West Sussex*, edited by M. B. Roberts and S. A. Parfitt, 395–415. London: English Heritage.

Pashkevich, G. A. 1977. "Palinologicheskoe Issledovanie Razreza Stoyanki Korman'

IV." In *Mnogosloinaya Paleoliticheskaya Stoyanka Korman' IV na Srednem Dnestre*, edited by G. I. Goretskii and S. M. Tseitlin, 105–111. Moscow: Nauka.

Păunescu, A. 1965. "Sur la Succession des Habitats Paléolithiques et Postpaléolithiques de Ripiceni-Izvor." *Dacia* 9: 5–31.

———. 1978. "Complexe de Locuire Musteriene Descoperite in Aşezarea de la Ripiceni-Izvor (Jud. Botoşani) şi Unele Consideraţii Privind Evoluţia Tipului de Locuinţă Paleolitică." *Studii şi Cercetări de Istorie Veche şi Arheologie* 29: 317–333.

———. 1987. "Inceputurile Paleoliticului Superior in Moldova." *Studii şi Cercetări de Istorie Veche şi Arheologie* 38 (2): 87–100.

Păunescu, A., A. Conea, M. Cârciumaru, V. Codaarea, A. V. Grossu, and R. Popovici. 1976. "Consideraţii Arheologice, Geocronologice şi Paleoclimatice Privind Aşezarea Ripiceni-Izvor." *Studii şi Cercetări de Istorie Veche şi Arheologie* 27: 5–21.

Pettitt, P. B., and N. O. Bader. 2000. "Direct AMS Radiocarbon Dates for the Sungir mid Upper Paleolithic Burials." *Antiquity* 74: 269–270.

Péwé, T. L. 1975. *Quaternary Geology of Alaska*. Geological Survey Professional Paper 835.

Pianka, E. R. 1978. *Evolutionary Ecology*. 2nd ed. New York: Harper and Row.

Pidoplichko, I. G. 1953. "Amvrosievskaya Paleoliticheskaya Stoyanka i Ee Osobennosti." *Kratkie Soobshcheniya Instituta Arkheologii* 2: 65–68.

———. 1969. *Pozdnepaleoliticheskie Zhilishcha iz Kostei Mamonta na Ukraine*. Kiev: Naukova Dumka.

———. 1976. *Mezhirichskie Zhilishcha iz Kostei Mamonta*. Kiev: Naukova Dumka.

Pike-Tay, A., V. Cabrera Valdés, and F. Bernaldo de Quirós. 1999. "Seasonal Variations of the Middle-Upper Paleolithic Transition at El Castillo, Cueva Morin and El Pendo (Cantabria, Spain)." *Journal of Human Evolution* 36: 283–317.

Polikarpovich, K. M. 1968. *Paleolit Verkhnego Podneprov'ya*. Minsk: Nauka i Tekhnika.

Potts, R. 1988. *Early Hominid Activities at Olduvai*. New York: Aldine de Gruyter.

Praslov, N. D. 1964. "Raboty po Issledovaniyu Paleoliticheskikh Pamyatnikov v Priazov'e i na Kubani v 1963 g." *Kratkie Soobshcheniya Instituta Arkheologii* 101: 74–76.

———. 1968. *Rannii Paleolit Severo-Vostochnogo Priazov'ya i Nizhnego Dona*. Leningrad: Nauka.

———. 1972. "Must'erskoe Poselenie Nosovo I v Priazov'e." *Materialy i Issledovaniya po Arkheologii SSSR* 185: 75–82.

———. 1982. "Istoriya Izucheniya Paleolita Kostenkovsko-Borshchevskogo Raiona i Slozhenie Kostenkovskoi Shkoly." In *Paleolit Kostenkovsko-Borshchevskogo Raiona na Donu 1879-1979*, edited by N. D. Praslov and A. N. Rogachev, 7–13. Leningrad: Nauka.

———. 1984a. "Razvitie Prirodnoi Sredy na Territorii SSSR v Antropogene i Problemy Khronologii i Periodizatsii Paleolita." In *Paleolit SSSR*, edited by P. I. Boriskovskii, 23–40. Moscow: Nauka.

———. 1984b. "Rannii Paleolit Russkoi Ravniny i Kryma." In *Paleolit SSSR*, edited by P. I. Boriskovskii, 94–134. Moscow: Nauka.

———. 1985. "Kostenkovskaya Gruppa Paleoliticheskikh Stoyanok." In *Kraevye*

Obrazovaniya Materikovykh Oledenenii: Putevoditel' Ekskursii VII Vsesoyuznogo Soveshchaniya, edited by S. M. Shik, 24–28. Moscow: Nauka.

———. 1995. "The Earliest Occupation of the Russian Plain: A Short Note." In *The Earliest Occupation of Europe*, edited by W. Roebroeks and T. van Kolfschoten, 61–66. Leiden: University of Leiden.

Praslov, N. D., and A. K. Filippov. 1967. "Pervaya Nakhodka Paleoliticheskogo Iskusstva v Yuzhnorusskikh Stepyakh." *Kratkie Soobshcheniya Instituta Arkheologii* 111: 24–30.

Praslov, N. D., and M. A. Ivanova. 1982. "Kostenki 21 (Gmelinskaya Stoyanka)." In *Paleolit Kostenkovsko-Borshchevskogo Raiona na Donu 1879–1979*, edited by N. D. Praslov and A. N. Rogachev, 198–210. Leningrad: Nauka.

Praslov, N. D., and A. N. Rogachev. eds. 1982. *Paleolit Kostenkovsko-Borshchevskogo Raiona na Donu 1879–1979*. Leningrad: Nauka.

Praslov, N. D., and S. A. Semenov. 1969. "O Funktsiyakh Must'erskikh Kremnevykh Orudii iz Stoyanok Priazov'ya." *Kratkie Soobshcheniya Instituta Arkheologii* 117: 13–21.

Praslov, N. D., and V. E. Shchelinskii. 1996. *Verkhnepaleoliticheskoe Poselenie Zolotovka I na Nizhnem Donu*. Saint Petersburg: Russian Academy of Sciences.

Puech, P. F., A. Prone, and R. Kraatz. 1980. "Microscopie de l'Usure Dentaire chez L'Homme Fossile: Bol Alimentaire et Environnement." *CRASP* 290: 1413–1416.

Richards, M. P., P. B. Pettitt, E. Trinkaus, F. H. Smith, M. Paunovic, and I. Karavanic. 2000. "Neanderthal Diet at Vindija and Neanderthal Predation: The Evidence from Stable Isotopes." *Proceedings of the National Academy of Sciences* 97 (13): 7663–7666.

Rightmire, G. P. 1986. "Africa and the Origins of Modern Humans." In *Variation, Culture and Evolution in African Populations*, edited by R. Singer and J. K. Lundy, 209–220. Johannesburg: Witwatersrand University Press.

———. 1998. "Human Evolution in the Middle Pleistocene: The Role of *Homo heidelbergensis*." *Evolutionary Anthropology* 6: 218–227.

Rink, W. J., H.-K. Lee, J. Rees-Jones, and K. A. Goodger. 1998. "Electron Spin Resonance (ESR) and Mass Spectrometric U-Series (MSUS) Dating of Teeth in Crimean Paleolithic Sites: Starosele, Kabazi II, and Kabazi V." In *The Middle Paleolithic of Western Crimea*. Vol. 1, edited by A. E. Marks and V. P. Chabai, 323–340. ERAUL 84.

Roebroeks, W. 1986. "Archaeology and Middle Pleistocene Stratigraphy: The Case of Maastricht-Belvédère (NL)." In *Chronostratigraphie et Faciès Culturels du Paléolithique Inférieur et Moyen dans l'Europe de Nord-Ouest*, edited by A. Trufeau and J. Somme, 81–86. Paris: Supplement au Bulletin de l'Association Française pour l'Étude du Quaternaire.

Roebroeks, W., N. J. Conard, and T. van Kolfschoten. 1992. "Dense Forests, Cold Steppes, and the Palaeolithic Settlement of Northern Europe." *Current Anthropology* 33: 551–586.

Roebroeks, W., J. Kolen, and E. Rensink. 1988. "Planning Depth, Anticipation and the Organization of Middle Palaeolithic Technology: the 'Archaic Natives' Meet Eve's Descendents." *Helinium* 28 (1): 17–34.

Roebroeks, W., and T. van Kolfschoten, eds. 1995. *The Earliest Occupation of Europe.* Leiden: University of Leiden.

Rogachev, A. N. 1953. "Issledovanie Ostatkov Pervobytno-Obshchinnogo Poseleniya Verkhnepaleoliticheskogo Vremeni u s. Avdeevo na r. Seim v 1949 g." *Materialy i Issledovaniya po Arkheologii SSSR* 39: 137–191.

———. 1955. *Aleksandrovskoe Poselenie Drevnekamennogo Veka u Sela Kostenki na Donu.* Materialy i Issledovaniya po Arkheologii SSSR 45.

———. 1957. "Mnogosloinye Stoyanki Kostenkovsko-Borshevskogo Raiona na Donu i Problema Razvitiya Kul'tury v Epokhy Verkhnego Paleolita na Russkoi Ravnine." *Materialy i Issledovaniya po Arkheologii SSSR* 59: 9–134.

Rogachev, A. N., and M. V. Anikovich. 1982a. "Kostenki 12 (Volkovskaya Stoyanka)." In *Paleolit Kostenkovsko-Borshchevskogo Raiona na Donu 1879–1979*, edited by N. D. Praslov and A. N. Rogachev, 132–140. Leningrad: Nauka.

———. 1982b. "Kostenki 6 (Streletskaya)." In *Paleolit Kostenkovsko-Borshchevskogo Raiona na Donu 1879–1979*, edited by N. D. Praslov and A. N. Rogachev, 88–91. Leningrad: Nauka.

———. 1982c. "Kostenki 4 (Aleksandrovskaya Stoyanka)." In *Paleolit Kostenkovsko-Borshchevskogo Raiona na Donu 1879–1979*, edited by N. D. Praslov and A. N. Rogachev, 76–85. Leningrad: Nauka.

———. 1984. "Pozdnii Paleolit Russkoi Ravniny i Kryma." In *Paleolit SSSR*, edited by P. I. Boriskowskii, 162–271. Moscow: Nauka.

Rogachev, A. N., M. V. Anikovich, and T. N. Dmitrieva. 1982. "Kostenki 8 (Tel'manskaya Stoyanka)." In *Paleolit Kostenkovsko-Borshchevskogo Raiona na Donu 1879–1979*, edited by N. D. Praslov and A. N. Rogachev, 92–109. Leningrad: Nauka.

Rogachev, A. N., and V. I. Belyaeva. 1982. "Kostenki 13 (Kel'sievskaya Stoyanka)." In *Paleolit Kostenkovsko-Borshchevskogo Raiona na Donu 1879–1979*, edited by N. D. Praslov and A. N. Rogachev, 140–145. Leningrad: Nauka.

Rogachev, A. N., and V. E. Kudryashov. 1982. "Borshchevo 1." In *Paleolit Kostenkovsko-Borshchevskogo Raiona na Donu 1879–1979*, edited by N. D. Praslov and A. N. Rogachev, 211–216. Leningrad: Nauka.

Rogachev, A. N., and V. V. Popov. 1982. "Kostenki 11 (Anosovka 2)." In *Paleolit Kostenkovsko-Borshchevskogo Raiona na Donu 1879–1979*, edited by N. D. Praslov and A. N. Rogachev, 116–132. Leningrad: Nauka.

Rogachev, A. N., N. D. Praslov, M. V. Anikovich, V. I. Belyaeva, and T. N. Dmitrieva. 1982. "Kostenki 1 (Stoyanka Polyakova)." In *Paleolit Kostenkovsko-Borshchevskogo Raiona na Donu 1879–1979*, edited by N. D. Praslov and A. N. Rogachev, 42–66. Leningrad: Nauka.

Rogachev, A. N., and A. A. Sinitsyn. 1982a. "Kostenki 14 (Markina Gora)." In *Paleolit Kostenkovsko-Borshchevskogo Raiona na Donu 1879–1979*, edited by N. D. Praslov and A. N. Rogachev, 145–162. Leningrad: Nauka.

———. 1982b. "Kostenki 15 (Gorodtsovskaya Stoyanka)." In *Paleolit Kostenkovsko-Borshchevskogo Raiona na Donu 1879–1979*, edited by N. D. Praslov and A. N. Rogachev, 162–171. Leningrad: Nauka.

———. 1982c. "Kostenki 16 (Uglyanka)." In *Paleolit Kostenkovsko-Borshchevskogo Raiona na Donu 1879–1979*, edited by N. D. Praslov and A. N. Rogachev, 171–181. Leningrad: Nauka.

Rolland, N. 1992. "The Palaeolithic Colonization of Europe: An Archaeological and Biogeographic Perspective." *Trabajos de Prehistoria* 49: 69–111.

Rolland, N., and H. L. Dibble. 1990. "A New Synthesis of Middle Paleolithic Variability." *American Antiquity* 55: 480–499.

Sablin, M. V. 1997. "Ostatki Mlekopitayushchikh iz Pozdnepaleoliticheskogo Poseleniya Pushkari I." In *Pushkarevskii Sbornik*, edited by V. I. Belyaeva, 31–34. Saint Petersburg: Obrazovanie Kul'tura.

Sahlins, M. 1976. *Culture and Practical Reason*. Chicago: University of Chicago Press.

Savich, V. P. 1975. "Pozdnepaleoliticheskie Poseleniya na Gore Kulichivka v g. Krements (Ternopol'skaya Oblast' USSR)." *Byulleten' Komissii po Izucheniyu Chetvertichnogo Perioda* 44: 41–51.

———. 1977. "Zhilishche Verkhnego Pozdnepaleoliticheskogo Sloya Stoyanki Kulichivka v Ternopol'skoi Oblasti." *Byulleten' Komissii po Izucheniyu Chetvertichnogo Perioda* 47: 121–127.

———. 1985. "Raboty Volyno-Podol'skoi Ekspeditsii." In *Arkheologicheskie Otkrytiya 1983 Goda*, 350–351. Moscow: Nauka.

———. 1986. "Issledovaniya Kulychivki v g. Krements." In *Arkheologicheskie Otkrytiya 1984 Goda*, 301–302. Moscow: Nauka.

———. 1987. "Pozdnyi Paleolit Volyni." In *Arkheologiya Prikarpat'ya, Volyni i Zakarpat'ya (Kamennyi Vek)*, edited by A. P. Chernysh, 43–65. Kiev: Naukova Dumka.

Schick, K. D., and Z. Dong. 1993. "Early Paleolithic of China and Eastern Asia." *Evolutionary Anthropology* 2: 22–35.

Schick, K. D., and N. Toth. 1993. *Making Silent Stones Speak: Human Evolution and the Dawn of Technology*. New York: Simon and Schuster.

Schledermann, P. 1978. "Preliminary Results of Archaeological Investigations in the Bache Peninsula Region, Ellesmere Island, N.W.T." *Arctic* 31 (4): 459–474.

Scott, K. 1989. "Mammoth Bones Modified by Humans: Evidence from La Cotte de St. Brelade, Jersey, Channel Islands." In *Bone Modification*, 335–346. Orono, Maine: Center for the Study of the First Americans.

Semenov, S. A. 1940. "Rezul'taty Issledovaniya Poverkhnosti Kamennykh Orudii." *Byulleten' Komissii po Izucheniyu Chetvertichnogo Perioda* 6–7: 110–113.

———. 1964. *Prehistoric Technology*. Translated by M. W. Thompson. New York: Barnes and Noble.

Sergeev, G. P. 1950. "Pozdneashel'skaya Stoyanka v Grote u c. Vykhvatintsy (Moldaviya)." *Sovetskaya Arkheologiya* 13: 203–212.

Sergin, V. Ya. 1974. "O Razmere Pervogo Paleoliticheskogo Zhilishcha v Yudinove." *Sovetskaya Arkheologiya* 3: 236–240.

———. 1975. "O Pervom Zhilishcho-Khozyaistvennom Komplekse Eliseevichei." *Kratkie Soobshcheniya Instituta Arkheologii* 141: 58–62.

———. 1977. "Raskopki v s. Yudinovo." In *Arkheologicheskie Otkrytiya 1976 Goda*, 70. Moscow: Nauka.

———. 1979. "K Faunisticheskoi Kharakteristike Suponevskogo Paleoliticheskogo Poseleniya." *Byulleten' Komissii po Izucheniyu Chetvertichnogo Perioda* 49: 134–136.

———. 1983. "Naznachenie Bol'shikh Yam na Paleoliticheskikh Poseleniyakh." *Kratkie Soobshcehniya Instituta Arkheologii* 173: 23–31.

Shchelinskii, V. E. 1974. *Proizvodstvo i Funktsii Must'erskikh Orudii (po Dannym Eksperimental'nogo i Trasologicheskogo Izucheniya)*. Kandidat Dissertation, Institute of Archaeology, Leningrad Branch, USSR Academy of Sciences.

———. 1981. "Vidy Ispol'zovaniya Kamennykh Orudii iz Must'erskoi Stoyanki Ketrosy." In *Ketrosy: Must'erskaya Stoyanka na Srednem Dnestre*, edited by N. D. Praslov, 53–58. Moscow: Nauka.

———. 1985. "Novye Dannye o Mnogosloinoi Rannepaleoliticheskoi Stoyanke Il'skaya 2 v Predgor'yakh Severo-Zapadnogo Kavkaza." In *Vsesoyuznaya Arkheologicheskaya Konferentsiya "Dostizheniya Arkheologii v XI Pyatilete." Tezisy Doklady*, 377–379. Baku: Nauka.

———. 1992. "Funktsional'nyi Analiz Orudii Truda Nizhnego Paleolita Prikuban'ya (Voprosy Metodiki). In *Voprosy Arkheologii Adygei*, edited by D. Kh. Mikulov, 194–209. Maikop: ANIIEYaLI.

———. 1994. "Terochnyi Kamen' iz Must'erskogo Kul'turnogo Sloya Barakaevskoi Peshchery." In *Neandertal'tsy Gupsskogo Ushchel'ya na Severnom Kavkaze*, edited by V. P. Lyubin, 148–150. Maikop: Meoty.

Shevyrev, L. T., and E. N. Khrisanfova. 1984. "Nakhodka Ostankov Iskopaemogo Cheloveka v Verkhnem Pleistotsene Tsentra Russkoi Ravniny (Mestonakhozhdenie Shkurlat III)." *Voprosy Antropologii* 73: 69–71.

Shevyrev, L. T., G. I. Raskatov, and L. I. Alekseeva. 1979. "Shkurlatskoe Mestonakhozhdenie Fauny Mlekopitayushchikh Mikulinskogo Vremeni." *Byulleten' Komissii po Izucheniyu Chetvertichnogo Perioda* 49: 39–48.

Shipman, P. 1986. "Scavenging or Hunting in Early Hominids: Theoretical Framework and Tests." *American Anthropologist* 88: 27–43.

Shipman, P., and J. Rose. 1983. "Evidence of Butchery and Hominid Activities at Torralba and Ambrona: An Evaluation Using Microscopic Techniques." *Journal of Archaeological Science* 10: 465–474.

Shovkoplyas, I. G. 1956. "Fastovskaya Pozdnepaleoliticheskaya Stoyanka." *Kratkie Soobshcheniya Instituta Istorii Material'noi Kul'tury* 65: 68–73.

———. 1965a. *Mezinskaya Stoyanka*. Kiev: Naukova Dumka.

———. 1965b. "Radomyshl'skaya Stoyanka – Pamyatnik Nachal'noi Pory Pozdnego Paleolita." In *Stratigrafiya i Periodizatsiya Paleolita Vostochnoi i Tsental'noi Evropy*, edited by O. N. Bader, 104–116. Moscow: Nauka.

———. 1972. "Dobranichevskaya Stoyanka na Kievshchine." *Materialy i Issledovaniya po Arkheologii SSSR* 185: 177–188.

Shovkoplyas, I. G., N. L. Korniets, and G. A. Pashkevich. 1981. "Dobranichevskaya Stoyanka." In *Arkheologiya i Paleogeografiya Pozdnego Paleolita Russkoi Ravniny*, edited by I. P. Gerasimov, 97–106. Moscow: Nauka.

Singer, R., B. Gladfelter, and J. Wymer. 1993. *The Lower Paleolithic Site at Hoxne, England*. Chicago: University of Chicago Press.

Sinitsyn, A. A. 1996. "Kostenki 14 (Markina Gora): Data, Problems, and Perspectives." *Préhistoire Européenne* 9: 273–313.

Sinitsyn, A. A., N. D. Praslov, Yu. S. Svezhentsev, and L. D. Sulerzhitskii. 1997. "Radiouglerodnaya Khronologiya Verkhnego Paleolita Vostochnoi Evropy." In *Radiouglerodnaya Khronologiya Paleolita Vostochnoi Evropy i Severnoi Azii: Problemy i*

Perspektivy, edited by A. A. Sinitsyn and N. D. Praslov, 21–66. Saint Petersburg: Russian Academy of Sciences.

Smirnov, S. V. 1979. "Must'erskaya Stoyanka Rikhta." *Kratkie Soobshcheniya Instituta Arkheologii* 157: 9–14.

Smith, F. H. 1982. "Upper Pleistocene Hominid Evolution in South-Central Europe." *Current Anthropology* 23: 667–703.

Soffer, O. 1985. *The Upper Paleolithic of the Central Russian Plain*. San Diego: Academic Press.

———. 1986. "Radiocarbon Accelerator Dates for Upper Paleolithic Sites in European U.S.S.R." *Oxford University Committee for Archaeology Monograph* 11: 109–115.

———. 1989. "The Middle to Upper Paleolithic Transition on the Russian Plain." In *The Human Revolution*, edited by P. Mellars and C. Stringer, 714–742. Princeton: Princeton University Press.

———. 1992. "Social Transformations at the Middle to Upper Paleolithic Transition: The Implications of the European Record." In *Continuity or Replacement: Controversies in Homo sapiens Evolution*, edited by G. Bräuer and F. H. Smith, 247–259. Rotterdam: A. A. Balkema.

———. 1993. "Upper Paleolithic Adaptations in Central and Eastern Europe and Man-Mammoth Interactions." In *From Kostenki to Clovis*, edited by O. Soffer and N. D. Praslov, 31–49. New York: Plenum Press.

Soffer, O., J. M. Adovasio, and D. C. Hyland. 2000. "The 'Venus' Figurines: Textiles, Basketry, Gender, and Status in the Upper Paleolithic." *Current Anthropology* 41 (4): 511–537.

Soffer, O., J. M. Adovasio, N. L. Kornietz, A. A. Velichko, Yu. N. Gribchenko, B. R. Lenz, and V. Yu. Suntsov. 1997. "Cultural Stratigraphy at Mezhirich, an Upper Palaeolithic Site in Ukraine with Multiple Occupations." *Antiquity* 71: 48–62.

Soffer, O., and C. Gamble. eds. 1990. *The World at 18 000 BP*. Vol. 1, *High Latitudes*. London: Unwin Hyman.

Soffer, O., P. Vandiver, B. Klima, and J. Svoboda. 1993. "The Pyrotechnology of Performance Art: Moravian Venuses and Wolverines." In *Before Lascaux: The Complex Record of the Early Upper Paleolithic*, edited by H. Knecht, A. Pike-Tay, and R. White, 259–275. Boca Raton, Fla.: CRC Press.

Stepanchuk, V. N. 1993. "Prolom II, a Middle Palaeolithic Cave Site in the Eastern Crimea with Non-Utilitarian Bone Artefacts." *Proceedings of the Prehistoric Society* 59: 17–37.

Stephens, D. W., and J. R. Krebs. 1986. *Foraging Theory*. Princeton: Princeton University Press.

Stiner, M. C. 1994. *Honor Among Thieves: A Zooarchaeological Study of Neandertal Ecology*. Princeton: Princeton University Press.

Stiner, M. C., N. D. Munro, T. A. Surovell, E. Tchernov, and O. Bar-Yosef. 1999. "Paleolithic Population Growth Pulses Evidenced by Small Animal Exploitation" *Science* 283: 190–194.

Straus, L. G. 1982. "Carnivores and Cave Sites in Cantabrian Spain." *Journal of Anthropological Research* 38: 75–96.

Stringer, C. 1993. "Secrets of the Pit of Bones." *Nature* 362: 501–502.

Stringer, C. B., and Andrews, P. 1988. "Genetic and Fossil Evidence for the Origin of Modern Humans." *Science* 239: 1263–1268.

Stringer, C., and C. Gamble. 1993. *In Search of the Neanderthals*. New York: Thames and Hudson.

Stringer, C., and R. McKie. 1996. *African Exodus: The Origins of Modern Humanity*. New York: Henry Holt and Co.

Sukhachev, V. N., V. I. Gromov, and O. N. Bader. 1966. *Verkhnepaleoliticheskaya Stoyanka Sungir'*. Moscow: Nauka.

Svezhentsev, Yu. S. 1993. "Radiocarbon Chronology for the Upper Paleolithic Sites on the East European Plain." In *From Kostenki to Clovis*, edited by O. Soffer and N. D. Praslov, 23–30. New York: Plenum Press.

Svoboda, J., V. Ložek, and E. Vlček. 1996. *Hunters between East and West: The Paleolithic of Moravia*. New York: Plenum Press.

Swisher, C.C.I., G. H. Curtis, T. Jacob, A. G. Getty, A. Suprijo, and Widiasmoro. 1994. "Age of the Earliest Known Hominids in Java, Indonesia." *Science* 263: 1118–1121.

Tarasov, L. M. 1961. "Uglyanskaya Paleoliticheskaya Stoyanka (Kostenki XVI)." *Kratkie Soobshcheniya Instituta Arkheologii* 85: 38–47.

———. 1977. "Must'erskaya Stoyanka Betovo." In *Paleoekologiya Drevnego Cheloveka*, edited by I. K. Ivanova and N. D. Praslov, 18–30. Moscow: Nauka.

———. 1979. *Gagarinskaya Stoyanka i Ee Mesto v Paleolite Evropy*. Leningrad: Nauka.

———. 1986. "Mnogosloinaya Stoyanka Korshevo I." In *Paleolit i Neolit*, edited by V. P. Lyubin, 46–53. Leningrad: Nauka.

———. 1989. "Periodizatsiya Paleolita Basseina Verkhnei Desny." In *Chetvertichnyi Period: Paleontologiya i Arkheologiya*, 166–175. Kishinev: Shtiintsa.

Tatarinov, K. A. 1977. "Fauna Pozvonochnykh Stoyanki Korman' IV." In *Mnogosloinaya Paleoliticheskaya Stoyanka Korman' IV na Srednem Dnestre*, edited by G. I. Goretskii and S. M. Tseitlin, 112–118. Moscow: Nauka.

Tatartsev, S. V. 1996. "K Voprosu o Verkhnem Sloe Pozdnepaleoliticheskoi Stoyanki Syuren'-I." *Arkheologicheskii Al'manakh* 5: 193–198.

Tattersall, I. 1986. "Species Recognition in Human Paleontology." *Journal of Human Evolution* 15: 165–175.

Thieme, H. 1997. "Lower Palaeolithic Hunting Spears from Germany." *Nature* 385: 807–810.

Torrence, R. 1983. "Time Budgeting and Hunter-Gatherer Technology." In *Hunter-Gatherer Economy in Prehistory: A European Perspective*, edited by G. Bailey, 11–22. Cambridge, U.K.: Cambridge University Press.

Trigger, B. G. 1989. *A History of Archaeological Thought*. Cambridge, U.K.: Cambridge University Press.

Trinkaus, E. 1981. "Neanderthal Limb Proportions and Cold Adaptation." In *Aspects of Human Evolution*, edited by C. Stringer. London: Taylor and Francis, 187–224.

———. 1983. *The Shanidar Neandertals*. New York: Academic Press.

Trinkaus, E., and W. W. Howells. 1979. "The Neanderthals." *Scientific American* 241: 94–105.

Trinkaus, E., and P. Shipman. 1992. *The Neandertals*. New York: Alfred A. Knopf.
Trivers, R. L. 1971. "The Evolution of Reciprocal Altruism." *Quarterly Review of Biology* 46: 35–57.
Tseitlin, S. M. 1965. "Geologiya Raiona Verkhnepaleoliticheskoi Stoyanki Sungir' vo Vladimirskoi Oblasti." In *Stratigrafiya i Periodizatsiya Paleolita Vostochnoi Evropy*, edited by O. N. Bader, I. K. Ivanova, and A. A. Velichko, 66–85. Moscow: Nauka.
Turner, A. 1984. "Hominids and Fellow Travellers: Human Migration into High Latitudes as Part of a Large Mammal Community." In *Hominid Evolution and Community Ecology*, edited by R. A. Foley, 193–217. London: Academic Press.
———. 1992. "Large Carnivores and Earliest European Hominids: Changing Determinants of Resource Availability during the Lower and Middle Pleistocene." *Journal of Human Evolution* 22: 109–126.
Turq, A. 1992. "Raw Material and Technological Studies of the Quina Mousterian in Perigord." In *The Middle Paleolithic: Adaptation, Behavior, and Variability*, edited by H. L. Dibble and P. A. Mellars, 75–85. Philadelphia: University of Pennsylvania.
Udartsev, V. P. 1980. "K Voprosu o Sootnoshenii Pokrovnykh i Lednilovykh Kompleksov Okso-Donskoi Ravniny." In *Vozrast i Rasprostranenie Maksimal'nogo Oledeneniya Vostochnoi Evropy*, edited by I. P. Gerasimov and A. A. Velichko, 20–72. Moscow: Nauka.
van der Hammen, T., T. A. Wijmstra, and W. H. Zagwijn.. 1971. "The Floral Record of the Late Cenozoic of Europe." In *Late Cenozoic Glacial Ages*, edited by K. Turekian, 391–424. New Haven: Yale University Press.
Van Peer, P. 1992. *The Levallois Reduction Strategy*. Monographs in World Archaeology, No. 13. Madison, Wisc.: Prehistory Press.
Vekilova, E. A. 1957. "Stoyanka Syuren' I i Ee Mesto Sredi Paleoliticheskikh Mestonakhozhdenii Kryma i Blizhaishikh Territorii." *Materialy i Issledovaniya po Arkheologii SSSR* 59: 235–323.
———. 1971. "Kamennyi Vek Kryma, Nekotorye Itogi i Problemy." *Materialy i Issledovaniya po Arkheologii SSSR* 173: 117–161.
———. 1979. "K Stoletiyu Otkrytiya Paleolita v Krymu (1879–1979 gg.)." In *Issledovanie Paleolita v Krymu*, edited by Yu. G. Kolosov, 5–15. Kiev: Naukova Dumka.
Velichko, A. A. 1961a. *Geologicheskii Vozrast Verkhnego Paleolita Tsentral'nikh Raionov Russkoi Ravniny*. Moscow: USSR Academy of Sciences.
———. 1961b. "O Vozmozhostyakh Geologicheskogo Sopostavleniya Raionov Paleoliticheskikh Stoyanok v Basseinakh Desny, Dona i na Territorii Chekhoslovakii." *Trudy Komissii po Izucheniyu Chetvertichnogo Perioda* 18: 50–61.
———. 1969. "Bassein Srednego Dona." In *Priroda i Razvitie Pervobytnogo Obshchestva na Territorii Evropeiskoi Chasti SSSR*, edited by I. P. Gerasimov, 20–22. Moscow: Nauka.
———. 1973. *Prirodnyi Protsess v Pleistotsene*. Moscow: Nauka.
———. 1980. "Voprosy Paleogeografii i Khronologii Rannego i Srednego Pleistotsena." In *Vozrast i Rasprostranenie Maksimal'nogo Oledeneniya Vostochnoi Evropy*, edited by I. P. Gerasimov and A. A. Velichko, 189–207. Moscow: Nauka.

———. 1984. "Late Pleistocene Spatial Paleoclimatic Reconstructions." In *Late Quaternary Environments of the Soviet Union*, edited by A. A. Velichko, 261–285. Minneapolis: University of Minnesota Press.

———. 1990. "Loess-Paleosol Formation on the Russian Plain." *Quaternary International* 7/8: 103–114.

Velichko, A. A., A. B. Bogucki, T. D. Morozova, V. P. Udartsev, T. A. Khalcheva, and A. I. Tsatkin. 1984. "Periglacial Landscapes of the East European Plain." In *Late Quaternary Environments of the Soviet Union*, edited by A. A. Velichko, 94–118. Minneapolis: University of Minnesota Press.

Velichko, A. A., L. V. Grekhova, and Z. P. Gubonina. 1977. *Sreda Obitaniya Pervobytnogo Cheloveka Timonovskikh Stoyanok*. Moscow: Nauka.

Velichko, A. A., Yu. N. Gribchenko, and E. I. Kurenkova. 1997. "Stratigrafischeskoe Polozhenie Stoyanok Pushkarevskoi Gruppy." In *Pushkarevskii Sbornik*, edited by V. I. Belyaeva, 19–30. Saint Petersburg: Obrazovanie Kul'tura.

Velichko, A. A., M. D. Gvozdover, G. P. Grigor'ev, Z. P. Gubonina, V. P. Udartsev, E. A. Vangengeim, and M. V. Sotnikova. 1981. "Avdeevo." In *Arkheologiya i Paleogeografiya Pozdnego Paleolita Russkoi Ravniny*, edited by I. P. Gerasimov, 48–56. Moscow: Nauka.

Velichko, A. A., and T. D. Morozova. 1982a. "Pochvennyi Pokrov Mikulinskogo Mezhlednikov'ya i Bryanskogo Intervala." In *Paleogeografiya Evropy za Poslednie Sto Tysyach Let*, edited by I. P. Gerasimov and A. A. Velichko, 81–91. Moscow: Nauka.

———. 1982b. "Izmeneniya Prirodnoi Sredy v Pozdnem Pleistotsene po Dannym Izucheniya Lessov, Kriogennykh Yavlenyi, Iskopaemykh Pochv, i Fauny." In *Paleogeografiya Evropy za Poslednie Sto Tysyach Let*, edited by I. P. Gerasimov and A. A. Velichko, 115–120. Moscow: Nauka.

Velichko, A. A., and V. P. Nechayev. 1984. "Late Pleistocene Permafrost in European USSR." In *Late Quaternary Environments of the Soviet Union*, edited by A. A. Velichko, 79–86. Minneapolis: University of Minnesota Press.

Velichko, A. A., F. M. Zavernyaev, Yu. N. Gribchenko, Z. P. Gubonina, E. M. Zelikson, A. K. Markova, and V. P. Udartsev. 1981. "Khotylevskie Stoyanki." In *Arkheologiya i Paleogeografiya Pozdnego Paleolita Russkoi Ravniny*, edited by I. P. Gerasimov, 57–69. Moscow: Nauka.

Vereshchagin, N. K. 1967. "Primitive Hunters and Pleistocene Extinction in the Soviet Union." In *Pleistocene Extinctions*, edited by P. Martin and H. E. Wright, 365–398. New Haven: Yale University Press.

Vereshchagin, N. K., and G. F. Baryshnikov. 1980. "Mlekopitayushchie Predgornogo Severnogo Kryma v Epokhu Paleolita." *Trudy Zoologicheskogo Instituta AN SSSR* 93: 26–49.

———. 1982. "Paleoecology of the Mammoth Fauna in the Eurasian Arctic." In *Paleoecology of Beringia*, edited by D. M. Hopkins, J. V. Matthews, C. E. Schweger, and S. B. Young, 267–279. New York: Academic Press.

———. 1984. "Quaternary Mammalian Extinctions in Northern Eurasia." In *Quaternary Extinctions*, edited by P. S. Martin and R. G. Klein, 483–516. Tucson: University of Arizona Press.

Vereshchagin, N. K., and A. D. Kolbutov. 1957. "Ostatki Zhivotnykh na Must'erskoi Stoyanke pod Stalingradom i Stratigraficheskoe Polozhenie Paleoliticheskogo Sloya." *Trudy Zoologicheskogo Instituta AN SSSR* 22: 75–89.

Vereshchagin, N. K., and I. E. Kuz'mina. 1977. "Ostatki Mlekopitayushchikh iz Paleoliticheskikh Stoyanok na Donu i Verkhnei Desne." *Trudy Zoologicheskogo Instituta AN SSSR* 72: 77–110.

Villa, P. 1990. "Torralba and Aridos: Elephant Exploitation in Middle Pleistocene Spain." *Journal of Human Evolution* 19: 299–309.

Vlček, E. 1975. "Morphology of a Neanderthal Child from Kiik-Koba in the Crimea." In *Palaeoanthropology, Morphology, and Palaeoecology*, edited by R. Tuttle, 409–417. The Hague: Mouton.

Voedvodskii, M. V. 1947. "Rezul'tati robit Desnyans'koi Ekspeditsii 1936–1938 rr." *Paleolit i Neolit Ukraini* 1: 41–59.

———. 1950. "Paleoliticheskaya Stoyanka Pogon." *Kratkie Soobshcheniya Instituta Istorii Material'noi Kul'tury* 31: 40–54.

———. 1952. "K Voprosu o Pushkarevskom Paleoliticheskom Zhilishche (Stoyanka 'Paseka')." *Uchenye Zapiski Moskovskogo Gosudarstvennogo Universiteta* 158: 71–74.

Walker, P. L. 1978. "Butchering and Stone Tool Function." *American Antiquity* 43: 710–715.

Wanpo, H., R. Ciochon, G. Yumin, R. Larick, F. Qiren, H. Schwarcz, C. Yonge, J. de Vos, and J. Rink. 1995. "Early *Homo* and Associated Artifacts from Asia." *Nature* 378: 275–278.

Waters, M. R., S. L. Forman, and J. M. Pierson. 1997. "Diring Yuriakh: A Lower Paleolithic Site in Central Siberia." *Science* 275: 1281–1284.

West, F. H., ed. 1996. *American Beginnings: The Prehistory and Paleoecology of Beringia*. Chicago: University of Chicago Press.

Whallon, R. 1989. "Elements of Cultural Change in the Later Paleolithic." In *The Human Revolution*, edited by P. Mellars and C. Stringer, 433–454. Princeton: Princeton University Press.

White, L. 1962. *Medieval Technology and Social Change*. Oxford, U.K.: Oxford University Press.

White, R. 1989. "Production Complexity and Standardization in Early Aurignacian Bead and Pendant Manufacture: Evolutionary Implications." In *The Human Revolution*, edited by P. Mellars and C. Stringer, 366–390. Princeton: Princeton University Press.

Whittaker, R. H. 1975. *Communities and Ecosystems*. 2nd ed. New York: Macmillan.

Williams, G. C. 1966. *Adaptation and Natural Selection: A Critique of Some Current Evolutionary Thought*. Princeton: Princeton University Press.

Wilson, A. C., and R. L. Cann. 1992. "The Recent African Genesis of Humans." *Scientific American* 266 (4): 22–27.

Wilson, E. O. 1975. *Sociobiology: The New Synthesis*. Cambridge, Mass.: Harvard University Press.

Winterhalder, B., and E. A. Smith. eds. 1981. *Hunter-Gatherer Foraging Strategies: Ethnographic and Archeological Analyses*. Chicago: University of Chicago Press.

Wobst, H. M. 1974. "Boundary Conditions for Paleolithic Social Systems: A Simulation Approach." *American Antiquity* 39 (2): 147–178.

———. 1976. "Locational Relationships in Paleolithic Society." *Journal of Human Evolution* 5: 49–58.

Wolpoff, M. H. 1999. *Paleoanthropology*. 2nd ed. Boston: McGraw-Hill.

Yakimov, V. P. 1957. "Pozdnepaleoliticheskii Rebenok iz Pogrebeniya na Gorodtsovskoi Stoyanke v Kostenkakh." *Sbornik Muzeya Antropologii i Etnografii* 17: 500–529.

———. 1961. "Naselenie Evropeiskoi Chasti SSSR v Pozdnem Paleolite i Mezolite." *Voprosy Antropologii* 7: 23–28.

Yanevich, A., V. N. Stepanchuk, and Yu. Cohen. 1996. "Buran-Kaya III and Skalisty Rockshelter: Two New Dated Late Pleistocene Sites in the Crimea." *Préhistoire Européene* 9: 315–324.

Yellen, J. E. 1977. *Archaeological Approaches to the Present*. New York: Academic Press.

Yevtushenko, A. I. 1998. "Kabazi V: Assemblages from Selected Levels." In *The Middle Paleolithic of Western Crimea*. Vol. 1, edited by A. E. Marks and V. P. Chabai, 287–322. ERAUL 84.

Zamyatnin, S. N. 1929. "Station Moustérienne a Ilskaja Province de Kouban (Caucase du Nord)." *Revue Anthropologique* 7–9: 282–295.

———. 1940. "Pervaya Nakhodka Paleolita v Doline Seima." *Kratkie Soobshcheniya Instituta Istorii Material'noi Kul'tury* 8: 96–101.

———. 1961. "Stalingradskaya Paleoliticheskaya Stoyanka." *Kratkie Soobshcheniya Instituta Arkheologii* 82 :5–36.

Zavernyaev, F. M. 1978. *Khotylevskoe Paleoliticheskoe Mestonakhozhdenie*. Leningrad: Nauka.

Zimmerman, M. R. 1985. "Paleopathology in Alaskan Mummies." *American Scientist* 73: 20–25.

Zubakov, V. A. 1988. "Climatostratigraphic Scheme of the Black Sea Pleistocene and Its Correlation with the Oxygen-Isotope Scale and Glacial Events." *Quaternary Research* 29: 1–24.

Index

Note: Page numbers in *italics* denote illustrations.

Abric Romani, cave at, 106, 111
Acheulean industry, 37, 53, 91, 93
adaptation, 3–4
Adzhi-Koba, 121
Africa, 1–2, 37, 140–142, 158, 188, 193–194, 249, 252–253
age (mortality) profile, 117, *117*, 125, 181, 241, 243
Ahmarian industry, 141
Alaska, 41, 158
alder, 33
Allen's rule, 7, 58, 132, 141, 158, 250
alliance networks, 10, 13–14, 125–126, 135, 138, 190, 248, 255
Alma River, 82
alpine chough (*Pyrrhocorax graculus*), 81
Altmühlian industry, 104
amber, 248
Ambrona, 39
Amvrosievka, 212–213, *213*, 228–229, 236, 242
Anetovka II, 212, 229, 242–243
anti-adaptationist backlash, 4–5
Arago, 38, 42, 56
arctic fox (*Alopex lagopus*), 24, 30–32, 74, 79, 81; cut marks on bones of, 242; nearly complete skeletons of, 225, 241–242; early Upper Paleolithic, 169, 171, 182–184, 189; late Upper Paleolithic, 202, 204, 206, 211, 225, 229, 241–242; trapping of, 189, 225
Arctic Ocean, 15, 17
art, 126, 141–142; decorative, 193; early Upper Paleolithic, 147, 153, 161, 167, 174–178; figurative, 142, 171, 174, 177, *178*, 191, 193, 208–209, 233–235, *234*; late Upper Paleolithic, 193, 201–202, 206–209, 219–220, 233–236. *See also* decorated objects; figurines, animal; Venus figurines

Artemisia, 23, 32
Ataki I, 146, 198, *199*, 228
Atapuerca: Gran Dolina, 38, 42; Sima de los Huesos, 57
Aurignacian industry, 141–142, 153, 160, 163, 165, 169, 173, 177, 190–191, 220
Aurignacoid industry, 167, 169, 172, 174–177, 190–191
Australia, 14, 141–142, 193–194
Autlev, P. U., 85
Avdeevo, 207–209, *208*, 220, 225–226, 233–235, 245
awls (bone), 107; early Upper Paleolithic, 160, 163, 169, 171–174, 189; late Upper Paleolithic, 197–199, 209, 215, 220, 225, 229
awls (stone), 107; microwear analysis of, 107

Babin I, 146, 163, 179, 187, 243
backed blades, 166, 219
backed bladelets, 173, 209, 214–215, 220–221, 223, 228–230, *231*
backed microblades, 173
backed points, 173, 223, 228
Bader, O. N., 151
Bailey, H. P., 8
Bakhchisarai, 82
Baku Transgression, 44
Balkans, 190–191, 254
Barakaevskaya Cave, 65, 85, 103, 110, 122–125, 129, 131; Neanderthal skeletal remains in, 88–91
"base camp" sites, 115, 119, 123, 211, 243–248, 255
beads. *See* ornaments
beech, 17, 28
Belarus, 15, 201–202
Belaya River, 153
Bel'bek River, 152
Belogorsk, 78

285

Belyaeva, E. V., 85
Berdyzh, 201–202, 220, 235
Bergmann's rule, 7, 58, 132, 158, 250
Beringia (or Bering Land Bridge), 1, 22, 25–26
Betovo, 66, 74, *75*, 116
Biache, 57
bifaces, 53; foliates, 76, 98, 103, 104–105; hafting of tools, relation to, 105; manufacture of, 91; Mousterian, 67, 69, 71–72, 74, 77, 79–82, 84–85, 93, 98, 104–105, *105*, 251; Upper Paleolithic, 147, 152–153, 166, 169, 174; uses of, 104–105. See also handaxes; points
Bilzingsleben, 38, 53
Binford, L. R., 111
birch, 33, 34
birds, 242
Biryuchya Balka, 171
bison. See steppe bison
Bison schoetensacki, 44, 46
Biyuk-Karasu River, 79
black grouse (*Lyrurus tetrix*), 242
Black Sea, 15, 28, 30, 33, 42, 82, 185, 248
blade technology, 78, 161, 165, 167, 169, 171, 174, 219; efficiency of, 159, 221. See also cores
bladelets, 174, 221, 228
Blombos Cave, 141
Bobuleshty VI, 166–167
Bodo, 140
body hair, human, 40
body size: carnivore, 83; human, 37
Bonch-Osmolovskii, G. A., 78, 80, 82, 152
bone, antler, and ivory artifacts, 106–107; early Upper Paleolithic, 139, 141–142, 151, 153, 159–160, 169; late Upper Paleolithic, 219, 225, 254
bone fuel, 24, 108–109, 192, 200, 226, 228, 237, 239, 241, 254
Bordes, F., 103
Boriskovskii, P. I., 204, 209
Borisovskoe Gorge, 85, 153, 215
Borshchevo I, 210, 236, 245
bows and arrows, 161, 193–194, 223, 229

Boxgrove, 38–39
Brace, C. L., 60
brachial index, 7, *7*, 58, 158, 195
brain volume: modern humans, 155; Neanderthals, 57–58
Bräuer, G., 140
Broken Hill, 140
Brunhes polarity epoch, 27
Bryansk, 72, 202–203
Bryansk Soil, *29*, 31, 143
Brynzeny Culture, 64, 166–167, *166*, 177, 190, 254
Brynzeny I, 143, 147, 161, 166–167, *166*, 176–177, *177*, 181
Bugorok, 204, 242
Buran-Kaya III, 152–153, 173–174, 214, 228
burials, 81, 88, 126–127, 150–152, 191, 210; effects of frost action on, 34–35, 155; ritual associated with, 174, 176. See also under modern humans; Neanderthals
burins: early Upper Paleolithic, 145–147, 149, 151, 153, 159, 162–171, 173–174; late Upper Paleolithic, 197–200, 202, 206–207, 209, 212–214, 219–220, 223, 228; microwear analysis of, 223; types of, 223; uses of, 159, 165, 223, 228
Burul'cha River, 153, 173
Buteshty, 71, 100
Buzduzhany I, 71, 103
Byzovaya, 152, 171

caloric: demand, 61, 109, 133; intake, 7–8
carnivore dens, 46, 116, 121, 181; shared with humans, 119, 123, 125
Carpathian Basin, 22, 54, 194, 220
Carpathian Mountains, 15, 36, 41, 89, 142, 249, 251
Caspian Sea, 15, 28, 30, 33, 42, 76
Caucasus Mountains, 15, 17, 20, 36, 42, 44, 190, 254
cave bear (*Ursus deningeri*), 46, 51, 85–86, 123–125
Central Asia, 190
central-place foraging, 46, 61, 112
Central Russian Upland, 15–16, 148, 185, 207–208, 220, 233, 241, 245

Ceprano, 38
ceramic technology, 161
Červený kopec, 41
Chatelperronian industry, 64, 139, 143, 161, 171, 174, 190, 254
Chaudian Transgression, 44
Chenopodiaceae, 32
chernozem soils, 19, 26, 30
Chernysh, A. P., 67, 145, 187
Chokurcha I, 81, 107, 121
Chokurcha II, 127
choppers, 38
Chulatovo III (Zarovskaya Krucha), 74
Chuntu rockshelter, 166, 199
Chutuleshty, 243–245
Clacton, 106
climatostratigraphy, 18
collared lemming (*Dicrostonyx torquatus*), 32, 74
competition between Neanderthals and modern humans, 2, 141, 178, 184, 253
competitive exclusion, 4–5
composite tools and weapons, 223, 225, 254
Coon, C. S., 58–61
cordage, 199, 207. *See also* textiles, woven
cores, 47–48, *50*, 71, 76, 79, 85–86, 146; centripetal flaking, 99; discoidal, 50, 81, 98–100, *99*, 166; double-platform, 163; Levallois, 50, 53–54, 60, 67, 69, 71, 74, 82, 84, *99*, 100; Levallois blade, 77, 86, 162; microblade, 174, 221; prismatic blade, 145, 162, 166, 169, 173–174, 221, 228; protoprismatic (or parallel), 50, 98–99, *99*; radial, 80–81, 84, 98–100, *99*; secondary, 221; subprismatic, 84. *See also* blade technology; Levallois techniques
clothing. *See* tailored fur clothing
Crimea. *See* Crimean Peninsula
Crimean Mountains, 17, 20, 78, 82
Crimean Peninsula, 17, 21, 30, 34–35; Mousterian, Neanderthals of, 63–66, 78–83, *78*, 88–89, 98, 100–104, 107, 109, 112–113, 119–123, 127, 129, 131, 188; Upper Paleolithic, modern humans of, 143, 152–156,
173, 176, 179, 183–184, 228, 243, 253
Cro-Magnon, 155, 156
crural index, 58, *92*, 158, 195
cut marks on bone from stone tools, 39, 61; Mousterian, 115, 117, 125; Upper Paleolithic, 178, 181, 183, 241–243

Dakhovskaya Cave, 123
Dar-es-Soltan, 140
Darwin, C., 5
Debets, G. F., 157
decorated objects, 169, 171–174, 176–177, *176*, 209, 235–236, *236*
Denekamp Interstadial, 31
Denticulate Mousterian of Levallois Type, 51, 54
denticulates, 50, 69, 80, 83, 86, 93, *102*, 103–104; microwear analysis of, 104; product of trampling and natural processes, 103; Upper Paleolithic, 146, 166–167, 172; uses of, 104
Desna River. *See* Middle Desna River
Dibble, H., 103
digging sticks, 111, 194
digging tools (bone), 197, 199. *See also* mattocks
dietary adaptations to cold climate, 7–8, 61, 109–110, 133, 178, 237
Diring Yuriakh, 41
Dmanisi, 37
DNA (fossil), 2, 39, 55–56, 89, 158
Dnepr Basin (or Dnepr Lowland), 15, 16–17, 20, 27, 72, 88, 143; late Upper Paleolithic, 196, 201, 206, 220, 229, 231, 235, 241, 245–246
Dnepr-Desna Basin. *See* Dnepr Basin
Dnepr Glaciation, 27, 42, 72, 74
Dnestr River, 15–16, 51, 65, 66, 110, 116, 166. *See also* Middle Dnestr Valley
Dobranichevka, 207, 228, 231, 242, 245–246, *247*
dolichocephalic skull, 217
Dolní Věstonice I and II, 220
domain-specific cognition, 12–13
Don Glaciation, 27, 42
Don hare (*Lepus tanaiticus*), 183. *See also* hare

288 INDEX

Don River, 17, 47, 88
Donets Ridge, 15, 17, 212, 242
Doronichev, V. B., 44
Dorset, Late, 246
Dragishte River, 71
drill, rotary, 161, 169, 253
Dubossary I, 47, 53
dwarf alder, 32
dwarf birch, 31–32
dwelling structures. *See* shelters, artificial

Early Glacial (OIS 5a–5d), 30, 65, 67, 69, 71, 74, 76, 79, 89, 100, 104, 251
Early Pleniglacial (OIS 4), 30–31, 65, 69, 71, 86, 89, 100, 104, 125, 136, 251–252
"earth-houses," 203
East Asia, 37, 93, 142, 252
"Eastern Cro-Magnon," 155
Eastern Gravettian, 173, 194, 209, 211, 214, 229, 248, 255; art and ornaments, 232–235; dating of, 200–202; occupation floors, 245–246; relation to loess-steppe, 220, 248, 254; technology and tools, 219–227. *See also* Kostenki-Avdeevo Culture; "longhouse" structures; Venus figurines
effective temperature, 8–10, 109
Efimenko, P. P., 206, 209–210
Ehringsdorf, 57
Electron Spin Resonance (ESR) dating, 44, 63, 83
Eliseevichi I, 204–205, *205*, 228–229, 231, 235–236, 248
elm, 17
"empty niche" hypothesis, 188, 253
end-scrapers: carinated, 169, 173; early Upper Paleolithic, 145–147, 149, 151–153, 159, 162–174, 191; late Upper Paleolithic, 197, 199–200, 212–214, 228; uses of, 159–160, 165, 225
Engels, F., 127–128
Epigravettian, 201, 227–233, 243, 254–255; art and ornaments, 235–236; dating of, 201; occupation floors, 246–248; technology and tools, 227–233

Equus altidens, 44
Ernst, N. L., 81
evolutionary ecology, 3–6, 12
Evolved Gravettian. *See* Epigravettian
exogamy, 128, 136
extinctions, 13; carnivore, 38, 249

Fastov, 241–242, 247
Feldhofer Cave, 158
Fennoscandian ice sheet, 30, 32, 192
figurines, animal, 177, *178*, 234
fire, use of controlled, 38–39, 52, 108, 250
fish, harvesting of, 193–194
Formozov, A. A., 82
Frolovo, 77
frost action, effects of, 35, 81, 151–152, 202, 215
frostbite, 7
frost-gley soil, 19, 24, 31
frost-heaving, 35. *See also* frost action
frozen mammal carcasses, 39
frozen ground, 34. *See also* permafrost
functional and seasonal differentiation of sites, 115, 123–125, 184, 241–243, 246–248

Gagarino, 210–211, 220, *224*, 225, 227, 233–235, 241, 245
Gamble, C., 39, 59
Garchi I, 171
Gavrilo River, 74
gene flow, 2, 56, 142, 252
genetic drift, 4, 39, 58, 59, 132
geographic speciation, 56
Gerasimovka, 43–44
giant deer (*Megaceros giganteus*), 39, 69, 77, 81, 115, 121, 152, 184
Gizel'don River, 86
Gmelin Soil, 195
goat (*Capra* sp.), 51, 85–86, 123–125
Golovanova, L. V., 44, 85
Gontsy, 207, 209, 228–229, 236, 241, 246, 248
goose (*Anser* sp.), 242
Goretskii, G. I., 47
Gorodtsov, V. A., 83, 203
Gorodtsov Culture, 172, *172*, 174–176, 191, 254
"grave goods," 127

"Great Interglacial," 44
"great leap" (*bol'shoi skachok*), 3, 12, 139
grinding stones, 110, 162–163, 193, 239; microwear analysis of, 110
Gubs River, 153
Gubs Shelter No. 1, 85, 153, 215
Gulf of Taganrog, 77

hafting of stone tools and weapons, 53, 60, 106, 109–111, 133, 136, 221, 250
hafts, 163, 197, 220
halophytic plants, 23
handaxes, 38, 72, 74, 76, 98. *See also* bifaces
handles. *See* hafts
hare (*Lepus* sp.): early Upper Paleolithic, 172, 181, 183–184, 189; late Upper Paleolithic, 199, 207, 209–210, 225, 229, 241–242; trapping of, 181, 189, 225
hazel, 31
hearths: debris concentrations, associated with, 129, 185, 197–198; early Upper Paleolithic, 187; late Upper Paleolithic, 226, 245; Mousterian, 108–109, 129; in shelters, 150, 163, 173, 227, 231, 233
Helix pomatia, 67. *See also* molluscs
Hengelo Interstadial, 31
hide burnishers (bone and ivory), 225; microwear analysis of, 225
hides, working and use of, 106–107, 135, 160, 165, 225, 229, 252
Hidjrati, N. I., 86
Holstein Interglacial, 44
Homo erectus, 38, 52–53, 93
Homo ergaster, 37, 52, 250
Homo heidelbergensis, 39, 52, 249
Homo neanderthalensis, 57, 250. *See also* Neanderthals
Homo sapiens neanderthalensis, 57. *See also* Neanderthals
Homo sapiens sapiens, 143. *See also* modern humans
hornbeam, 17, 31
horse (*Equus* sp.), 24, 28, 31–32, 39, 44, 50; hunted in groups, 183; Mousterian, 67, 69, 72, 80–81, 112, 115–116, 121; Upper Paleolithic, 145–147, 149, 152, 179–183, 185, 197–199, 206, 215, 239, 241
Howell, F. C., 56, 58
Hoxne, 38–39
Hoxnian Interglacial, 44
hunter-gatherer adaptations to northern environments, 7–12, *11*
hunting: Middle Pleistocene, 39, 46, 52, 61; by modern humans, 179–184, 239, 241–243; by Neanderthals, 61, 111–125, 133
hyena (*Crocuta* sp.), 80–81, 119, 121
"hyperpolar" morphology. *See* Neanderthals

ice cellars, 161, 226
ice wedges (and ice-wedge casts), 203–204
Ikva River, 146
Il' River, 83
Il'skaya I and II, 83–85, 115–117, *117*, 123–125, 129
intelligence, general, 12
Interpleniglacial. *See* Middle Pleniglacial
Inuit, 161

Jersey, island of, 115
juniper, 81

Kabazi I, 131
Kabazi II, 66, 82–83, 100, 104
Kabazi V, 123
Kacha River, 82
Kama River, 171
Kamenkutsa River, 71
Kamennaya Balka II, 213, 228, 242
Kamennomostskaya Cave, 153
Kapovaya Cave, 233
Kärlich, 38, 41
katabatic wind, 22
Katanda, 141
Ketrosy, 67, 100, 107–108, 110, 113, 115, 128,
Khotylevo I (Mousterian), 65, 72–74, *73*, *99*, 100, 104, 116, 119, 201, 251
Khotylevo II (Upper Paleolithic), 196, 202, 220, 225–226, 233–234, 245
Khryashchi, 43, 47–51, *49*, 53, 64
Khvalyn Transgression, 76

Khvoiko, V. V., 206, 209
Kiev, 206
Kiik-Koba, 78, 80–81, 102–103, 107, 123; Neanderthal burial at, 88–91, 127
kill-butchery sites, 183, 213, 241–243
kilns, 161
kin selection, 5–6, 131, 190
Kirillovskaya, 206, 246
Klasies River Mouth, 140
Klein, R. G., 141
knives (stone), 67, 77, 102–103, *102*; backing of, 103; Upper Paleolithic, 145–146, 162–163, 197
Kodak, 116
Kodiak Island, 158
Kolosov, Yu. G., 79, 82
Konstantinovsk, 213
Korman' IV, *21*; Mousterian, 66–67, 108, 113, 115; Upper Paleolithic, 144, 146, 162–163, 184, 197–199, 225, 228
Korolevo, 38, 54
Korpach, 147, 167
Korshevo I and II, 65, 74, 104
Kosoutsy, 197–199, 215, 228–229, 235
Kostenki, 66; early Upper Paleolithic, 143, 167, 170, 172–176, 181, 185, 187, 189–191, 253; late Upper Paleolithic, 128, 208–210; modern human skeletal remains at, 153, 157, 215, 217–218; stratigraphy and topographic setting, 148–151, *149–150*. *See also* Efimenko, P. P.; Gorodtsov Culture; Kostenki-Avdeevo Culture; Lower Humic Bed; Spitsyn Culture; Strelets Culture; Upper Humic Bed
Kostenki I: early Upper Paleolithic, 148–150, 160, 169–170, *170*, 175, 177, 185, 187; late Upper Paleolithic, 207–211, 220, *224*, 225–227, *227*, 233–235, *234*, *244*, 245
Kostenki II, 108, 197, 210, 228, 235; modern human skeletal remains at, 215, 217–218
Kostenki IV, 210, 225, 227, 234–235, 241, 245
Kostenki VI (Streletskaya), 169–170, *170*, 181, 185

Kostenki VIII, 149, 173, 176–177, 183–184, 187, 220, 225
Kostenki XI (Anosovka II), 170, 210, 229, 234–236, 246
Kostenki XII, 149–150, 161, 167, 169–170, *170*, 172, 175–176, 181
Kostenki XIII, 220, 225, 233, 242
Kostenki XIV (Markina Gora): early Upper Paleolithic, 149–150, 155–157, 161, 172, *172*, 175–176, *176*, 181–183, 187; late Upper Paleolithic, 220; modern human skeletal remains at, 155–157
Kostenki XV, 149–150, 155, 172, 175–176, 181, 183, 187
Kostenki XVI, 172, 181, 183, 187
Kostenki XVII, 149, 155, 167–169, *170*, 172, *175*, 181, 185
Kostenki XVIII, 197, 215, 217, 220
Kostenki XXI, 209–210, 235
Kostenki-Avdeevo Culture, 209, 248. *See also* Eastern Gravettian
Kostenki knives, 221–223, *223*; uses of, 221–223. *See also* truncations
Kostenki points, 221, 223, *223*; microwear analysis of, 221; uses of, 221. *See also* points (stone): shouldered
Krasnaya Balka, 79, 123
Krasnodar, 83
Kremenets, 146
Krutitsa Phase or soil (Mezin Soil Complex), 30–31
Kuban' River, 83
Kulichivka, 146, *147*; early Upper Paleolithic, 162–163, 165, 176, 179, 184, 187, 189; late Upper Paleolithic, 199–200, 227, 235, 245
Kurdzhips River, 85, 174, 214
Kursk, city of, 17, 207
Kursk Soil, 28

La Cotte de St. Brelade, 115
language, 1, 12–13, 138, 188, 253; Neanderthals and, 125–127; organization and, 13–14, 125, 188; technology and, 12–13, 158, 160–161, 188. *See also* symbols
Last Glacial Maximum (OIS 2), 32–33, 66, 115–116, 136, 138, 143, 148,

159, 161, 192–248, 254; environments of, 24–25, *33*
Last Interglacial (OIS 5e), 20, 28–30, *29*, 33, 47, 59, 64–65, 67, 69, 71–76, 81–84, 86, 88–89, 100, 104, 148, 250–251; environments of, 28, *29*. *See also* Mezin Soil Complex
Late Glacial (OIS 2), 33, 195, 198, 208, 210, 214–215, 219, 228–229, 239
Late Pleniglacial. *See* Last Glacial Maximum
Late Stone Age, 141, 193–194
Lehringen, 61, 106
leopard, 119
Levallois cores. *See* cores: Levallois
Levallois techniques, 93, 98, 100, 104–105, 133, 135, 251; hafting of tools, relation to, 60, 101, 133
Le Vallonet, 41
Likhvin Interglacial, 47
limace, 46–47, 50
linden, 28
lion (*Panthera* sp.), 121, 123
Lipa Culture, 163
Lipa VI, 163
loess, 19–23, 25–34, *29*, 47, 51, 65–67, 69, 74, 113, 128, 146; Last Glacial Maximum, 32, 192, 195, 199, 203–204, 242; stratigraphy, 19, *20*. *See also* Valdai Loess
loess-steppe, 22–26, 28, 31–32, 188, 192, 211, 254; as environment without modern analogue, 25, 192; late Upper Paleolithic adaptations to, 192–195, 211, 217, 226, 228, 248; Neanderthals unable to occupy, 55, 136, 251–252; productivity of, 23–25, 237, 252; reconstruction of, 22–26; role of loess deposition and weathering on, 22–23
loessic colluvium (or loess-derived colluvium), 19, *21*, 67, 144, 148, 197–198, 206–208
long-distance movement of raw materials, 11, 112, 135, 189–190, 252; early Upper Paleolithic, 137–138, 167, 169, 172, 178, 184–185, 190; late Upper Paleolithic, 237, 243, 248, 253–255
Longgupo, 37

"longhouse" structures, 207–209, *244*, 245–246; Late Dorset examples, 246
Lower Bug River, 212, 243
Lower Dnepr River, 116, 233, 248
Lower Dnestr River, 47–48, 53, 243–245
Lower Don River, 77
Lower Humic Bed (Kostenki), 31, 148, 167, 169–170, 175, 181, 187, 190
Lower Paleolithic, 36–54, 93, 110
Lubny, 207
Lycopodium, 23
lynx (*Felis* sp.), 121
Lyubin, V. P., 85

Maastricht-Belvédère, 57
Magdalenian industry, 194
Maikop, 214
mammal bone, natural concentrations of, 26, 113, 179, 211–212, 226, 232
mammoth. *See Mammuthus trogontherii*; woolly mammoth
mammoth-bone structures. *See* shelters, artificial
Mammuthus primigenius. *See* woolly mammoth
Mammuthus trogontherii, 44, 47
Mariin Terrace (Don River), 47
Marxist-Leninist ideas, application to archaeology of, 127–128, 209
mating network, 10
matrilineal clan organization of Upper Paleolithic, 128, 233
mattocks, 209, 220, 226, *227*, 231; use of, 226, 231; wear patterns on, 231
Matuzka Cave, 84–85, 123–125
Mauer, 38, 52
meat, proportion of diet of, 8, 109–110, 178, 193–194, 237, 250
Merezhkovskii, K. S., 78
Mezhirich, 206–207, 211, 228–231, *232*, 235–236, 245–246, 248
Mezin, 206, 228–229, *231*, 235–236, *236*, 241–242, 245–246, 248
Mezin Soil Complex, 28, *29*, 30, 64, 74
Mezmaiskaya Cave: Mousterian, 66, 85, *86*, 109, 117, 123–125; Neanderthal infant burial in, 88–91, 127, 157–158; Upper Paleolithic, 152, 174, 214–215
Micoquian industry, 104

microblades, 221, 228
microliths, 201, 210, 212, 214, 223, 229; geometric segments, 214–215; as inset blades and points, 223, 229; trapezoidal, 174
micropoints, 174, 221
microwear patterns on stone tools, 60, 102–104, 106, 110, 133, 221, 223, 225
Middle Desna River: Mousterian of, 65, 72–75, 116, 119; Upper Paleolithic of, 195, 202–204, *203*, 229, 233, 241–242, 245, 251
Middle Dnestr Valley, 30–31; Mousterian of, 67–71, *68, 70*, 103, 107, 127–128; early Upper Paleolithic of, 143–146, 150, 162–163, 176, 179, 184, 187; late Upper Paleolithic of, 197–200, 215, 228–229, 233, 239, 243
Middle Don River, 31, 65–66, 74; early Upper Paleolithic of, 143, 148, 167, 174–177, 181, 184, 187, 189, 253; late Upper Paleolithic of, 197, 208, 210, 215, 220, 227, 229, 231, 233, 241, 245–246
Middle Paleolithic. *See* Mousterian
Middle Pleniglacial (OIS 3), 31, 66–67, 69, 74, 77, 83, 85–86, 89, 100, 104, 125, 148
Mikhailovskoe, 43, 47–50, *49, 50*, 53, 64
mineral salts, 22, 24–25
minimum equilibrium size of hunter-gatherer population, 10
Minsk, 195
Mius River, 212
modern humans (*Homo sapiens sapiens*), 106–107, 111, 115–116, 128–129, 136–137, 153, 253–255; adaptation to warm climates, 2, 139–141, 158, 189, 253; African origins, 1, 139–142, 188; brachial index, 155; burials, 82–83, 150, 215–218; cranial capacity, 155; crural index, 155, 217; dispersal of, 1–2, 13–14, 141–142, 158, 188, 193–195, 253; distal limb segments, 140, 155, 217; East European skeletal remains, 153–158, 215–218; frontal bone, 140, 155; mandible, 140, 155; occipital, 140; phalanges, 140–141; planning and scheduling of resource use, 184; scapula, 140; socioecology, 12–14, 138, 142, 178, 185–188, 243–248, 253–255; technological complexity, 138, 158–161, 189, 253–255; transition to, 1–2, 138–139, 157–158, 253–254; variability of East European samples, 155, 190, 217
Moldova, 15, 199
molluscs, 24, 26, 30, 67, 73–76
Molodova Culture, 162–163, 190, 254
Molodova I: Mousterian, 67, 71, 78, 84, 100, 107–108, 113, 115, 127–129; Upper Paleolithic, 144, 146, 179, 184
Molodova V: early Upper Paleolithic, 143–146, *145*, 162, *165*, 176, 179, 184, 187; late Upper Paleolithic, 197–198, 200, 225, 228–229, 233, 235–236, 243, 245; Mousterian, 65–67, 71, 78, 84, 100, 108, 113, 115, 128–129
Monasheskaya Cave, 85, 100, 102, 119, 123–125; Neanderthal isolated teeth in, 89–91
Morgan, L. H., 127
morphological adaptations to cold, 7, 11, 39–40, 52, 58–59, 251. *See also* Neanderthals
Moskva Glaciation, 28
Mount Elbrus, 17
Mousterian, 53–54, 63, 88–89, 144, 152, 185, 202, 212; of Acheulean Tradition, 104; "cultures" defined, 98; Crimea, 78–83, 173–174; East European Plain: central, 71–75; East European Plain: south, 75–78; East European Plain: southwest, 66–71; Northern Caucasus, 83–86, 173–174; occupation floors, 127–131; ornaments in, 174; technology and tool types in Upper Paleolithic sites, 146–147, 151, 161–163, 166–169, 172–173, 191; topographic setting of sites, 67–69, 72–74, *76*, 76–77

Mousteroid assemblages, 190
Muralovka, 235
musk-ox (*Ovibos moschatus*), 24, 28, 200, 206, 241
Myshtulagty lagat (Weasel Cave), 42, 51, 53–54, 64–65, 78, 85–86, 88–89, 100, 123

Ndutu, 140
Neanderthals, 2–3, 13–14, 36, 39, 48, 52–53, 138–139, 147, 150, 152–153, 155, 157, 159, 173, 188–189, 193, 197, 219, 249–254; adaptations to cold climates, 58–62, 109–110, 132–133, 250–252; brachial index, 58, 91; burials, 81, 88, 126–127; clavicle, 58; cranium, 57, 59; crural index, 58, 91; diet, 109–110, 132, *134*; distal limb segments, 58, 91, 132; East European skeletal remains, 88–91; foraging strategy, 110–125; front teeth, 57, 60, 91; group size, 112–113, 126, 131, 136; hands and feet, 58–59, 91; "hyperpolar" morphology, 52, 58, 132, 250; infraorbital foramina, 58–59; lithic technology and tools, 91–105; mandible, 57; muscle attachments, 57, 60; nasal cavity, 57; nonlithic technology, 106–109; occipital bun, 57; origins, 55–57; paramastoid process, 91; physiology, 58, 132–133; planning and scheduling of resource use, 111–112; pubic bone, 58, 60; "radiator nose," 58–59; retromolar space, 57, 59; scapula, 58, 60, 91; socioecology, 125–131, 136; speech, 126; suprainiac fossa, 57, 91; tailored fur clothing, lack of evidence for, 107, 109; technological complexity, measurement of, 109–111, 132–135, *134*; teeth, 60, 91; Upper Paleolithic industries produced by, 64, 66, 143, 159, 161–162, 167, 171, 190; vertebrae, 57–58; windbreaks, use of, 107–108, 115, 129. *See also* Mousterian
Near East: Aurignacian of, 160; colonization of Eastern Europe and, 53–54, 157–158, 190, 250; *Homo ergaster* in, 37; modern humans of, 140, 158, 252; Neanderthals of, 59, 88–89, 101, 133, 141, 251–253
needle case (bone), *224*, 225, 254
needles (bone and ivory), 107; early Upper Paleolithic, 160, 171–172, 189; late Upper Paleolithic, 194, 199, 209, 220, *224*, 225, 229
Negotino, 74
Negotino-Rudnyanka, 74
Neolithic, 72
nets, fishing, 194
neural hypothesis, 141, 253
niche, 2, 4–5; modern human, 13, 192, Neanderthal, 111; overlap, 178, 252
Nihewan Basin, 37
Nile Valley, 193
nonadaptive characters, 4, 59
North American Plains, 213
North Atlantic Ocean, effect on climate in Europe of, 3, 17
Northern Caucasus, 21, 26, 34–35, 42, 48, 51, 53, 250; Mousterian, Neanderthals of, 63, 65, 66, 83–86, *84*, 88–89, 100–103, 109–110, 112–113, 115–117, 119, 122–125, 127, 129, 131, 157, 188; Upper Paleolithic, modern humans of, 143, 153, 173–174, 176, 179, 183, 214–215, 221, 228, 237, 243, 253
Northern Donets River, 43, 47–48, 53, 64, 171, 212
Nosovo I, 77, 104, 113, 129
notched bladelets, 162
notches, 93, 103–104, 147, 167; product of trampling and natural processes, 103–104
Novgorod-Severskii, 74, 206, 241

oak, 17, 28, 31
occupation floors: early Upper Paleolithic, 185–188, *186*; late Upper Paleolithic, 243–248; Mousterian, 125–131, *130*
Oka-Don Lowland, 16, 27, 153
Oka River, 151, 171, 177, 187, 211, 245
Oktyabr'skoe, 241

Oldowan industry, 37
Omo, 140
opportunistic encounter strategy, 111
Optical Stimulated Luminescence (OSL) dating, 63
optimality models, 5; optimal foraging, 5
Ordzhonikidze, 85
Origin of Races, The (Coon), 58
Origin of the Family, Private Property, and the State, The (Engels), 128
ornaments, 141–142; bracelets, 177, 236; brooches, 236; early Upper Paleolithic, 145, 147, 151, 153, 161–162, 169–173, 174–177, *175*, *177*; headbands, 235; late Upper Paleolithic, 209, 212, 219–220, 235; Neanderthal, 126, 139, 174; rings, 177
Oselivka I and II, 146, 162–163, 187
Osokorovka, 233, 242, 248
Oswalt, W., 110
owls as bone collectors in sites, 119
oxygen-isotope stages, 18–19, *20*

paleomagnetism, 19, 27
Pavlovsk, 75
pebble tools, 47, 53, 93
Pechora River, 152, 171
Pedokompleks I, II, and III (Central Europe), 28, 30–31
pendants. *See* ornaments
perforated teeth and shell (ornaments), 175–176, 235
perforators. *See* awls
periglacial steppe. *See* loess-steppe
permafrost, 33, 161, 192, 215, 226
Petralona, 42, 56
physiological responses to temperature, 7, 11, 58, 60, 132–133
pine, 17, 23, 30, 33–34
pits, 152, 171, 200–202, 207, 211, 214, 220, 226–227, 242, 245–246; bone-filled, 202–206, 210, 226, 229; hearth, 187; permafrost-refrigerated ("ice cellars"), 161, 192, 226, 228–229, 254; storage of food in, 226, 254; storage of fuel in, 192, 226, 228–229, 254

plant foods, proportion of diet of, 52, 109–110, 237, 239, 250–251
pleiotropy, 4
Pliocene, 193, 250
Pogon, 204, 242
Pogrebya, 47, 53
points (bone), 160, 169, 171–174, 197–199, 209, 212, 215, 220; laterally grooved, 212
points (stone), 50, 67, 77, 80–81, 84–85, 93, 100, 102–103, *102*; bifacial, 74, 147, 151, 153, 167, 170–171; Levallois, 162; microwear analysis of, 103, 221; shouldered, 200, 202, 208, 220–221, *223*; triangular, 169; Upper Paleolithic, 147, 153, 166–167, 170–173, 199, 202; uses of, 103, 221
Poles'e, 72, 103
polar desert, 65, 192
polar fox. *See* arctic fox
Polish Plain, 15
pollen-spores, 22–24, 26, 28, 34, 51, 69, 74, 77, 81, 86, 110
postmold arrangements, 107, 233, 248. *See also* shelters, artificial
Praslov, N. D., 43, 47, 77, 83
primary productivity, 6, 8, 23, 24–25, 251
Pripyat' Marshes (or Basin), 16, 146, 202
productivity paradox of loess-steppe, 25. *See also* loess-steppe
proglacial lakes, 32, 192, 195
Prolom I, 79–80
Prolom II, 79–80, 107, 121, 127
"promiscuous horde" of Neanderthals, 128
protobiface, 46
Prut River, 15, 51; Mousterian of, 65, 69, 71, 103, 107, 116, 128; Upper Paleolithic of, 146, 166–167, 176, 181, 199
Pshekha River, 84
Pushkari I, 203–204, 220, 225–226, 235, 245
Pushkari sites, 203–204

Qafzeh, 140

"radiator nose." *See under* Neanderthals
radiocarbon dating, 63, 143, 196, 200, *201*
Radomyshl', 219
Rakovets River, 71, 147
reciprocal altruism, 6
red deer (*Cervus elaphus*), 28, 46, 50; Mousterian, 69, 72, 112, 115–116, 121, 250; Upper Paleolithic, 147, 181, 194, 199, 239, 241, 243
Reilingen, 57
reindeer (*Rangifer tarandus*), 24, 26, 28, 30–32; hunted in groups, 183; Mousterian, 69, 71, 79, 80–81, 115; Upper Paleolithic, 145–147, 152, 179, 181, 183, 194, 197–200, 206, 212, 239, 241–242
retouched blades, 146–147, 149, 151, 159, 163, 169, 173, 220
retouched bladelets, 174
retouched flakes, 102
retouched microblades, 153, 169, 173, 212–214, 228
Reut River, 166
Rikhta, 72, 113
Rikhta River, 72
Ripiceni-Izvor: Mousterian, 69, 71, 100, 107–108, *108*, 113, 115, 128–129; Upper Paleolithic, 146–147, 167, 181, 187
Riwat, 37
rods, bone, 160, 171, 174
roe deer (*Capreolus capreolus*), 24
Rogachev, A. N., 207–210
Romankovo, 88
Romania, 69, 199
Ros' River, 206
Rosava River, 206
Rovno, 163
Rozhok I and II, 65, 77, 88, 107–108, 113, 116–119, *118*, 129

saiga (*Saiga tatarica*), 24, 28; Mousterian, 79–81, 112, 121–123, 133; Upper Paleolithic, 153, 183–184, 214, 243, 251
Salyn Phase (Mezin Soil Complex), 28, 30, 64
Sary-Kaya I, 79, 123
Satanai, 215–217, 228, 243

Scandinavian glacier. *See* Fennoscandian ice sheet
scavenging, 46, 111–112, 115, 117, 121, 249–250
scrapers. *See* side-scrapers
Schöningen, 61, 106
sea level, changes in, 28, 30, 33, 42, 51
Sea of Azov, 43, 65, 119, 129, 235
seasonality, 125, 183, 213, 241–242, 246
secondary productivity, 6, 24–25
Seim River, 207
settlement hiatus: during Early Pleniglacial (OIS 4), 55, 65, 71, 132, 136; during Last Glacial Maximum cold peak (20,000–18,000 years BP), 192, 200–201, 212, 219, 227, 254; in Siberia, 194
Sevastopol', 152, 173, 214
sexual dimorphism, 37, 195
sexual division of labor, 8, 131
Shaitan-Koba, 82–83, 100, 121, 123
Shan-Koba, 214
sheep (*Ovis* sp.), 85, 122–125
shellfish, 194
shelters, artificial, 128, 137; early Upper Paleolithic, 149, 152, 162–163, 165, 167, 169–171, 173, 184, 189, 253; late Upper Paleolithic, 197–198, 200–203, 214, 220, 226–228, 231–233, 241–242, 245–248; "longhouse" structures (Eastern Gravettian), 204, 207–209, 226–227, *244*, 245–246, 255; mammoth bone structures (Epigravettian), 192, 201–207, 210, 227–228, 231–232, *232*, 241–242, 246–247, *247*, 254; Mousterian, lack of evidence for, 107–109, 135, 252; multiple structures on one occupation floor, 187–188, 205–207, 210, 231, 246–248, *247*, 255; postmold arrangements representing, 107, 233, 248; semi-subterranean, 203, 227, 254; simultaneous occupation of multiple structures, 246, 255
Shkurlat III, 65, 74, 88–90
Shlyakh, 77–78
shouldered points. *See* points (stone): shouldered

shovel-shaped tools (bone), 160, 172
Shovkoplyas, I. G., 206
Siberia, 3, 22, 36, 41, 52, 107; modern humans of, 14, 142, 158, 160, 174, 188, 193–195, 233, 251
side-scrapers, 47, 50, 67, 69, 76–77, 79–86, 93, 101–103, *102*; canted, 102; microwear analysis of, 102, 107; resharpening of, 101; Upper Paleolithic, 147, 151–153, 162–163, 166, 169, 171–172; uses of, 102–103, 107
Simferopol', 78, 80, 214
Singil' fauna, 27
single-crystal-laser-fusion dating, 51
Skhul, 140
Smolensk, 32, 192
Smolensk Cryogenic Horizon, 30–31
snares. *See* untended facilities
socioecology, 5–6. *See also under* modern humans; Neanderthals
Soffer, O., 242, 246
solar energy, 6, 22, 25
Solutrean industry, 194
Soroki, 198
South Africa, 193–194
Sozh River, 201
spear-throwers, 161, 193, 223, 229
Spitsyn Culture, 167–169, *170*, 171, 175, 190, 254
spruce, 17
stable isotope analysis of human bone, 11, 52, 109, 133, 178, 250
Starosel'e, 66, 82–83, 88, 102, 104, 121, 123; dating of modern human burial at, 63, 82–83, 88, 157
Stephanorhinus hundsheimensis, 44
steppe bison (*Bison priscus*), 24, 28, 31–32, 50; cut marks on bones of, 125, 243; hunted in groups, 117, 213, 242–243; Mousterian, 69, 72, 76–77, 81, 84–85, 112, 115–117, *117*, 119, 122–125, 133, 136, 250–251; Upper Paleolithic, 179–181, 192, 198–199, 212–215, 239, 242–243
Stinka, 69, 71, 104, 113, 115
storage: food, 10, 109, 111, 135, 137, 252; fuel, 192, 226; pits used for, 161, 227. *See also* pits

Strelets Culture, 169–171, *170*, 175–176, 190, 254
Styr' River, 146, 199
Sudost' River, 204
Sukhaya Mechetka (Volgograd), 65, 75–77, *76*, 102, 104, 108, 113, 116–117, 119, 128, *130*
Sungir', 151–152, *151*, 171, 177, *178*, 183–184, 187–189; modern human skeletal remains at, 153, 155–158
Supoi River, 207
Suponevo, 202–203, 235–236
Svinoluzhka River, 72
Swanscombe, 57
symbols, 1–2, 12–14, 142, 174, 189–190, 234, 253; archaeological evidence for, 125, 138, 158, 160, 188; organization and, 13–14, 138, 142, 188–190, 253–255; technology and, 12–14, 138, 142, 158–161, 188–189, 253, 255. *See also* art; burials; language
Syuren' I, 152–153, 155–156, 160, 173, 183–184, 214
Szeletian industry, 64, 143, 162, 167, 171, 190–191

tailored fur clothing, 10, 137, 160, 165, 172, 189, 194, 225, 229, 253; Mousterian, absence of evidence for, 107, 109, 135, 252; sequence of steps for production of, 160–161. *See also* needles; trapping, fur
taphonomy, 52, 61, 111–112, 121, 179, 181, 237, 239, 241
Tarasov, L. M., 210
teal (*Anas* sp.), 242
technology: cold climates and, 10–11, *10*, 133–135, 158–159, 189; complexity of, 10, *10*, 12–13, 109–111, 133–135, *134*, 253–254; innovations in, 13, 138, 142, 253–254. *See also* symbols
"technounits," 10, *10*, 109–111, *134*. *See also* technology
Terek River, 86
territory size, 8, 11, 126, 132, *134*, 135
textiles, woven, 161
throwing sticks. *See* spear-throwers

Timonovka I and II, 202–203, 235, 248
Tiraspolian faunal complex, 44
Torralba-Ambrona, 38
trapping, fur: early Upper Paleolithic, 161, 169, 171–172, 181, 183, 189; late Upper Paleolithic, 192, 237, 242, 253
traps and snares. *See* untended facilities
Treugol'naya Cave, 42, 44–46, *45*, *46*, 53
Trinka I, 71, 116
trophic levels, 6
Trubchevsk Horizon, *29*, 32, 195, 202
truncations, 202, 208–209, 221–223, *223*; uses of, 221–223. *See also* Kostenki knives
tundra-steppe. *See* loess-steppe
Turner, A., 38

Udai River, 207
Ukraine, 15, 66, 72, 206
Upper Humic Bed (Kostenki), 148–149, 170, 172–173, 175–176, 181, 183, 187, 191
Upper Paleolithic, 66, 72, 88, 107–109, 127–129, 139, 141–143, *144*, 159–162, 193–195, *196*; early Upper Paleolithic sites of Crimea and Northern Caucasus, 152–153; early Upper Paleolithic sites of East European Plain: central and north, 148–152, 167–173; early Upper Paleolithic sites of East European Plain: southwest, 144–148, 162–167; late Upper Paleolithic sites of Crimea and Northern Caucasus, 214–215; late Upper Paleolithic sites of East European Plain: central, 200–212; late Upper Paleolithic sites of East European Plain: south, 212–214; late Upper Paleolithic sites of East European Plain: southwest, 197–200; topographic setting of sites, 147–148, 150–151, *150*, 211–212; variability of early Upper Paleolithic, 159. *See also* Aurignacian industry; Brynzeny Culture; Chatelperronian industry; Eastern Gravettian; Epigravettian;

Gorodtsov Culture; Lipa Culture; Molodova Culture; Spitsyn Culture; Strelets Culture; Szeletian industry
untended facilities, 10, 13, 109, 135, 137, 161, 171, 189, 225, 252–254; trap components, 161, 229, 254
Ural Mountains, 15, 16, 41, 171, 233, 249
uranium series (U-series) dating, 83

Valdai Glaciation, 30, 32
Valdai Loess, *29*, 31–32, 65
Velichko, A. A., 203
Venus figurines, 200, 209, 211, 220–221, 233–234, *234*, 237, 248, 255
Vértesszöllös, 38, 39, 53
Vladimir, 151, 171
Vladimir Cryogenic Horizon, 32
Voedvodskii, M. V., 206
Vogelherd, 177
volcanic ash, 86, 148, 167
Volchii Grot, 78
vole (*Arvicola chosaricus*), 51
Volga River, 16, 27–28, 65, 75, 119, 129, 195
Volga Upland, 15
Volgograd, 17
Volgograd (Mousterian site). *See* Sukhaya Mechetka
Volkov, F. K., 206
Volyn-Podolian Upland, 15–16, 24, 72; early Upper Paleolithic of, 146, 162–163, 176, 179, 187, 189; late Upper Paleolithic of, 199, 227, 241, 245
Voronezh, 148
Voronovitsa I, 146, 163, 179, 187
Vykhvatintsy, 69, 116

Weasel Cave. *See* Myshtulagty lagat
wild ass (*Equus hydruntinus*), 77, 81–83, 121, 123, 133
Willandra Lakes, 194
willow ptarmigan (*Lagopus lagopus*), 242
wolf (*Canis lupus*), 46, 119, 121, 123, 180–183, 206, 209, 225, 241; nearly complete skeletons of, 225, 241–242; trapping of, 181, 225
wooden tools and weapons, 106, 61–62, 109–111, 133

woolly mammoth (*Mammuthus primigenius*), 24, 28, 31–32; hunting of, 113–115, 241; Mousterian, 67, 69, 72, 81, 112–113, 116, 120–123; taphonomy of, 239–241; Upper Paleolithic, 146, 152, 179–183, 185, 201–202, 204, 206–207, 209–211, 239–241; use of bones and tusks for raw material, 115, 239

woolly rhinoceros (*Coelodonta antiquitatis*), 32, 115–116, 179

workshop sites, 119

Yami, 212
Yaroslavl' Cryogenic Horizon, 32–33
Younger Dryas event, 33
Yudinovo, 205–206, 228, 231, 235–236, 242, 246, 248
Yukon, 41
Yurovichi, 202, 220

Zaraisk, 209, 211, 221, 245
Zamyatnin, S. N., 75, 83, 210
Zaskal'naya V, 65, 79, *80*, 81, 84, 104, 109, 123, 129; Neanderthal skeletal remains at, 88, 90–91
Zaskal'naya VI, 79, 81, 84, 104, 123; Neanderthal skeletal remains at, 88, 90–91
Zavernyaev, F. M., 72
Zhitomir, 72
Zolotovka I, 213–214, 228
Zuya River, 80

About the Author

John F. Hoffecker received a B.A. from Yale (1975), M.A. from the University of Alaska in Fairbanks (1979), and a Ph.D. from the University of Chicago (1986), and is currently a research associate at the Institute of Arctic and Alpine Research at the University of Colorado in Boulder. He has focused his research primarily on human paleoecology in cold environments, conducting archaeological and zooarchaeological studies in Russia (especially the Northern Caucasus) and central and northern Alaska.

CPSIA information can be obtained
at www.ICGtesting.com
Printed in the USA
LVHW011148291118
598455LV00003B/22/P

9 780813 529929